Controlled Particle, Droplet and Bubble Formation

Series Editor
Professor R. A. Williams
University of Exeter, Camborne School of Mines, UK

Series Advisors
Professor J. Gregory
University College, London, UK

Dr. K. Nishinari
National Food Research Laboratory, Japan

Professor D. C. Prieve
Carnegie-Mellon University, USA

Professor T. Ring
Ecole Polytechnic, Lausanne, Switzerland

Professor H. N. Stein
Technical University of Eindhoven, The Netherlands

Dr. D. J. Wedlock
Shell Research Ltd, UK

Forthcoming titles
Particle Deposition and Aggregation: Measurement Modelling and Simulation
Biocolloid Engineering

Controlled Particle, Droplet and Bubble Formation

Edited by
David J. Wedlock

o 6 177864

CHEMISTRY

Butterworth-Heinemann Ltd
Linacre House, Jordan Hill, Oxford OX2 8DP

℞ A member of the Reed Elsevier group

OXFORD   LONDON   BOSTON
MUNICH   NEW DELHI   SINGAPORE   SYDNEY
TOKYO   TORONTO   WELLINGTON

First published 1994

British Library Cataloguing in Publication Data
Controlled Particle, Droplet and Bubble
Formation.–(Colloid and Surface
Engineering Series)
  I. Wedlock, David J.   II. Series
  660

ISBN 0 7506 1494 3

Library of Congress Cataloging in Publication Data
Controlled particle, droplet, and bubble formation/David J. Wedlock.
  p. cm.–(Colloid and surface engineering series)
  includes bibliographical references and index.
  ISBN 0 7506 1494 3
  1. Particles.   2. Drops.   3. Bubbles.   I. Wedlock, David J.
  II. Series.
  TP156.P3C66   1994
660'.2945–dc20   93–32069
                                                                          CIP

Composition by Genesis Typesetting, Laser Quay, Rochester, Kent
Printed in Great Britain by Redwood Books, Trowbridge, Wiltshire

Contents

Preface

In industrial chemical processing (chemical in the broadest sense) there is rarely a scheme which does not involve at least one stage where a new phase is created (condensation) or an existing phase is subject to a size reduction process (homogenization/comminution). I have sought in this book to collect together contributions from recognized authorities in their field to cover most aspects of 'particle' formation and control likely to be encountered by the research worker, the industrial chemist, chemical engineer and process engineer.

Whether the continuous phase or the discontinuous phase is solid, liquid or gaseous, many common, usually thermodynamic, themes dominate the discussion. In general terms, particulates below about 1 μm in diameter fall into a class termed 'colloidal particles'. Colloidal particulates in dispersion have very many of their properties dictated by the fact that the ratio of surface area to mass is very high and therefore interfacial/surface tensions are of paramount importance. Furthermore, their motions and colloisional encounters are determined primarily by Brownian considerations.

This book does not seek purely to cover the subject of formation of colloidal particulates, but does inevitably return to the theme of surface effects on particulate properties where appropriate. In chapters on emulsion formation, aerosol formation, bubble growth and suspension stability, the recurring theme, for example, of the Laplace pressure in relation to particle radius and interfacial tension and, furthermore, its relationship to particle growth comes through.

Two chapters deal specifically with the formation of inorganic particulates. The chapter on sol–gel processing particularly covers characterization of the particulates once formed as well as the formation processes. The other chapter on inorganic particulates deals especially with control of the size distribution during particle formation. The so-called 'monodisperse particles' to which this control can lead, has a significant place in subsequent processing, for example to form advanced materials. Many advanced materials require controlled particle packing to make fault-free ceramic green body precursors for sintering. This requires both the ability to control particle-size distributions and of course the ability to characterize the particle-size distributions once formed. Advances in the characterization of particle-size distributions with a particular emphasis on the process industries is covered in a chapter by itself.

Bubbles are one discontinuous phase that does not spring immediately to mind under the general heading of 'particles'. However, their nucleation, growth and collapse are central to foam formation and stability, gas release at electrolytic surfaces and a number of related process problems. There are striking parallels in the behaviour of bubbles compared to more dense discontinuous phases and it is hoped that those highlighted (together with the differences) will aid understanding.

Probably the single most common processing operation in the chemical industry is the precipitation of organic and inorganic compounds in industrial crystallizers. These processes are covered in the book in two chapters, but from quite different approaches. The first is the concept of particle formation by agglomerative precipitation processes. This discusses particle birth, death, associations and general population balance concepts in 'mixed-suspension mixed-product removal reactors' (MSMPR), with the associated differential equations that would be invoked in a chemical engineer's characterization of such processes. The second approach is that of crystal growth as a summation of a number of two-dimensional (surface) growth processes leading to a three-dimensional array of molecules or ions that may be characterized by structural and symmetry parameters. That is, crystal forms. Crystal habit (the external manifestation of the constrained growth processes dictated by particular crystal forms and face growth kinetics) and how it arises is explained. Modern computational approaches for the calculation of likely crystal habit from structural and surface chemistry considerations are outlined. These computational approaches are quite novel and are only now becoming available to the scientific community at large. It should allow the development of predictive capability in crystal engineering and processing which will remove the hitherto hit-and-miss approach to attempting control of precipitation and crystal-growth processes.

Most researchers tackling the study of suspended particles resort first to use of what are generally referred to as emulsion polymers. This type of particle formation by condensation is of great industrial relevance also since it is the basis of most modern paint formulations *inter alia*. In either event, whether seeking emulsion polymers for fundamental research or for surface coatings, the control of particle size is critical. A significant chapter is dedicated to this technology with the emphasis on factors allowing control of the polymerization process.

Similarly, a large amount of chemical processing requires the formation of well-defined particulates by a size-reduction process (comminution). This is dealt with in a dedicated chapter outlining the general considerations in size reduction by milling/crushing of both dry particulates and in liquid-based suspension. The factors determining size-reduction rates, energy inputs, final particle stability and associated regrowth processes are discussed.

On occasions there is no alternative in control of particulate performance but to resort to encapsulation of (particularly) biologically active molecules. To this end, two chapters deal with this problem, again, from quite different approaches. The first might be referred to as the 'synthetic approach' which involves the formation of a chemically synthesized distinct barrier, to microencapsulate the agent and give rise to a controlled release of the active agent. Thus, the book discusses microencapsulation; techniques of formation. The second approach is consideration of the formation of natural barriers to diffusion using surface active agents, resulting in well-defined particulate suspensions which may be utilized to microencapsulate; namely liposomal or vesicular preparations. The former case involves the use of naturally occurring phospholipids formed into more or less concentric bilayer preparations which may or may not contain a biologically active molecule at various different sites, to permit controlled release. In the latter case the considerations are largely similar but do not involve the use of natural phospholipid surfactants, although any synthetic surfactant capable of forming a bilayer that may be turned into a (concentric) bilayer sphere(s) may be involved.

Thus, it is hoped that the book will prove a useful companion to anyone brave enough to attempt to work with particulates and conceited enough to think he or she can really totally control what is going on!

David J. Wedlock
Shell Research Ltd
Thornton Research Centre

Contributors

G. Clydesdale
Department of Pure and Applied Chemistry, University of Strathclyde, Glasgow G1 1Xl, UK.

Ian Colbeck
Institute for Environmental Research, University of Essex, Wivenhoe Park, Colchester CO4 3SQ, UK.

Eric Dickinson
Procter Department of Food Science, University of Leeds, Leeds LS2 9JT, UK.

R. Docherty
ICI Specialities, Hexagon House, Blackley, Manchester M9 3DA, UK.

Ken R. Geddes
Crown Berger Ltd, Hollins Road, Darwen, Lancs BB3 0BG, UK.

Alan G. Jones
Department of Chemical and Biochemical Engineering, University College London, Torrington Place, London WC1E 7JE, UK.

M. Jayne Lawrence
Department of Pharmacy, King's College London, University of London, London SW3 6LX, UK.

S. D. Lubetkin
DowElanco Europe, Letcombe Laboratories, Letcombe Regis, Wantage, Oxon OX12 9JT, UK.

Paul F. Luckham
Department of Chemical Engineering and Chemical Technology, Imperial College of Science and Technology, Prince Consort Road, London SW7 2BY, UK.

Egon Matijević
Center for Advanced Materials Processing, Clarkson University, Potsdam, New York 13699, USA.

S. J. Peng
Department of Chemical Engineering, UMIST, P.O. Box 88, Sackville Street, Manchester M60 1QD, UK.

John D. F. Ramsey
CNRS, Institut de Recherches sur la Catalyse, 2 Avenue Albert Einstein, 69626 Villeurbanne Cedex, France.

K. J. Roberts
Department of Pure and Applied Chemistry, University of Strathclyde, Glasgow G1 1XL; and SERC Daresbury Laboratory, Warrington WA4 4AD, UK.

David J. Wedlock
Shell Research Ltd, Thornton Research Centre, Chester CH1 3SH, UK.

Richard A. Williams
Department of Chemical Engineering, UMIST, P.O. Box 88, Sackville Street, Manchester M60 1QD, UK.

Acknowledgements

Chapter 3
The author is indebted to Dr. J. Hostomsky (Prague) for many helpful discussions during the preparation of the manuscript.

Chapter 4
Research on morphological modelling has been supported for a number of years through collaborative programmes involving the SERC, Exxon Chemicals and ICI Chemicals and Polymers, whom we gratefully acknowledge.

Chapter 11
The author acknowledges helpful discussions with Mr P. Hobin.

Chapter 1

Sol–gel processing

J. D. F. Ramsay

1.1 Introduction

Over the past decade there has been a remarkable increase in interest in sol–gel processes. This has mainly stemmed from the development of technological applications in the fabrication of materials such as oxide ceramics and glasses, catalysts and porous adsorbents (Brinker *et al.*, 1988; Klein, 1988; Brinker and Scherer, 1990; Guizard *et al.*, 1990). For these applications, sol–gel processes have inherent advantages because the microstructure of the material produced can be controlled together with its bulk form (e.g. spherical particles, fibres and thin films).

In parallel with these technological applications there has been a rapid advance in the understanding of the characteristics of concentrated colloidal dispersions and gels, particularly in aqueous media (see e.g. Goodwin and Ottewill, 1991). This progress has mainly arisen from the introduction of several advanced techniques in colloid science, such as small angle neutron scattering (SANS), dynamic light scattering and nuclear magnetic resonance (NMR) spectroscopy. These developments have recently been reviewed extensively (see e.g. Ramsay, 1986a; Goodwin and Ottewill, 1991) and, therefore, will not be described in detail here although, as will be discussed, their significance in the exploitation of sol–gel processes is very important. Perhaps at this stage we should consider a phenomenological description of gelation in the context of gel processes, particularly because these can encompass several somewhat different methods.

Gelation may occur with a wide range of inorganic colloidal dispersions and is an important characteristic requiring control in numerous commercial products, such as paints, pigments, lubricants and drilling fluids. Gel formation is typified by a change from a free-flowing state to one where the system becomes viscoelastic and eventually rigid, while retaining homogeneity and, frequently, an unchanged appearance. Gelation may be induced by a variety of processes, such as increases in concentration, changes in temperature or the addition of other components, e.g. electrolytes, polymers and surfactants. Evidently the mechanism may be complex and different.

We can distinguish three general classes of process which have been developed to produce gel particles with controlled properties. These are illustrated schematically in Figure 1.1. In process 1 an aqueous colloidal dispersion of oxide particles, called a 'sol', is produced. This is achieved either by the peptization of oxide or hydrous oxide powders or by the controlled hydrolysis, nucleation and growth of particles in an aqueous salt solution. Such

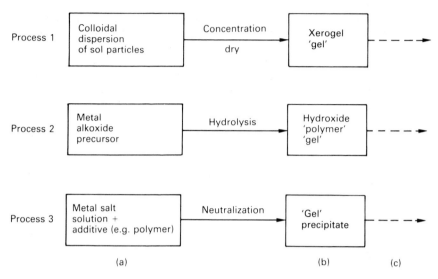

Figure 1.1   *Three general processes for forming an oxide/hydroxide gel. After stage (b), subsequent treatment involves drying and controlled calcination*

processes have been developed to produce stable concentrated (volume fraction of more than 10%) sols of numerous oxides and are used on a commercial scale to prepare, for example, silica and alumina sols. When sols of this type are sufficiently concentrated, by evaporation for example, a hydrated gel and eventually a porous xerogel is formed. Other more specialized dehydration techniques (solvent displacement and spray drying) can be used to produce gel particles with controlled shape and size.

In process 2 a gel is formed by the controlled hydrolysis and condensation of a metal alkoxide solution. This method can be considered as a process in which an 'inorganic polymer network' is eventually formed through the cross-linking of metal–oxygen bonds.

An enormous effort has been made in the development and application of alkoxide sol–gel processes over the past 10 years. These processes are well suited to the fabrication of thin films and fibres, for example, and have been used extensively in the development of ceramics and glasses for optical and electronic applications, where controlled chemical composition and purity are paramount (see e.g. Brinker *et al.*, 1988; Brinker and Scherer, 1990).

There are also gel processes that differ from the previous two and which have not been extensively employed. These include 'gel precipitation' and 'internal gelation' processes. These processes were originally developed over 20 years ago for the production of oxide ceramic nuclear fuel in the form of highly uniform spherical particles with narrow and controlled size distribution.

In gel precipitation, droplets of a metal salt solution, containing a water soluble polymer and a 'modifying' agent, such as formamide, are 'gelled' by neutralizing with a base such as ammonia in, for example, a long column (Stringer *et al.*, 1984). The diameter of the particles so formed can range from about 50 μm to 1 mm. The particles may have a very open structure composed of a cross-linked polymer network within which is contained a hydrous oxide precipitate.

The second of these variants on the traditional sol–gel process, known as 'internal gelation', involves the controlled hydrolysis of an aqueous metal salt solution in the form of droplets in an oil bath. Here hydrolysis is achieved 'internally' by the thermal decomposition of an organic precursor, such as urea or tetramine, dissolved in the salt solution. During the decomposition ammonia is produced gradually, and controlled hydrolysis to give a hydrous oxide precipitate in the form of droplets is achieved.

In this brief review the scientific and technical basis of process 1 will be described in more detail, since this route is well adapted for the controlled production of ceramic particles covering a range of sizes. Furthermore, as will be illustrated, the internal microstructure (surface area and porosity) of these particles is determined by the organization of the primary particles derived from the sol. This latter feature will be considered in some depth here.

1.2 Mechanisms of sol formation

A schematic mechanism in which hydrous oxide or hydroxide colloids are formed from an aqueous solution of metal cations is illustrated in Figure 1.2. This stepwise process involves the hydrolysis of a solvated aquo-ion followed by condensation to give a polynuclear ion. Such a sequence of hydrolytic and condensation reactions leads, under conditions of oversaturation of the metal

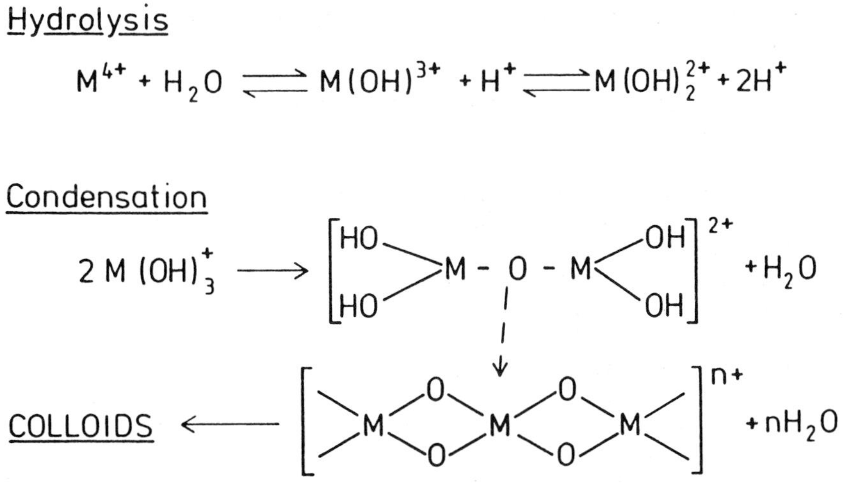

<u>Hydrolysis</u>

$$M^{4+} + H_2O \rightleftharpoons M(OH)^{3+} + H^+ \rightleftharpoons M(OH)_2^{2+} + 2H^+$$

<u>Condensation</u>

$$2\,M(OH)_3^+ \longrightarrow \begin{bmatrix} HO \\ HO \end{bmatrix} M - O - M \begin{matrix} OH \\ OH \end{matrix} \end{bmatrix}^{2+} + H_2O$$

<u>COLLOIDS</u> $\longleftarrow$ $\begin{bmatrix} M \begin{matrix} O \\ O \end{matrix} M \begin{matrix} O \\ O \end{matrix} M \end{bmatrix}^{n+} + nH_2O$

Figure 1.2 *Hydrolysis and polymerization of polyvalent metal ions in aqueous solution*

hydroxide, to the formation of colloidal hydroxy polymers which can eventually form hydroxide or hydrous oxide precipitates. This tendency for cation hydrolysis, involving the coordination of hydroxyl ligands (which act as electron-pair donors) increases with the acidity or charge density of the metal cation. Consequently, ions of small ionic radius and high charge are extensively hydrolysed in solution, as illustrated by silicic acid ($Si(OH)_4$), which can exist as a monomer in aqueous solutions of high pH (Iler, 1979). The tendency for hydrolysis can furthermore be generalized for a wide range of cations, as demonstrated by the dependence of the first hydrolysis constant K_1, on the charge to the M–O distance (z/d) shown in Figure 1.3. The tendency for cation hydrolysis is also paralleled by the ease of polynuclear ion formation. Thus most trivalent and quadrivalent ions form polynuclear species, whereas the process only occurs with the smallest of the divalent ions (Be^{2+}) (Schwarzenback and Wenger, 1969). Such polynuclear species are formed by oversaturating a salt

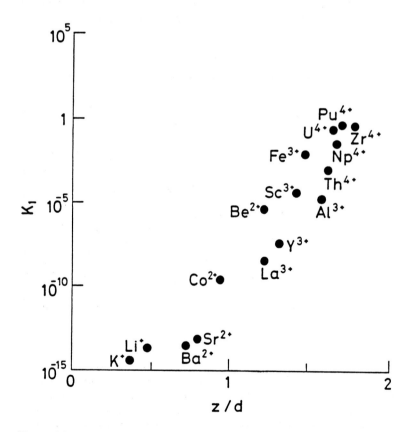

Figure 1.3   *Dependence of the first hydrolysis constant K_1 on the ratio of the charge to M–O distance* d *for different cations*

solution with respect to the solid hydroxide or hydrous oxide by adding, for example, a base.

This process can be illustrated by the behaviour of aqueous solutions of Fe^{3+} for which the following equilibria have been reported (Stumm and Morgan, 1981).

$$Fe^{3+} + H_2O \rightleftharpoons FeOH^{2+} + H^+ \tag{1.1}$$
$$Fe^{3+} + 2H_2O \rightleftharpoons Fe(OH)_2{}^+ + 2H^+ \tag{1.2}$$
$$2Fe^{3+} + 2H_2O \rightleftharpoons Fe_2(OH)_2{}^{4+} + 2H^+ \tag{1.3}$$
$$Fe(OH)_3(s) + 3H^+ \rightleftharpoons Fe^{3+} + 3H_2O \tag{1.4}$$
$$Fe(OH)_3(s) + H_2O \rightleftharpoons Fe(OH)_4^- + H^+ \tag{1.5}$$

Using the representative equilibrium constants, which have been determined by a variety of techniques, including potentiometric measurements and spectrophotometry, it is possible to construct distribution diagrams for the different species as depicted in Figure 1.4. In the hatched areas in the figure, the solutions become oversaturated with respect to $Fe(OH)_3(s)$ and, in consequence, polynuclear hydrolysis species will form as intermediates in the formation of

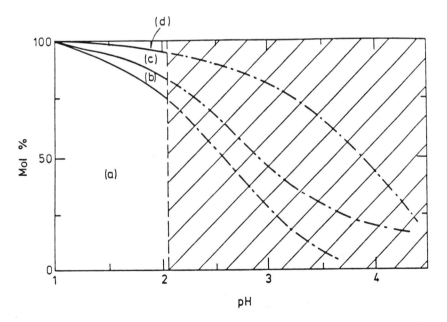

Figure 1.4 *Distribution diagram for iron(III) species in aqueous solution at a total iron(III) concentration of 10^{-2} mol dm^{-3} as a function of pH. (a)–(d) correspond to proportions of Fe^{3+}, $FeOH^+$, $Fe(OH)_2^{4+}$ and $Fe(OH)_2^+$, respectively. The vertical dashed line corresponds to the saturation limit with respect to $Fe(OH)_3(s)$. $(- \cdot - \cdot -)$ Proportions of metastable species in the hatched area where polynuclear hydrolysis species will form as intermediates before the eventual precipitation of the hydroxide occurs*

$Fe(OH)_3(s)$ which occurs in this range of relatively low pH. If neutralized completely, solutions of Fe^{3+} ions will rapidly precipitate $Fe(OH)_3$ which may age to form crystalline β-FeOOH. Furthermore, in alkaline solution (pH >10) the formation of ferrate ions $(Fe(OH)_4^-)$ becomes increasingly important.

Although there is extensive evidence for the existence of small polynuclear ions (Baes and Messmer, 1976) containing a few metal atoms (e.g. $Be_3(OH)_3^{3+}$, $Zr(OH)_8^{8+}$, $Th_2(OH)_2^{6+}$ and SiO_8^{8-}) the structure and stoichiometry of larger thermodynamically stable species still remain obscure. This is partly due to a lack of suitable techniques for characterizing such small macromolecules, and to the time-dependent and heterogeneous nature of many aqueous systems which may contain a variety of species simultaneously. Hydrolysed aluminium salt solutions are among the systems containing polycations which have been most widely studied (Akitt *et al.*, 1972; Waters and Henty, 1977; Bottero *et al.*, 1982) and here there is evidence of both $Al_8(OH)_{20}^{4+}$ and $Al_{13}O_4(OH)_{24}^{7+}$ species in solution. Furthermore, there is a vast amount of data on the nature of different oligomeric silicate anions which may exist in aqueous solutions which has been obtained from ^{29}Si NMR studies (Engelhardt and Michel, 1987).

Novel information which can be obtained from light scattering and incoherent neutron scattering measurements on reasonably well-defined and relatively concentrated (>0.1 mol dm^{-3}) polynuclear ion solutions containing Zr(IV) and Al(III) have also been reported (Ramsay, 1986b, 1990).

A vast number of specialized procedures are described in the patent literature for the preparation of oxide sols. The basis of these is sometimes obscure but, in general, colloid formation is achieved either by controlled nucleation and growth processes in solution or by the peptization of powders or precipitates of metal oxides and hydroxides. The mechanism of the first process is frequently based on the controlled growth from a seed nucleus, first described by Zsigmondy and Hückel (1925) and later refined and developed by LaMer (LaMer and Barnes, 1946; LaMer and Dinegar, 1950) for the preparation of monodispersed colloids. This is illustrated simply in Figure 1.5 by the behaviour which occurs when a reaction proceeds and produces a partially soluble product, such as an oxide or hydroxide, for example, having a solute concentration C_p. In this process C_p will rise, passing the saturation concentration C_s until the critical supersaturation or nucleation concentration C_n is reached. At this point nuclei will form, and proceed to grow at a rate determined by the diffusion of neighbouring solute species, which are consumed. The growth process will continue until the solute concentration falls to C_s. To obtain a monodispersed system, nucleation should occur in a short time interval or 'burst'; growth must then proceed by the slow production of solute species which have adequate time to diffuse to the seed nuclei, thus avoiding a recurrence of supersaturation and further nucleation. These principles have been applied to numerous systems, particularly by Matijević (1981) and others, to produce sols such as chromium hydroxide (Demchak and Matijević, 1969), aluminium hydroxide (Brace and Matijević, 1973), and silica (Bechtold, 1951; Stöber *et al.*, 1968).

The stability and fate of particles so formed depends on their surface properties and the nature of the aqueous medium. Two distinct situations can arise, depending on the interparticle interactions. In the first, where there is strong

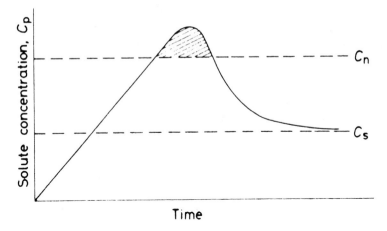

Figure 1.5 *Diagram illustrating the formation of a colloidal dispersion by controlled nucleation and growth. Initially, the solute concentration C_p rises, passing the saturation limit C_s, and continues until the critical nucleation concentration C_n is reached, at which precipitation occurs in the hatched zone. Further controlled growth of these nuclei results in a decrease in C_p, as depicted*

interparticle repulsion, discrete particles remain as a stable colloidal dispersion. This situation occurs where the particles have a high surface charge and the concentration of electrolyte in the solution is low. For monovalent counterions this is typically $<10^{-3}$ mol dm^{-3}. Where these conditions are not fulfilled the particles will tend to aggregate and may eventually form a precipitate. These extreme situations are illustrated by the behaviour of silica dispersions as depicted in Figure 1.6 (Iler, 1979).

The production of stable concentrated sols of other oxides has been achieved by nucleation and growth processes, by maintaining control of pH and electrolyte concentration. It is inevitably more demanding to obtain concentrated dispersions by these processes, because of the high concentration of destabilizing counterions (normally anions) in the system. This problem has been overcome by partially neutralizing salt solutions with amines in a separate immiscible phase. Thus, using long-chain tertiary amines (e.g. $(C_nH_{2n+1})_3NH_2$, where $n \approx 18$), the extraction of anions from concentrated solutions has been achieved in a controllable manner (Woodhead, 1984). An alternative, but less efficient, method involves controlled dialysis of dilute salt solutions followed by a concentration stage using membrane filtration (Ramsay *et al.*, 1983).

Another more general route to concentrated sols is by the peptization of hydrous oxide precipitates which have been prepared under controlled conditions (Dell, 1972; Ramsay *et al.*, 1978). Referring to Figure 1.5, it can be seen that rapid neutralization of a concentrated metal salt solution will cause the solute supersaturation limit to be far exceeded, resulting in the sudden production of a vast number of nuclei which have only a limited possibility of further growth.

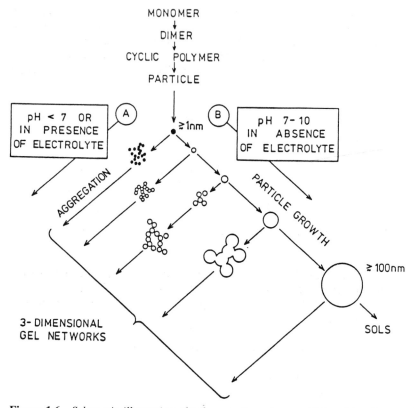

Figure 1.6 *Schematic illustration of polymerization of monosilicic acid to form seed particles. In basic solution (B) these can grow to give sols of different particle size (Ostwald ripening), whereas at lower pH or in the presence of electrolyte (A) aggregation occurs, which can lead to the formation of three-dimensional networks (after Iler (1979))*

Such a process will yield a precipitate containing very small amorphous or poorly crystalline particles, typically about 10 nm in size. If washed, to remove excess counterions, such precipitates can be peptized by controlled addition of either a dilute acid or alkali to give a stable concentrated sol containing particles which have a relatively narrow size distribution.

1.3 Properties and characterization of sols

1.3.1 Properties of oxide–water interfaces

The surface charge and nature of oxide–water interfaces have a dominant effect on the stability and properties of oxide sols. A hydroxylated oxide surface in contact with water can develop either a positive or negative surface charge, as shown schematically in Figure 1.7. This demonstrates that oxide surfaces have amphoteric properties and that the charge-determining ions are either H^+ or OH^-. The surface charge will consequently be dependent on pH and will vary depending on the type of oxide surface studied. The pH of zero point of charge

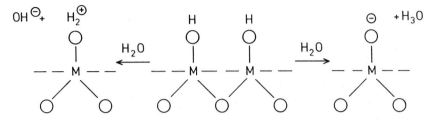

Figure 1.7 *Schematic illustration of surface charge generation on an oxide surface*

(ZPC), reflects the acid–base character of the surface; a low ZPC. corresponds to a more acidic nature, as is the case for silica. Values of ZPC for a number of oxides are given in Table 1.1. These are typical values which may vary by one or even two pH units depending on the method of preparation of the oxide and its pretreatment. These variations probably reflect different surface structures and the presence of adsorbed foreign ions and contaminants.

Table 1.1 *Typical pH values for oxides at the zero point of charge (ZPC)*

Oxide	ZPC	Oxide	ZPC
WO_3	~0.5	ZrO_2	~5
SiO_2	~2	γ-Al_2O_3	7.5
SnO_2	4.3	α-Fe_2O_3	8.5
TiO_2	~6	CeO_2	6.8
		Co_3O_4	11.4

As a result of the charged groups an electrostatic potential ϕ_0 is developed at the surface relative to the bulk solution. Furthermore, to retain electroneutrality, ions of opposite charge (counterions) are concentrated close to the surface, whereas the charge-determining ions are depleted in this zone. By assuming a Boltzmann distribution of ions, such a description leads to the familiar diffuse double-layer model (Verwey and Overbeek, 1948), in which the potential decreases with the distance from the surface r as

$$\phi(r) = \phi_0 \exp(-\kappa r) \qquad (1.6)$$

where κ is the Debye–Hückel inverse screening length given by

$$\kappa = \left(\frac{2c_iNe^2v^2}{\epsilon_0\epsilon kT} \right)^{1/2} \qquad (1.7)$$

Here c_i is the ionic strength of the solution, v is the valency of the ions, ϵ is the dielectric constant of the solvent, ϵ_0 is the permittivity of vacuum, N is Avogadro's number, and k is the Boltzmann constant. For spherical particles, of radius R, say, the surface charge z_p is related to ϕ_0 by the approximation

$$\phi_0 = z_p/4\pi\epsilon\epsilon_0R(1 + \kappa R) \qquad (1.8)$$

Colloid dispersions of oxide particles are in general stabilized by a mutual repulsion resulting from the interaction between their electrical double layers, which overcomes the van der Waals attractive interaction. Since details of lyophobic colloid stability theory have been treated extensively elsewhere (Verwey and Overbeek, 1948) this aspect will not be developed further here. However, it will be noted that stability is enhanced when ϕ_0 is large and $\phi(r)$ falls away gradually from the surface. The latter situation is favoured when κ^{-1} is large, i.e. where c_i is small (e.g. $<10^{-3}$ mol dm^{-3}) and ϵ large.

This simple description of the electrical double layer, although satisfactory for many colloid systems in which the ionic strength is relatively low ($<10^{-2}$ mol dm^{-3}) and where the surface charge density is modest (<10 μC cm^{-2}), is particularly inadequate for oxide interfaces. This has been widely demonstrated (Lyklema, 1971) by the very large values of surface charge density obtained by titration, which far exceed those from electrokinetic and other measurements (Penfold and Ramsay, 1985). These differences are partly reconciled by the Stern theory (Stern, 1924) which assumes that a layer of counterions of finite size is located within a few angstroms of the surface, across which the electrical potential falls linearly.

Even this description often cannot completely explain the very large surface charges and low zeta potentials of oxides and has led to further refinements such as the site binding model (James and Parks, 1980) and the 'porous gel' model (Lyklema, 1971).

1.3.2 Particle interactions and structure in concentrated sols

In sol–gel processes a concentrated sol is converted to a solid hydrous oxide gel and finally a ceramic oxide, as depicted schematically in Figure 1.8. Recent work has shown that the structure and interactions in the sol are of importance, since these may predetermine the rheological behaviour during the conversion process and ultimately the final microstructure of the gel – in particular its surface area and porous properties (Ramsay and Booth, 1983; Ramsay and Avery, 1986; Ramsay, 1989). This feature is illustrated schematically in Figure 1.9, showing unaggregated and aggregated sols which yield gels of low and high porosity respectively. These contrasting types of sol will be referred to subsequently.

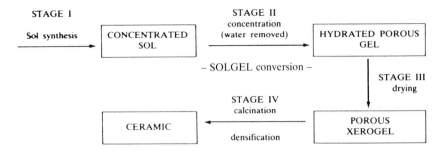

Figure 1.8   *The different stages in the formation of a ceramic from an oxide sol*

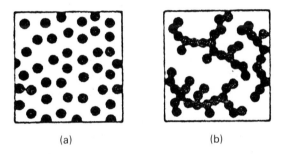

(a)   (b)

Figure 1.9   *Schematic illustration of two types of sol system which have been studied using small angle neutron scattering. In* (a) *the sol particles are nearly monodispersed and interact strongly in concentrated dispersions. In* (b) *the sol is composed of aggregates of small primary particles*

Some of the techniques which have been applied and the information obtainable on sol–gel systems are summarized in Table 1.2. Many of these are familiar techniques which have been described extensively elsewhere (Goodwin *et al.*, 1982) and need not be dealt with here. The recent application of scattering techniques (light, X-ray and neutrons) has, however, led to considerable advances in the understanding of concentrated sols and gels (Ramsay, 1986a and b, 1990). In particular, these techniques have provided direct microscopic information of the structure and interparticle interactions in colloidal dispersions which influence the stability and rheology of colloidal dispersions and the microstructure of gels (Ramsay, *et al.* 1991). The application of neutron scattering has been unique in this respect, particularly when used in parallel with other classical techniques for determining the macroscopic properties of concentrated sols and gels, as we briefly illustrate here.

Neutron coherent scattering has its counterpart in small angle X-ray (SAXS) scattering, and diffraction, for which the theory is very similar (Guinier and Fournet, 1955), although the possibilities afforded by contrast variation are

Table 1.2   *Techniques for characterization of structure and interactions in oxide sols and gels*

	Technique for study	Information
Sol interactions	Rheology – measurements under steady and oscillatory shear (viscoelastic behaviour)	Flow behaviour of dispersions – particle interactions at high volume fraction
	Electrophoresis – electrophoretic light scattering	Electrophoretic mobility – surface charge, colloid stability
	Inelastic and quasielastic neutron scattering	Properties of interfacial water layers – particle interaction, polynuclear hydroxy ions
	IR spectroscopy neutron diffraction	
Sol structure	Static light scattering	Particle size of sol MW and size of sol aggregates
	Quasielastic light scattering (photon correlation spectroscopy)	Diffusional and rotational motion of colloid particles (surface–solvent interactions)
	Ultracentrifugation	Sedimentation coefficient
	Small angle X-ray scattering	Particle size/shape of sol; particle ordering (radial distribution function); interparticle forces
Gel structure	Gas adsorption isotherms	Size and arrangement of sol particles on removal of dispersion medium (water) (pore size, pore volume, specific surface area); interparticle and interfacial chemical structure
	Electron microscopy	
	X-ray line broadening	
	Small angle neutron scattering	
	Solid-state NMR (e.g. ^{29}Si, ^{27}Al, ^{31}P), EXAFS	

unique to neutron scattering. Small angle scattering (SAS) arises from variations of scattering length density which occur over distances d_{SAS} (where $d_{SAS} \approx \lambda/2\theta$, corresponding to a scattering angle 2θ for radiation with a wavelength λ) exceeding the normal interatomic spacings in solids and liquids. Such an effect thus occurs with: (i) colloidal dispersions of particles (Ottewill, 1982) and polymers in liquids; (ii) assemblies of small particles in air or vacuum, comprising a porous material (Kostorz, 1979; Booth and Ramsay, 1985); and (iii) solid solutions such as alloys. One of the important features of neutron scattering is its ability to probe both the structure and dynamic properties of materials over a wide spatial range extending from about 1 to 10^3 Å. For structural studies of colloids and gels the upper part of this range is particularly important and overlaps with that covered by SAXS and static light scattering (SLS).

Typical systems where SANS can provide information on structure and interactions are illustrated schematically in Figure 1.9. These represent the type of structure frequently occurring in concentrated oxide sols such as silica, which have been mentioned previously. In the first the concentrated sol is composed of small discrete spherical particles, whereas in the second it is composed of aggregates of primary particles having a similar diameter. As we will see, these two different structures result in gels with contrasting surface and porous structures.

Discrete particle systems

For the case of a two-phase system, composed of particles having a homogeneous scattering length density ρ_p, dispersed in a liquid of scattering density ρ_s, the intensity distribution of scattered neutrons can be simplified to the following:

$$I(Q) = V_p^2 n_p (\rho_p - \rho_s)^2 P(Q) S(Q) \qquad (1.9)$$

where Q is the momentum transfer and is given by $Q = 4\pi \sin \theta / \lambda$, V_p is the particle volume, n_p is the particle number density, $P(Q)$ is a particle form factor, and $S(Q)$ is a particle structure factor. $P(Q)$ is dependent on the particle shape. For spheres of radius R, it is given by

$$P(Q) = \left(\frac{3[\sin (QR) - QR \cos (QR)]}{Q^3 R^3} \right)^2 \qquad (1.10)$$

In concentrated dispersions, the intensity distribution is modified by the effects of interference, which depend on the spatial ordering of the particles, which is described by the function $S(Q)$. Thus $S(Q)$ is determined by the nature of the particle interaction potential; for non-interacting systems, $S(Q) = 1$. The spatial distribution of the particles as a function of the mean interparticle separation r is given by the particle pair-distribution function $g(r)$, and is related to $S(Q)$ by the Fourier transform.

$$g(r) - 1 = \frac{1}{2\pi^2 n_p} \int_0^\infty (S(Q) - 1) \, Q^2 \, \frac{\sin (QR)}{QR} \, dQ \qquad (1.11)$$

Using this type of analysis the structural changes which occur during the progressive concentration of sols into gels has been investigated for several oxide systems (see e.g. Ramsay, 1986). Results for silica sols (Figure 1.10) illustrate the dependence of scattered neutron intensity $I(Q)$ on momentum transfer Q for a range of concentrations. At the lowest concentration there is a gradual decrease

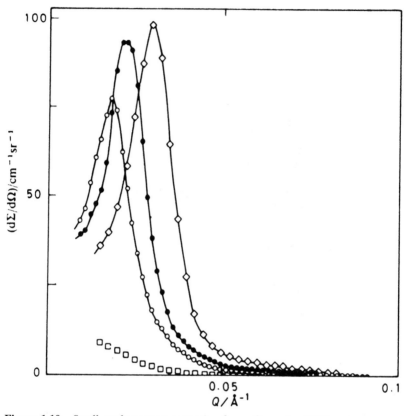

Figure 1.10   *Small angle neutron scattering from silica sols of different concentrations* c *(g cm^{-3}): (□) 0.014; (○) 0.14; (●) 0.27; (◇) 0.55*

in $I(Q)$, which has a form expected (i.e. eqn (1.10)) for discrete non-interacting particles having a diameter in close agreement with those determined from electron microscopy, i.e. about 16 nm.

The development of the maxima in $I(Q)$ at higher concentrations is caused by interference effects and indicates that the particles are not arranged at random but have some short-range ordering due to interparticle repulsion. Thus the movement of the maxima to higher values of Q with increasing sol concentration reflects a reduction in the equilibrium separation distance $r_{g(r)max}$ $(= r^*)$ between the particles. The form of $g(r)$, which defines the probability that the centres of a pair of particles will be separated by a distance r, is typified by results for silica sols (see Figure 1.11). An insight into the structural changes that occur in converting sols into gels can be obtained from the dependence of r^* on the sol concentration c.

In general, it is found (Ramsay and Booth, 1983; Ramsay, 1986a) that as c increases the interparticle separation decreases inversely as $c^{1/3}$, and as the solid gel is formed approaches that of the particle diameter $2R$.

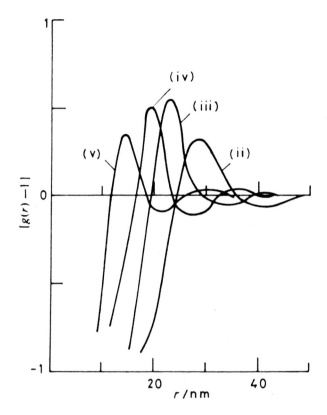

Figure 1.11   *Radial distribution functions g(r) for silical sols and gels of different concentrations (g cm⁻³): (ii) 0.16; (iii) 0.41; (iv) 0.65; (v) ca. 1.2. Particle diameter ca. 16 nm*

An indication of the arrangement of the particles in both sols and gel can also be obtained from the plots of $g(r)$ against r. Thus the average number of nearest neighbours Z is obtained from the relationship

$$Z = 8\pi n_{\mathrm{p}} \int_0^{r^*} r^2 g(r) \mathrm{d}r \qquad\qquad (1.12)$$

where r^* is again the position of the first maximum in $g(r)$. From such a procedure it has been shown that values of Z are, in general, close to 8. Such a large value implies that the particles are packed relatively efficiently when the gel is formed, as is depicted schematically in Figure 1.12. A dense particle packing is also demonstrated by the low porosities and pore sizes of the dehydrated gels, which will be discussed in more detail later. These are in close

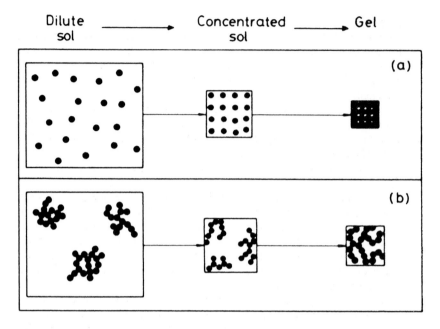

Figure 1.12   *Diagram depicting the formation of gels of low and high porosity from* (a) *unaggregated and* (b) *aggregated sols*

accord with a random close-packed structure having a porosity ϵ of approximately 0.36 (see Table 1.3).

A direct insight into the interparticle interaction behaviour can also be obtained from the form of $S(Q)$ for sols containing discrete particles. This aspect has recently become a topic of wide interest for a range of concentrated colloidal dispersions, such as polymer latex particles (Cebula *et al.*, 1983), surfactant micelles (Hayter and Penfold) and oxide sol (Penfold and Ramsay, 1985)

Table 1.3   *Porosities of regular and random packings of uniform spheres*

Type	Z	ϵ
Hexagonal	12	0.260
Body centred cubic	8	0.320
Simple cubic	6	0.476
Tetrahedral	4	0.660
Trihedral	3	0.815
Random packing	~7.5	~0.36

systems. Thus a knowledge of multibody interactions at a microscopic level can lead to a more confident prediction and control of many macroscopic properties, such as the rheology and stability of dispersions. Here the behaviour of such dispersions has been modelled as a one component fluid of colloidal particles using established liquid-state models that involve either computer simulation or, more frequently, integral equation methods. Using the latter approach, the behaviour of oxide sol systems has been modelled using the hard-sphere potential and, more recently, a screened Coulombic potential using the solution of the rescaled mean spherical approximation (MSA), as developed by Hayter and coworkers (Hansen and Hayter, 1982).

Thus it has been shown (Penfold and Ramsay, 1985) that the experimental $S(Q)$ for sols of different concentration can be quite closely simulated to that of an assembly of hard spheres, in which σ, the hard-sphere diameter, is dependent on the concentration of both the sol and electrolyte. Thus, for dilute sols of low electrolyte concentration, σ, is considerably larger than the real particle diameter, but as the sol concentration is increased during the conversion into the gel σ progressively decreases. Furthermore, the effective volume fraction ϕ_{HS} of the more dilute sols is relatively large (about 0.3) and does not increase greatly as the sol concentration is increased.

Such microscopic properties involving the interparticle interactions and structure are in accord with the bulk behaviour of sols containing discrete particles and contrast with those containing particle aggregates. A notable difference arises in the rheological properties, during the sol-to-gel conversion. Here discrete particle systems are predominantly viscous, showing a progressive increase in viscosity with increase in volume fraction, which only begins to rise markedly when ϕ exceeds about 0.5, in accord with the predicted behaviour of hard-sphere systems. This behaviour contrasts markedly with that observed for aggregated particle systems, as will be described.

Aggregated particle systems – fractal structures

Up to now we have considered the application of SANS in the study of structure and interactions in sol systems containing discrete, almost spherical particles which are nearly monodispersed. Another more complex system occurs where the sol is composed of aggregates of primary particles. A colloidal dispersion of this type can be prepared either by aggregating discrete sol particles (Ramsay, 1978) or by dispersing pyrogenic oxide (e.g. silica or alumina) powders in water (Ramsay and Scanlon, 1986); this is depicted schematically in Figure 1.9b.

Dispersions of this type are typical of a wide range of colloid systems occurring as flocs and particle clusters, which are important in a variety of natural and commercial processes. The importance of such systems has indeed stimulated considerable efforts to define floc structure, particularly in terms of computer simulations which model flocculation processes (Vold, 1963; Sutherland, 1967; Witten and Sander, 1983; Meakin, 1983). The assessment of these models has, however, been difficult, due mainly to a lack of suitable techniques which can probe floc structure. It has, however, been demonstrated recently that the properties of such aggregates can be determined from scattering measurements and thence defined in terms of fractal structure.

This approach for defining the structure of particle aggregates is based on the concept of the fractal dimension D (Mandelbrot, 1977) and has been developed theoretically using computer simulations of aggregate formation (Meakin, 1983; Witten and Sander, 1983; Kolb et al., 1983). Aggregates so formed have the property of self-similarity. That is, the gross structure of the aggregate is the same on length scales greater than that of the range of interaction between the individual particles within the aggregate.

The Hausdorff or fractal dimension D of an object can be defined as

$$N(r) = N_0 r^D \tag{1.13}$$

where $N(r)$ is the mass contained in a radius r about any point in the object, and N_0 is a constant. For the case of a solid mass, for example, $D = 3$ and $N_0 = 4\pi/3$. Furthermore, for this simple case, D and the Euclidean dimension d of free space will be identical (i.e. $D = d$). However, for fractal objects, such as aggregates of spherical particles, $D < d$. For these, the density–density correlation function is scale invariant and obeys the relationship

$$<\rho(r' + r)\rho(r')> \approx r^{-A} \tag{1.14}$$

where the exponent A is given by

$$D = d - A \tag{1.15}$$

This non-integral dimensionality d thus implies that the average density of particles in an aggregate will decrease as the volume sampled is increased. In effect it can be shown that the average number of particles $N(a)$ within a volume of radius a, say, will vary as a^D, which is expressed as

$$a^D \approx N(a) = \int_0^a d^3r < \rho(r)\rho(0) >/< \rho_0 > \tag{1.16}$$

This expression defines the fractal dimension D in terms of the ensemble average density function given by

$$g(r) = <\rho(r')\rho(r' + r)/\rho(r) \tag{1.17}$$

where $\rho(r')$ is the density at position r'.

It also follows that

$$g(r) \approx r^{(D-d)} \tag{1.18}$$

Furthermore, it can be shown that the scattered amplitude (or intensity $I(Q)$) is given by the general expression

$$I(Q) \approx 4\pi r^2 g(r) \frac{\sin (Qr)}{Qr} d^3r \tag{1.19}$$

Such that $I(Q)$ and $g(r)$ are related by a Fourier transform.

It thus follows that $I(Q)$ will also obey a corresponding power law relation, i.e.

$$I(Q) \approx Q^{-D} \tag{1.20}$$

Evidently, eqn (1.20) will only apply for a certain range of length r, which for aggregates will be approximately between the diameter of the individual particles and the overall size of the aggregate. Correspondingly, the fractal power law observed from scattering measurements (i.e. eqn (1.20)) will hold for a range of Q, which is determined in reciprocal space by these limiting values of r. For small r, comparable with the primary particle size (i.e. high Q), the power law expected would then be dominated by surface fractal structure. Thus for scattering samples composed of two components (such as aggregates dispersed in a liquid, and porous solids) which are separated by a smooth interface it can be shown that (Bale and Schmidt, 1983)

$$I(Q) \approx S \cdot Q^{-(6-D)} \tag{1.21}$$

where S is the surface area of the interface and D is the surface fractal dimension of 2. This leads to the familiar Porod law relation (Porod, 1951)

$$I(Q) \approx S \cdot Q^{-4} \tag{1.22}$$

In extreme cases (Kjems and Schofield, 1985), however, where the surface is irregular or has curvature on a scale smaller than reciprocal space Q, D may become greater than 2. This can lead to power-law exponents in eqn (1.22) which are smaller than –4. Such a situation may arise with very small colloidal particles (diameter $<100\,\text{Å}$), as has been demonstrated. A schematic representation of a particle aggregate having range of self-similarity between approximately a_1 and a_2 and the form of the scattering expected is as depicted in Figure 1.13.

Experimental measurements of scattering from colloidal aggregates or clusters of particles in aqueous dispersions may thus in principle provide an insight into the uniformity, size and openness of their structure. The latter feature is implicit in eqn (1.16) which leads to the relationship between the radius of gyration R_g of a cluster containing N_c particles:

$$N_c \approx R_g^D \tag{1.23}$$

The processes of aggregation and gelation have been modelled extensively by computer simulation. These models are defined by the laws governing the kinetics of diffusion of the elementary particles or subunits of the particle clusters on the one hand, and also by the probability of aggregation or 'sticking' resulting from particle–particle or particle–cluster contact. The latter is determined by the particle interaction potential. If this is repulsive this will lead to a sticking probability less than 1 (i.e. reaction limited).

As an explanation, in Figure 1.14 six models of growth are illustrated. These differ in terms of the mechanism which controls growth (reaction limited or self-diffusion) of colloidal species, and the nature of the interacting colloidal species (single particle–cluster and cluster–cluster). These different mechanisms result in different values of D, as illustrated. We note that in the reaction-limited case

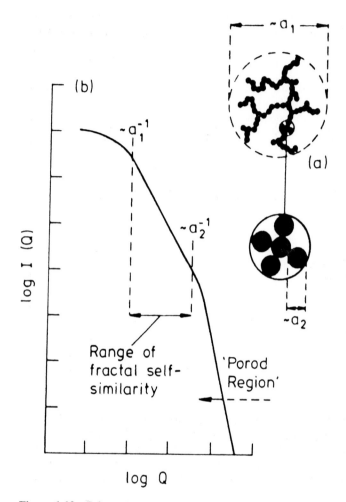

Figure 1.13 *Schematic representation of a particle aggregate* (a) *having a range of self-similarity between approximately* a_1 *and* a_2. *The form of the scattering is depicted in* (b)

the values of D will be smaller and the aggregates formed will have a more compact structure. Two idealized models have been considered extensively; these involve the diffusion-limited aggregation (DLA) of single particles, where Meakin (1983) has derived a value of D of 2.5, and that of the aggregation of clusters of comparable size (DLCA) where a dimensionality of about 1.80 is obtained (Kolb *et al.*, 1983).

A combination of light scattering and SANS has been used to investigate colloidal dispersions of pyrogenic silica and alumina (Ramsay and Scanlon, 1986). The scattering behaviour of different concentrations of pyrogenic silica

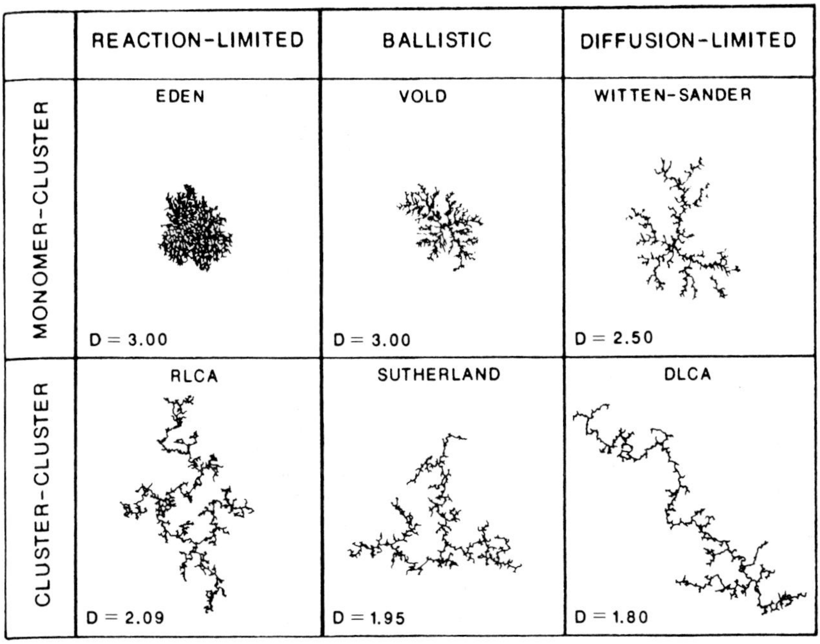

	REACTION-LIMITED	BALLISTIC	DIFFUSION-LIMITED
MONOMER–CLUSTER	EDEN D = 3.00	VOLD D = 3.00	WITTEN–SANDER D = 2.50
CLUSTER–CLUSTER	RLCA D = 2.09	SUTHERLAND D = 1.95	DLCA D = 1.80

Figure 1.14 *Simulated structures resulting from various kinetic growth models. Fractal dimensions are listed for three-dimensional clusters (after Brinker and Scherer (1990))*

(Aerosil 200) is illustrated in Figure 1.15 (i) and (ii). Above a $Q \approx 2.5 \times 10^{-2}\,\text{Å}^{-1}$ it will be noted that $I(Q)$ decays with a power law of $Q^{-3.9}$, which corresponds closely to that of the Porod law. This demonstrates that the scattering behaviour is dominated by the total surface area of the particles in the dispersion, when Q begins to exceed the inverse size of the primary particles, which here is about 10 nm.

Light scattering results obtained at lower Q are also shown in Figure 1.15 for several dispersion concentrations of the same aggregated silica. On account of multiple scattering and attenuation effects these measurements are restricted to considerably more dilute dispersions. Note that the results all show a power law increase in $I(Q)$ with an exponent between −1.7 and −1.8, which is in accord with the limiting behaviour of $I(Q)$ at low Q observed in SANS.

From similar studies made with other Aerosil silica dispersions (Ramsay and Scanlon, 1986) it has been shown that the fractal dimensionality falls within the approximate range 1.6–2.0. This suggests that the aggregates have a relatively open structure, which is common to all the grades of Aerosil which have been studied. Furthermore, the value of D is similar to that predicted (about 1.8) by computer simulation for the process of DLCA, where clusters are formed by homogeneous aggregation of a collection of particles: these small clusters subsequently diffuse and stick together to give larger aggregates.

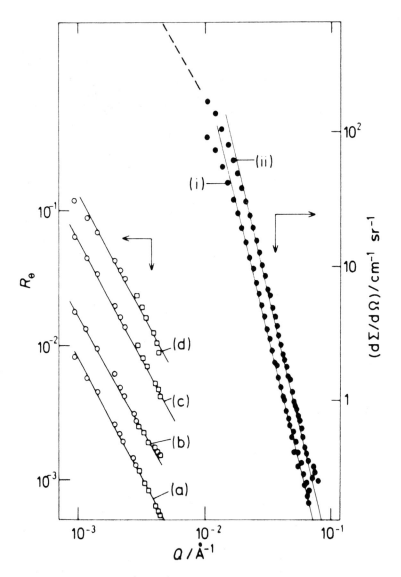

Figure 1.15 *Static light scattering and small angle neutron scattering (SANS) of colloidal dispersions of silica particle aggregates of different concentrations. Concentrations for light scattering (g cm^{-3}): (a) 3.4 × 10^{-4}; (b) 6.8 × 10^{-4}; (c) 1.7 × 10^{-3}; (d) 3.4 × 10^{-4}. Measurements were made using wavelenghts of (○) 546 nm and (□) 365 nm. Concentrations (in D$_2$O) for SANS (g cm^{-3}): (i) 0.10; (ii) 0.23. Slope: (—) ca. –3.9; (– – –) –1.7*

1.4 The sol-to-gel conversion

1.4.1 Gelation and percolation processes

We will now consider the mechanism of gelation of sols that occurs on concentration, with particular reference to systems containing discrete particles or particle aggregates. These two systems show marked differences in the porous properties of the xerogels formed. For discrete-particle systems such as silica, where there is strong interparticle repulsion, the porosity ϵ obtained is relatively low (typically about 0.36). The effective pore size of the xerogels scales with the size of the primary particles as demonstrated in Figure 1.16, and the specific surface area decreases inversely. For gels produced by cluster aggregation, shrinkage of the network is constrained, as will be discussed, and residual gel porosities are much higher, i.e. $\epsilon > 0.7$ (Ramsay, 1989). Correspondingly, the effective pore size is much larger for an equivalent primary particle size as illustrated in Figures 1.17 and 1.18. The nature of the forces resisting the deformation of the elastic network can be inferred from measurements of bulk elastic moduli, and depend on the structure of the particle network (see e.g. Clerq *et al.*, 1983; Brown and Ball, 1985; van Voorst Vader and Groenweg, 1989).

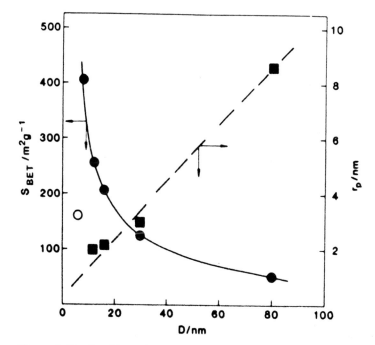

Figure 1.16   *Specific surface areas* S_{BET} *and mean pore radii* r_p *of silica and ceria gels derived from non-aggregated sols having particle diameters* D. (●) S_{BET} *of silica;* (○) S_{BET} *of ceria* (■) r_p *of silica*

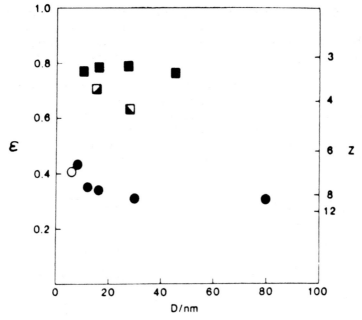

Figure 1.17 *Porosities ε of oxide gels derived from sols having different primary particle sizes D. Non-aggregated sols of: (●) silica; (○) ceria. Aggregated sols of: (■) silica; (◪) alumina; (◪) titania. Z is the coordination number of sphere packing having the corresponding porosity*

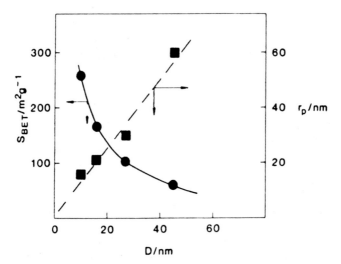

Figure 1.18 *Specific areas S_{BET} and mean pore radii r_p of silica gels derived from aggregated sols having particle diameters D. (●) S_{BET}; (■) r_p*

Recently, stochastic models have been developed to simulate the formation of such gel networks. Some of these are illustrated in Figure 1.19. In percolation networks, the available space is filled with a grid, and network elements are placed on sites of this grid in a random manner up to a certain volume fraction (Clerq *et al.*, 1983).

Another approach is based on the diffusional contact between a large number of initial clusters, i.e. cluster–cluster aggregation (CCA). Such initial clusters may form either by diffusion limited aggregation (DLCA) or reaction limited aggregation (RLCA) mechanisms. For the latter there is an energy barrier resisting particle contact which results in denser flocs, with a lower fractal dimension (see Figure 1.14). Gels formed by such statistical processes have

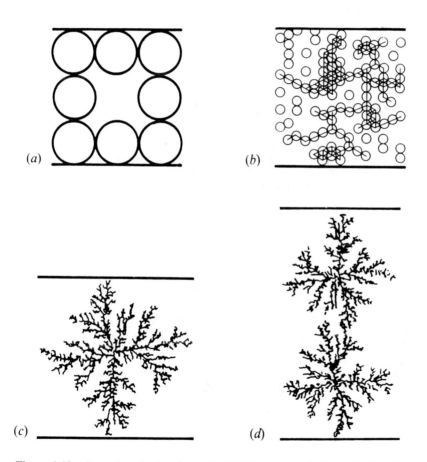

Figure 1.19 *Examples of network models. (a) Linear network. (b–d) Stochastic networks: (b) Percolation network – infinite aggregates exist only if $\phi > \phi_c$ (where ϕ_c is the percolation limit); (c) diffusion-limited aggregate or reaction-limited aggregate; (d) cluster – cluster aggregate (after van Voorst Vader and Groeneweg (1989))*

fractal properties, and hence show symmetry on dilatation. This implies that physical properties can be described by exponential laws because they must preserve their form under a change of scale. This approach has been applied to describe the storage moduli of such flocculated networks. External forces exerted on such a network are resisted by its 'backbone', i.e. the shortest continuous path across the network between opposite sides (Figure 1.20). For percolation

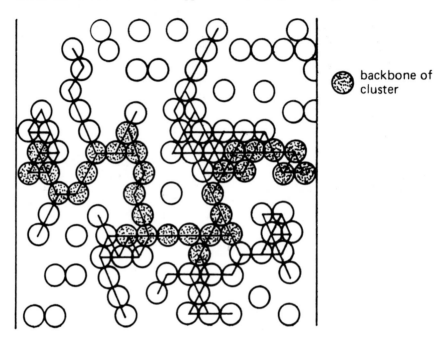

backbone of cluster

Figure 1.20   *Backbone of a stochastic cluster (after van Voorst Vader and Groeneweg (1989))*

clusters, the volume fraction of filled sites required to form an infinite network ϕ_c which spans the whole volume depends only on the assumed grid geometry; ϕ_c is defined as the percolation threshold. The following scaling equation for the storage modulus G' can be derived (Kantor and Webman, 1984) for such a percolation network:

$$G'(\phi) = G(\phi - \phi_c)^{\tau_P} \tag{1.24}$$

where G'_0 is a scaling constant. Close to the percolation threshold, the theoretical value of the exponent τ_p is 3.55 and tends to unity at the limit of closest packing. For such a situation, $\phi_c > 0.144$. For gels formed by CCA the following scaling equation is obeyed:

$$G'(\phi) = G'_0\phi^{\tau} \tag{1.25}$$

where $\tau = 3.5$ for DLCA and 4.5 for RLCA.

Thus the theoretical relationship between G' and ϕ for a gel network depends on the stochastic process chosen to generate the network. Typically ϕ_c is in the range 0–0.15 and τ is in the range 1–4.5.

This theoretical description is also in accord with experimental measurements, as illustrated in Figure 1.21 by the relationship between G' and ϕ during the conversion of an aggregated alumina sol (Degussa C) to a rigid gel. Here the exponent τ is approximately 4.0 and ϕ_c, obtained from extrapolation, is approximately 0.03 (≈ 0.1 g ml^{-1}). Recent SANS measurements (Ramsay, 1988) on this system have been extended to much lower Q, thus allowing the

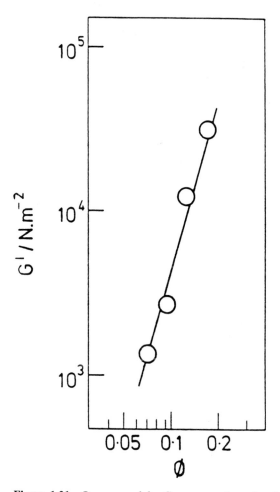

Figure 1.21   *Storage modulus G$_l$ versus volume fraction ϕ for colloidal dispersions of pyrogenic alumina*

investigation of concentrated dispersions and gels (Figure 1.22). Such measurements are important because they give insight into the interactions and interpercolation between aggregates. Here the size of the primary particles of the alumina is about 15 nm. The scattering from the dry powder (Figure 1.22a) is typical of a mass fractal with $D = 1.76$, which corresponds to a (CCA) process of formation. For $Q > 3 \times 10^{-2}$ Å^{-1}, the power law shown is –4. Results in Figure 1.22b and 1.22c correspond to concentrated dispersions in water of 0.2 and 0.5 (w/w) Al_2O_3 and H_2O, respectively, where as noted above the more concentrated dispersion has marked viscoelastic properties. Compared with the SANS of the powder and dilute dispersions, these show striking changes by the appearance of interference features at about 7×10^{-3} and 2×10^{-2} Å^{-1} (note the shifts in abscissae). These features can be ascribed to interparticle correlations arising

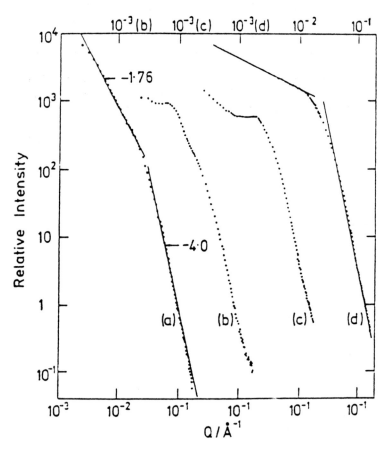

Figure 1.22 *Small angle neutron scattering for* (a) *pyrogenic alumina powder,* (b, c) *dispersions in water (0.2 and 0.5 (w/w) Al_2O_3/H_2O) and* (d) *the resultant dehydrated gel*

from interpercolation of aggregates. Such interference effects are marked with the alumina system here, due to strong interparticle repulsion arising from a high surface charge at the low pH (<4) of the system. Such interference features begin to become apparent at a concentration close to the percolation threshold where separate aggregates begin to interact and form a continuous network. This behaviour is linked to the dehydration and shrinkage behaviour of the gels and favours the formation of crack-free xerogel structures.

1.4.2 Drying of gels

The foregoing analysis is important for understanding the porous texture formed in the dehydrated xerogel and its mechanical properties. Indeed, considerable advances have been made recently in the theoretical understanding of drying processes, particularly by Scherer and coworkers (see e.g. Brinker and Scherer, 1990). This aspect is very important for obtaining crack-free coherent gel films and also porous particles. We will only briefly indicate some aspects of this topic here.

We can represent the evaporation of a liquid from a porous medium as shown in Figure 1.23. Here a meniscus of radius r forms at the liquid–vapour interface at the surface of the body. The tension P in the liquid is related to r by

$$P = -2\gamma_{LV}/r \tag{1.26}$$

where γ_{LV} is the interfacial energy.

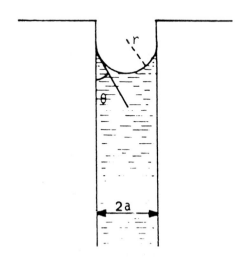

Figure 1.23 *Meniscus formation at a liquid – vapour interface in a porous medium for the case of a wetting liquid*

The capillary pressure reaches a maximum P_R when the radius of the meniscus becomes equal to the pore radius. For a cylindrical pore of radius a, r is given by

$$r = -a/\cos \theta \qquad (1.27)$$

where θ is the contact angle. This gives the familiar Laplace relationship

$$P_R = 2\gamma_{LV} \cos \theta/a \qquad (1.28)$$

The capillary pressure (or tension) in the liquid is supported by the solid phase which is then under compression. This compressive stress causes the gel to contract. The solid structure becomes increasingly rigid during this process and contraction continues until a critical value is reached which can resist the capillary tension P_R. For gels containing very small pores the capillary tension can reach very large values, of the order of hundreds of megapascals. This analysis will not be detailed further here, although it can be shown that uniform and slow evaporation of liquid from the porous structure is important to avoid solid fragmentation. This is particularly important with large microporous bodies which have a very low permeability.

Several technical approaches have been applied to minimize fragmentation. These include:

(a) slow drying
(b) control of porosity
(c) reduction of the solid/liquid interfacial tension, γ_{LV}
(d) enhancement of the solid rigidity.

For example, the displacement of water by partially miscible solvents, such as long-chain alcohols (butanol and hexanol) can have an effect in (c). This is illustrated in Figure 1.24 by the isotherms of spherical xerogel particles of zirconia produced by gel precipitation (Ramsay et al., 1991). In Figure 1.24a where the xerogel has been dried in air the particles have contracted to give a microporous structure of low porosity, whereas in Figure 1.24b where the water has been exchanged with butanol (which has subsequently been evaporated) the porosity is very high and the corresponding pore size is considerably larger (Table 1.4). It should be noted that the microporous structure is more prone to cracking on subsequent calcination and sintering.

Table 1.4 *Surface and porous properties of zirconia gels dehydrated by different routes from nitrogen adsorption isotherms*

Drying method	S_{BET} $(m^2\ g^{-1})$	Mean pore radius (nm)	Pore volume $(cm^3\ g^{-1})$
Air dried	130	$\leqslant 2$	0.09
Solvent displacement dried	320	22	1.76

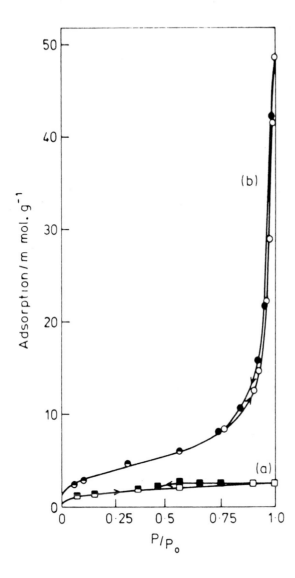

Figure 1.24   *Nitrogen adsorption isotherms at 77 K for zirconia gels. (a) Gel dehydrated by evaporation in air. (b) Gel dried after water displacement with butanol. Open and filled symbols represent adsorption and desorption, respectively. Note the much greater uptake for (b), showing that the gel structure has contracted much less and therefore contains a higher pore volume*

Finally, perhaps mention should be made of the method of supercritical drying of gels, first developed by Kistler (1932). The principle of this method is illustrated in Figure 1.25, which shows the pressure–temperature relationship for a fluid, extending beyond the critical state. If a wet gel is heated under pressure in an autoclave to condition B where the liquid state no longer exists and

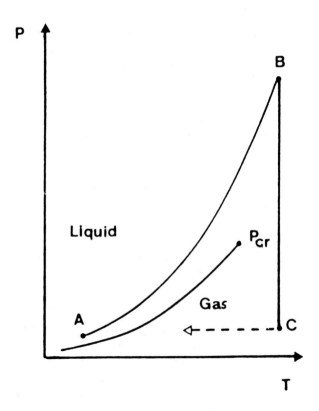

Figure 1.25 *Pressure–temperature diagram depicting the development of pressure and temperature during drying by evaporation of a solvent under supercritical conditions. Note that the curve ABC is traced; drying occurs suddenly in the transition B→C*

suddenly evacuated to condition C, an extremely porous structure termed an 'aerogel' is formed. In the process of evacuation there is no compressive stress due to the liquid–vapour tension. This process has been applied extensively in the production of porous catalysts by Pajonk (1991) and developed for other applications, notably by Fricke (1986) (see also Vacher *et al.*, 1989).

1.5 Formation of gel particles

Until now, emphasis has been given to aspects of the sol-to-gel conversion process which are important in controlling the microstructure and texture of the gel. As illustrated, the structure and interactions between sol particles can have a profound effect on the surface and porous properties of xerogels. Another important feature of sol–gel processes is the range of methods available for converting the sol to the gel and, thereby, controlling the bulk form of the xerogel, and subsequently the size and shape of ceramic monoliths. In this respect considerable emphasis has been given to the production of particles using methods which involve droplet formation. We will outline some of these here, particularly because they relate to the context of the present volume.

Methods for producing droplets include spray-drying techniques, emulsion-drying processes and others where precipitation occurs within a liquid droplet produced from a nozzle or spinnarette device. Each of these methods has its particular advantage and application, as summarized in Table 1.5.

Table 1.5 *Methods of droplet formation used in sol–gel processes*

Method	Size range	Comments
Spray drying	≈ 2–$50\,\mu m$	Size control depends on spray formation parameters and sol rheology (particles polydispersed)
Nozzle (gravity)	≈ 0.5 to 2–$3\,mm$	Monodispersed particles. Size depends on orifice diameter and sol rheology
Nozzle (spinarette – high-frequency vibration)	$\approx 50\,\mu m$ to $0.5\,mm$	Monosized particles. Control depends on orifice diameter and vibration frequency
Water/oil emulsion	$\geqslant 0.5\,\mu m$ to $\leqslant 30\,\mu m$	Good size control. Depends on water/oil interfacial tension, stirring rate and surfactant addition

In general, the spray drying and atomization of sols is suitable for producing spherical gel particles with a size in the range from about 3 μm to greater than 50 μm. The process is normally optimized to produce particles with the desired size, although in general the size distribution is relatively broad. The properties of the sol (concentration and rheology) are important, as are the dryer operation conditions (rotor velocity, nozzle characteristics, temperature, humidity etc.). The method is cheap, but may yield non-uniform particles (e.g. hollow or deformed spheres). The lower size limit is determined by the surface tension of the liquid. The spray-drying method has also been adapted to produce thin (approximately 10 μm fibres; here sols of high viscosity and low viscoelastic characteristics are required. Gel particles have also been formed using emulsion-drying techniques; here aqueous sol droplets are formed in a partially miscible solvent (e.g. trichlorethylene). Water is progressively removed from the sol by transfer to the organic phase, and rigid gel particles eventually form. The size of the emulsion droplets can be controlled using appropriate surfactant additions to the sol and by the stirring conditions used to disperse the aqueous phase in the organic solvent. This method generally yields highly uniform particles which can have a size in the range >1 μm to about 30 μm. In both of the above methods, narrower size distributions can be achieved by subsequent centrifugal classification.

Larger spherical particles (typically in the range >100 μm to <5 mm) can be prepared by gelation of individual droplets formed by flowing sols from an orifice into another liquid medium. These methods usually involve the injection of droplets into the top of a column and during the descent dehydration or precipitation occurs. Such techniques involving water extraction into a counter current flow of 2-ethylhexanol, for example, have been described (Matthews and Swanson, 1979). Usually surfactants are added to the organic phase to prevent coalescence and clustering of gel spheres. In gel precipitation (see Figure 1.1) droplets of an aqueous solution of metal ions are injected into a column containing an ammonia solution. In the preparation of larger spheres the droplets are allowed to fall initially through an emulsion of ammonia gas. Here a protective skin forms around the droplets, thus preventing particle deformation. To obtain smaller particle sizes, droplets are injected from vibrating spinnarettes, usually with a multihead device.

Other methods have been developed where sol droplets are injected into hot paraffin to induce gelation (van der Grift *et al.*, 1991). A somewhat similar procedure was used in the internal gelation process where the alkali precipitant was derived from the hydrothermal decomposition of tetramine, contained in the aqueous droplets dispersed in hot oil. The size of the particles formed in such processes is generally very uniform and controllable.

In these methods involving droplet formation there are many variables which control particle size, in particular the diameter of the orifice. Thus the size of the sphere formed, when a drop of one fluid (sol) flows through an orifice into another immiscible fluid, depends on the balance between the falling and restraining forces. When the forces acting to separate the drop from the orifice (i.e. gravity, due to the density difference between the liquids, and kinetic effects resulting from fluid flowing from the orifice) exceed the forces acting to keep the drop in the orifice (interfacial tension and the drag force of the suspended medium), the drop breaks away. Harkins and Brown (1919) equated buoyancy and interfacial tension, and derived the following expression for the drop volume of static bubbles at low flow rates:

$$V_s = \frac{F\pi\gamma D}{g\Delta\rho} \tag{1.29}$$

where V_s is the volume of the drop, γ is the interfacial tension, D_o is the diameter of the orifice, $\Delta\rho$ is the density difference between the two fluids, g is the acceleration due to gravity, and *F* is a correction factor for the portion of the drop that stays in the orifice.

This equation can be used to predict conditions for controlling sphere size. In general, larger spheres are obtained by increasing D_o. The sol density and interfacial tension between the two liquids are also other important variables. Thus increases in γ yield larger droplets. If the kinetic and drag forces are considered, it can be shown that the drop size increases with increasing flow rate of the fluid through the orifice and increasing viscosity of the continuous phase. The rheological properties (viscosity and viscoelasticity) of the liquid flowing from the jet are also very important in determining the size, uniformity and break-up of the droplets. Such properties are generally complex, since concentrated colloidal dispersions and liquids containing dissolved polymers, which are frequently employed in these processes, exhibit non-Newtonian behaviour.

1.6 Applications

Although it is not within the scope of this chapter to detail the wide technological potential of sol–gel processes, mention should be made of some applications where an awareness of the fundamental basis of the process as described here has been exploited.

Here a distinction can be made between applications requiring porous and non-porous materials. The former includes oxide adsorbents, and catalyst supports and the latter various dense ceramic artefacts. In some of these the ability to produce spherical particles with high uniformity and controlled size has been a particular advantage (Harata and Delmon, 1986). This is particularly the case in the production of chromatographic adsorbents for high pressure liquid chromatography (HPLC), for example, where particles with a diameter in the approximate range 3–20 μm are used as column packings. Furthermore, in this application the porous texture and surface properties are of inherent importance in a range of separation processes (Ramsay, 1978).

In the case of densified ceramics, there are frequently requirements for small free-flowing powders for plasma spraying and other applications. Furthermore, as mentioned above, the production of larger ceramic reactor fuel particles has been another development in the nuclear industry. Here the fuel is generally a mixed actinide oxide (Pu/UO_2) ceramic which may be produced as uniform spherical particles in the range of millimetres for vibrocompaction into a metal tube or fuel pin. The flow and packing of such material is critically dependent on the size and uniformity of the particles, and the performance is controlled by the density and ceramic microstructure (Matthews and Swanson, 1979).

1.7 Conclusion

In this chapter an outline of the basis of some sol–gel processes has been described. Particular emphasis has been given to some of the fundamental aspects which involve the nucleation, controlled growth and association of colloidal particles in sols. These processes are directly related to the microstructure of the gel formed which, as has been shown, requires control in many technological applications. We have also indicated how a control of some of the interfacial processes involved in the drying and formation of xerogels, particularly in droplet form, is crucial to many of these applications.

These developments have arisen largely from advances in understanding in several areas of colloid and interface science and have benefited from the application of recent advanced physicochemical techniques, which have been used in research investigations on sol–gel processes.

References

Akitt, J. W., Greenwood, N. N., Khandelwal, B. L. and Lester, G. D. (1972) *J. Chem. Soc., Dalton Trans.*, 604

Baes, C. F. and Messmer, R. E. (1976) *The Hydrolysis of Cations*, Wiley Interscience, New York

Bale, H. D. and Schmidt, P. W. (1983) *Phys. Rev. Lett.*, **53**, 596

Bechtold, M. F. and Synder, O. E. (1975) *U.S. Patent 2574902*

Booth, B. O. and Ramsay, J. D. F. (1985) in *Principles and Applications of Pore Structural Characterization* (ed. Haynes, J. M. and Rossi-Doria, P.), J. W. Arrowsmith, Bristol, 97

Bottero, J. Y., Tchoubar, D., Cases, J. M. and Flessinger, F. (1982) *J. Phys. Chem.*, **86**, 3367

Brace, R. and Matijević, E. (1973) *J. Inorg. Nucl. Chem.*, **35**, 3691

Brinker, C. J. and Scherer, G. W. (1990) *Sol–gel Science*, Academic Press, New York, 453

Brinker, C. J., Clark, D. E. and Ulrich, D. R. (eds) (1988) *Better Ceramics Through Chemistry*, Vol. III, Materials Research Society, Pittsburgh, PA

Brown, W. D. and Ball, R. C. (1985) *J. Phys. A.*, **18**, L517

Cebula, D. J., Goodwin, J. W., Jeffrey, G. C., Ottewill, R. H., Parentich, A. and Richardson, R. A. (1983) *Faraday Discuss. Chem. Soc.*, **71**, 71

Clerq, J. P., Giraud, G., Rousseuq, J., Blanc, R., Carton, J. P., Guyon, E., Ottavi, H. and Stauffer, D. (1983) *Ann. Phys. (Paris)*, **8**, 5

Dell, R. M. (1972) New preparative processes for mixed oxide and alloy powders, in *Proceedings of the 7th International Symposium on the Reactivity of Solids* (ed. Anderson, J. S.) Chapman & Hall, London, 553

Demchak, R. and Matijević, E. (1969) *J. Colloid Interface Sci.*, **31**, 257

Engelhardt, G. and Michel, D. (1987) *High Resolution Solid-State NMR of Silicates and Zeolites*, Wiley, New York, 75

Fricke, J. (ed.) (1986) *Aerogels*, Springer-Verlag, Berlin

Goodwin, J. W. and Ottewill, R. H. (1991) *J. Chem. Soc., Faraday Trans.*, **87**, 357

Goodwin, J. W. *et al.* (1982) in *Colloidal Dispersions (Special Publication No. 43)* Royal Society of Chemistry, London

Guinier, A. and Fournet, G. (1955) *Small Angle Scattering of X-rays*, Wiley, New York

Guizard, C., Larbot, A., Cot, L., Perez, S. and Rouviere, J. (1990) *J. Chim. Phys.*, **87**, 1901

Hansen, J. P. and Hayter, J. B. (1982) *Mol. Phys.*, **46**, 651

Harata, M. and Delmon, B. (1986) *J. de Chim. Phys.*, **83**, 859

Harkins, W. D. and Brown, F. E. (1919) *J. Am. Chem. Soc.*, **41**, 499

Hayter, J. B. and Penfold, J. (1851) *J. Chem. Soc.*, Faraday Trans, 1981, **77**

Iler, R. K. (1979) *The Chemistry of Silica*, Wiley, New York, 312

James, R. O. and Parks, G. A. (1980) in *Surface and Colloid Science*, Vol. II (ed. Matijević, E.) Wiley Interscience, New York

Kantor, Y. and Webman, I. (1984) *Phys. Rev. Lett.*, **52**, 1891

Kistler, S. S. (1932) *J. Phys. Chem.*, **36**, 52

Kjems, J. K. and Schofield, P. (1985) in *Scaling Phenomena in Disordered Systems* (ed. Pynn, R. and Skjelthorpe, A., Plenum Press, New York, 141

Klein, L. C. (ed.) (1988) *Sol–gel Technology for Thin Films, Fibers, Preforms, Electronics, and Speciality Shapes*, Noyes, New Jersey

Kolb, M., Botet, R. and Julien, R. (1983) *Phys. Rev. Lett.*, **51**, 1123

Kostorz, G. (1979) in *A Treatise on Materials Science and Technology* (ed. Herman, H.) Academic Press, New York, 227

LaMer, V. K. and Barnes, M. D. (1946) *J. Colloid Sci.*, **1**, 71

LaMer, V. K. and Dinegar, R. H. (1950) *J. Am. Chem. Soc.*, **72**, 4847

Livage, J., Henry, M. and Sanchez, C. (1988) Sol–gel chemistry of transition metal oxides. *Prog. Solid State Chem.*, **18**, 259

Lyklema, J. (1971) *Croat. Chem. Acta*, **43**, 249

Mandelbrot, B. B. (1977) *Fractals, Forms, Chance and Dimension*, W.H. Freeman, San Francisco

Matijević, E. (1981) *Acc. Chem. Res.*, **14**, 22

Matthews, R. B. and Swanson, M.L. (1979) *Am. Ceram. Soc. Bull.*, **58**, 223

Meakin, P. (1983) *Phys. Rev. A.*, **27**, 1495

Ottewill, R. H. (1982) in *Colloidal Dispersions* (ed. Goodwin, J. W.) Royal Society of Chemistry, London, 143

Pajonk, G. M. (1991) *Appl. Catal.*, **72**, 217

Penfold, J. and Ramsay, J. D. F. (1985) *J. Chem. Soc., Faraday Trans. 1*, **81**, 117

Porod, G. (1951) *Kolloidn. Zh.*, **124**, 83

Ramsay, J.D.F. (1978) in *Chromatography of Synthetic and Biological Polymers*, Vol. I (ed. Epton, R.) Ellis Horwood, Chichester, 339

Ramsay, J. D. F. (1986a) *Chem. Soc. Rev.*, **15**, 335

Ramsay, J. D. F. (1986b) in *Water and Aqueous Solutions* (ed. Neilson, G. W. and Enderby, J. E.) Adam Hilger, Bristol, 207

Ramsay, J. D. F. (1988) *Mat. Res. Symp. Proc.*, **121**, 293

Ramsay, J. D. F. (1989) Synthesis of metal oxide adsorbents with controlled surface and porous properties, in *Proceedings of the 3rd International Conference on Fundamentals of Adsorption* (ed. Mersmann, A. B. and Scholl, S. E.), 701–714

Ramsay, J. D. F. (1990) in *Structure, Dynamics and Equilibrium Properties of Colloidal Systems* (ed. Bloor, D. M. and Wyn-Jones, E.) Kluwer, Dordrecht, 635–651

Ramsay, J. D. F. (1992) Characterisation of inorganic colloids. *Pure Appl. Chem.*, **64**, 1709–1713

Ramsay, J. D. F. and Avery, R. G. (1986) *Br. Ceram. Proc.*, **38**, 275

Ramsay, J. D. F. and Booth, B. O. (1983) *J. Chem. Soc., Faraday Trans. 1*, **79**, 173

Ramsay, J. D. F. and Scanlon, M. (1986) *Colloids and Surf.*, **18**, 207

Ramsay, J. D. F., Daish, S. R. and Wright, C. J. (1978) *Faraday Discuss. Chem. Soc.*, **65**, 65

Ramsay, J. D. F., Avery, R. G. and Benest, L. (1983) *Faraday Discuss. Chem. Soc.*, **76**, 53

Ramsay, J. D. F., Russell, P. J. and Swanton, S. W. (1991) in *Characterisation of Porous Solids* (ed. Reinoso, F. R., Rouquerol, J. and Sing, K. S. W.) Elsevier, Amsterdam, 257

Schwarzenback, G. and Wenger, H. (1969) *Helv. Chim. Acta*, **52**, 644

Stern, O. (1924) *Z. Electrochem.*, **30**, 508

Stöber, W., Fink, A. and Bohn, E. (1968) *J. Colloid Interface Sci.*, **26**, 62

Stringer, B., Russell, P. J., Davies, B.W. and Danso, K.A. (1984) *Radiochim. Acta*, **36**, 31

Stumm, W. and Morgan, J. J. (1981) *Aquatic Chemistry*, Wiley Interscience, New York

Sutherland, D. N. (1967) *J. Colloid Interface Sci.*, **25**, 373

Vacher, R., Phalippou, J., Pellous, J. and Woigner, T. (eds) (1989) *Rev. Phys. Appl.*, 24

van der Grift, Mulder, A. and Geus, J. W. (1991) *Colloids Surf.*, **53**, 223

van Voorst Vader, F. and Groeneweg, F. (1989) in *Food Colloids* (Special Publication No. 75) (ed. Bee, R. D., Richmond, P. and Mingins, J.) Royal Society of Chemistry, London, 218

Verwey, E. W. J. and Overbeek, J. Th. G. (1948) *Theory of Stability of Lyophobic Colloids*, Elsevier, Amsterdam
Vold, M. J. (1963) *J. Colloid Sci.*, **18**, 684
Waters, D. N. and Henty, M. S. (1977) *J. Chem. Soc., Dalton Trans.*, 1977
Witten, T. A. and Sander, L. M. (1981) *Phys. Rev. Lett.*, **47**, 1400
Witten, T. A. and Sander, L. M. (1983) *Phys. Rev. B.*, **27**, 5686
Woodhead, J. L. and Segal, D. L. (1984) *Chem. Br.*, **20**, 310
Zsigmondy, R. and Hückel, Z. (1925) *Z. Physik. Chem.*, **116**, 291

Chapter 2

Formation of monodisperse inorganic particulates

E. Matijević

2.1 Introduction

The secrets of the formation of finely dispersed matter consisting of particles uniform in size and shape have mystified scientists in general, and colloid chemists in particular, for well over a century. The challenge of synthesizing a countless number of identical tiny solids seemed overwhelming. Even Nature offers only a few examples of monodispersed systems, which have been discovered or recognized as such only quite recently. The best examples are opals, which are composed of perfect silica spheres. Figure 2.1 illustrates

4 μm

Figure 2.1 *Scanning electron micrograph of colloidal iron sulphide (pyrite) embedded in black shale (courtesy of Professor Neal R. O'Brien, State University of New York, Potsdam, New York)*

octahedral pyrite particles found in shale (O'Brien and Slatt, 1990). Some 15 years ago magnetotactic bacteria were identified (Blakemore, 1975), which were shown to contain uniform magnetite particles (Towe and Moench, 1981).

Several reports of synthetic monodispersed particulates can be found in older literature. Probably, the best known are the gold sols, first prepared by Faraday (1857). Other uniform dispersions of elements are those of sulphur (LaMer and Barnes, 1946) and selenium (Watillon *et al.*, 1958). Examples of inorganic compounds include barium sulphate (Andreasen, 1943; Miura *et al.*, 1956; Petres *et al.*, 1966) or lead iodate (Herak *et al.*, 1958). It is interesting to note that only very few successes had been recorded with the most important well-defined finely dispersed inorganics, i.e. metal (hydrous) oxides. Indeed, the only truly uniform of such synthetic materials were akageneite (β-FeOOH) (Watson *et al.*, 1960), and tungstic acid (Watson *et al.*, 1948), which were extensively studied for their optical properties. It should be also mentioned that most of the cited preparations were essentially empirical.

The situation has changed dramatically over the past 20 years or so, during which time several techniques have been developed, based on certain fundamental principles, which have given us a large variety of uniform inorganic particulates of diversified chemical composition (simple or composite), of various shapes, and of modal sizes ranging from a few nanometres to a few micrometres. The impetus for these activities came both from academic and applied areas. The monodispersed materials are essential if one wishes to relate optical, magnetic, electric, adsorptive and other properties to particle shape, size and structure, in addition to the chemistry of the solids. On the other hand, it has been recognized that uniform particles are useful in many applications, especially when stringent specifications and reproducibility are of importance, such as in pigments, recording materials, catalysis, medical diagnostics, and various aspects of ceramics.

Several recent reviews (Matijević, 1985, 1993; Haruta and Delmon, 1986; Sugimoto, 1987) deal extensively with the preparation of monodispersed solids. Another series of articles describing various aspects of the subject have been published in two special issues of the *MRS Bulletin* (December 1989 and January 1990).

It is not possible to review all techniques presently developed for the synthesis of uniform powders. Instead, this chapter deals essentially with the precipitation in homogeneous solutions, with some references to phase transformation of precipitated solids, and to particle formation in surfactant solutions.

2.2 Precipitation from homogeneous solutions

The thermodynamic condition required to produce a solid phase in a homogeneous solution is to exceed the solubility product of the constituent species. As a rule, the precipitates so obtained are morphologically ill defined, unless the kinetics of the process involved is properly controlled. The guiding principle for the formation of uniform particles was first expressed by LaMer (1952), according to which a short-lived single nucleation stage should be followed by the diffusional growth of the so-generated initial solids to larger, identical particles. Although it is now generally recognized that this idealized mechanism applies only in a limited number of cases, the concept has given an impetus to extensive studies and many successful syntheses of monodispersed colloids. Considerable evidence is now available documenting the fact that the mechanisms of the generation of uniform particles can be much more involved, e.g. in many cases a huge number of tiny primary particles aggregate into monosized products (Hsu *et al.*, 1988; Look *et al.*, 1990). In other instances initial, very finely dispersed matter may undergo phase transformation into different, but well-defined particles, most likely through a dissolution/recrystallization process (Bailey *et al.*, 1993).

2.2.1 Spontaneous decomposition

In some systems the kinetics of precipitation can be controlled by spontaneous decomposition of solute species. One such process, which has led to many monodispersed metal (hydrous) oxides, is the so-called 'forced hydrolysis', in which hydrated metal ions keep deprotonating on ageing solutions at appropriate electrolyte concentration, pH and temperature, until the solid phase is formed. An alternative approach is to decompose organometallic solutes in a given reactive environment.

Forced hydrolysis

The hydrolysis of cations, especially those bearing higher charges, is greatly promoted with an increase in temperature, which under certain conditions may result in the formation of uniform particles. The nature of the products depends on the number of parameters, including the obvious ones (pH and temperature), as well as the concentration of the metal ion in a given solution and, especially, on the anion of the dissolved electrolyte. In practice, a metal salt solution of an appropriate concentration and pH is heated to an elevated temperature (usually <100°C) until the solid phase appears. It is easily understood that the hydrolysability of the metal ion will dictate the optimum pH. Thus, in some instances it is necessary to acidify solutions to rather low pH values in order to control the rate of hydrolysis, which would ultimately result in uniform particles.

The forced hydrolysis method has been successfully employed in the preparation of monodispersed metal (hydrous) oxides of aluminum (Brace and Matijević, 1973; Scott and Matijević, 1978), chromium (Demchak and Matijević, 1969; Matijević *et al.*, 1971), iron(III) (Matijević *et al.*, 1975; Matijević and Scheiner, 1978; Hamada and Matijević, 1981), titanium (Matijević *et al.*, 1977), thorium (Milić and Matijević, 1982), cerium(IV) (Hsu *et al.*, 1988), gallium (Hamada *et al.*, 1986), indium (Hamada *et al.*, 1990; Yura *et al.*, 1990), and hafnium (Ocaña *et al.*, 1991).

The actual composition, structure and shape of the particulates depend strongly on the experimental conditions. The anions are frequently incorporated either as contaminants or in a stoichiometric form. Thus, chromium hydroxide particles obtained by forced hydrolysis of chrom alum solutions contain sulphate ions, removable by repeated washings (Zettlemoyer *et al.*, 1978), and haematite precipitated in $FeCl_3$ solutions contains chloride, which is leachable (Hesleitner *et al.*, 1987). In contrast, uniform particles generated by heating ferric sulphate solutions consist of well-defined and stable ferric basic sulphate colloidal crystals of alunite (Matijević *et al.*, 1975).

Many of the dispersions prepared by forced hydrolysis consist of spherical particles, some of which are clearly amorphous (e.g. aluminium or chromium hydroxide), while others show X-ray characteristics of known minerals (e.g. cerium(IV) oxide). In the latter case it has been established, that the resulting spheres actually consist of a large number of much smaller (30–50 Å) subunits. Figure 2.2a illustrates amorphous chromium hydroxide particles and Figure 2.2b shows the electron micrograph of a 'crystalline' cerium oxide powder. The structural variations point to different mechanisms of the processes involved.

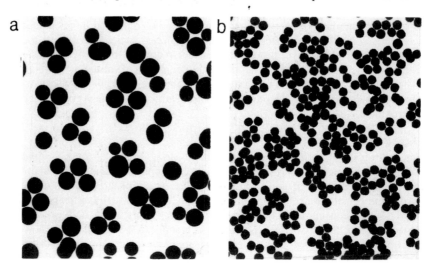

Figure 2.2   (a) *Transmission electron micrograph of (TEM) amorphous chromium hydroxide particles obtained by ageing at 75°C for 24 h a 4 × 10⁻⁴ mol dm⁻³ chrom alum solution (Demchak and Matijević, 1969). (b) TEM of cerium oxide particles obtained by ageing at 90°C for 48 h a 1.2 × 10⁻³ mol dm⁻³ Ce(SO₄)₂ solution (Hsu et al., 1987)*

Particles of shapes other than spheres are often generated, but at present it is difficult, if not impossible, to predict *a priori* different morphologies from any fundamental principles, although some efforts in this direction are now being made (Livage and Henry, 1988; Livage *et al.*, 1989). Two parameters especially may affect particle shape: the concentration of reactants and the nature of the anion present in the metal salt solution that is being aged. In some instances the particle composition remains the same, while the morphology differs, while in

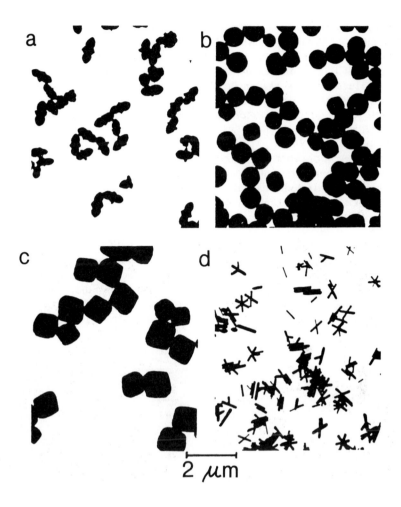

Figure 2.3   *Transmission electron micrographs of particles obtained in solutions of FeCl₃ + HCl under the following conditions (Matijević, 1985):*

	Fe^{3+} (mol dm^{-3})	Cl^- (mol dm^{-3})	Initial pH	Final pH	Temp. of ageing (°C)	Time of ageing
(a)	0.018	0.104	1.3	1.1	100	24 h
(b)	0.315	0.995	2.0	1.0	100	9 days
(c)	0.09	0.28	1.65	0.88	100	24 h
(d)	0.09	0.28	1.65	0.70	150	6 h

other cases the shape is associated with the change in the chemical composition, when either of the two parameters is altered. Figure 2.3 exemplifies different shapes of particles obtained by ageing ferric chloride solutions of varying acidities (HCl). In three cases haematite (Fe_2O_3) is formed, while in the last system (Figure 2.3d) akageneite (β-FeOOH) precipitates (Matijević and Scheiner, 1978). A small amount of phosphate ions added to a ferric chloride solution yields, on ageing, exceedingly uniform ellipsoidal haematite particles (Ozaki *et al.*, 1984).

Decomposition of organometallic compounds

Controlled decomposition of organometallic compounds of different stability in appropriate liquid media can also lead to the formation of monodispersed particles. The classical example is the hydrolytic break-up of tetraethylorthosilicate (TEOS) in aqueous ammonia solutions (Stöber *et al.*, 1968; Bogush *et al.*, 1988; Hsu *et al.*, 1993), which yields exceedingly uniform spherical silica particles, as exemplified in Figure 2.4.

Different metal alkoxides are especially convenient starting materials for the preparation of metal (hydrous) oxides because of their reactivity with water. Thus, spherical aluminum hydroxide was obtained by the hydrolysis of aluminium(III) sec-butoxide (Catone and Matijević, 1974). The procedure was proven useful in the synthesis of pure or doped titania from alcoholic solutions

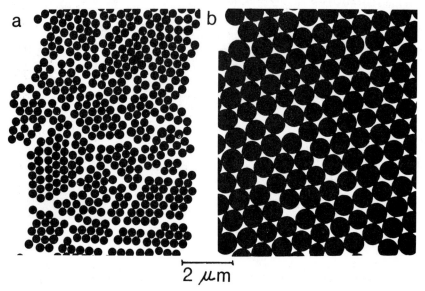

Figure 2.4 *Transmission electron micrographs of silica particles obtained by ageing at 40°C for 1 h solutions of* (a) *0.25 mol dm^{-3} tetraethylorthosilicate (TEOS), 0.90 mol dm^{-3} NH$_3$, and 8.0 mol dm^{-3} H$_2$O in ethanol;* (b) *0.45 mol dm^{-3} TEOS, 1.2 mol dm^{-3} NH$_3$, and 3.0 mol dm^{-3} H$_2$O in isopropanol*

of titanium(IV) alkoxides (Barringer and Bowen, 1982; Fegley *et al.*, 1984; Jean and Ring, 1986). All solids so prepared are amorphous. The present trend is to affect the nature of the resulting particles by modification of the molecular composition or structure of the alkoxide (and similar) precursors and by the use of different solvents (Livage and Henry, 1988; Livage *et al.*, 1990).

Dispersions of various chemical compositions (mostly metal oxides or metal basic compounds) and particle shapes can also be obtained by decomposition of metal chelate complexes in alkaline solutions. This method usually requires somewhat higher temperatures (150–250°C) and the particles are mostly crystalline and larger (10–20 μm) than those prepared by reactions in acidic media. The scanning electron micrographs shown in Figure 2.5 illustrate alunite precipitated from the iron(III) triethanolamine (TEA) complex in an NaOH solution (Sapieszko and Matijević, 1980a) and of metallic nickel particles obtained by decomposition of the corresponding nickel complex (Sapieszko and Matijević, 1980b). In the latter case, hydrazine was also added in order simultaneously to reduce the solid product. Obviously, this procedure offers numerous variations by the choice of the chelating agents which yield metal complexes of different stability.

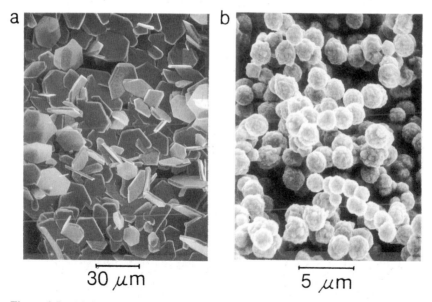

a

b

30 μm   5 μm

Figure 2.5   (a) *Scanning electron micrograph (SEM) of haematite (α-Fe_2O_3) platelets obtained by ageing (ClO_4)$_3$ at 250°C for 1 h a solution 0.18 mol dm^{-3} in Fe(ClO_4)$_3$ and 0.2 mol dm^{-3} in triethanolamine (TEA), and 1.2 mol dm^{-3} in NaOH (Sapieszko and Matijević, 1980a). (b) SEM of nickle particles obtained by ageing at 250°C for 1 h a solution 0.04 mol dm^{-3} in nickel acetate, 0.2 mol dm^{-3} in TEA, 1.2 mol dm^{-3} in NaOH, and 0.85 mol dm^{-3} in hydrazine (Sapieszko and Matijević, 1980b)*

2.2.2 Precipitation by continuous introduction of a reactant

The kinetic control of the precipitation process can also be achieved by the slow introduction of a reactant (especially anionic) into a solution of a coreactant, most commonly an electrolyte of the desired cation. The oldest such procedure consists of heating a salt solution in the presence of urea. The latter decomposes at elevated temperatures according to

$$CO(NH_2)_2 \rightleftarrows NH_3 + HNCO \rightleftarrows NH_4^+ + NCO^-$$

The cyanate ion hydrolyses rapidly either in acidic solutions

$$NCO^- + 2H^+ + H_2O \rightarrow NH_4^+ + CO_2$$

or in basic solutions

$$NCO^- + OH^- + H_2O \rightleftarrows CO_3^{2-}$$

Thus, when heated at moderate temperatures, the pH of such systems rises and the carbonate ion is formed. As a result, depending on the solubility of the products, either metal hydroxides or basic carbonates are precipitated in the presence of urea, which under a proper decomposition rate yields uniform particles, many of spherical shape. Using this procedure, monodispersed colloidal lanthanides (Matijević and Hsu, 1987; Akinc et al., 1988), yttrium (Akinc and Sordelet, 1987; Aiken et al., 1988), zirconium (Aiken et al., 1990), cadmium (Janeković and Matijević, 1985), zinc (Castellano and Matijević, 1989), copper (Kratohvil and Matijević, 1991), and aluminium (Blendell et al., 1984) basic carbonates or hydroxides have been obtained. Figure 2.6a illustrates gadolinium basic carbonate prepared in this way and Figure 2.6b is the scanning electron micrograph of $CdCO_3$. Recently, it has been shown that such processes can be carried out *continuously*, resulting in uniform particles in high yields (Her et al., 1990).

An analogous principle can be applied in order to synthesize other inorganic compounds. Thus, metal sulphides were obtained by ageing at elevated temperatures solutions of metal salts in the presence of thioacetamide (TAA), which on heating releases HS– anions. Under proper conditions, exceedingly uniform spheres of CdS (Matijević and Murphy-Wilhelmy, 1982) and ZnS (Murphy-Wilhelmy and Matijević, 1984; Celikkaya and Akinc, 1990) (Figure 2.6c), and prisms of PbS (Murphy-Wilhelmy and Matijević, 1985) (Figure 2.6d) were precipitated. It is noteworthy that, despite the sphericity, the X-ray pattern of the CdS powder corresponded to greenockite and that of ZnS to sphalerite, indicating the composite structure of these particles. Alternatively, colloidal ZnS (Chiu, 1981) and CuS (Chiu, 1977) were obtained by passing hydrogen sulphide through metal salt solutions containing a chelating agent (EDTA). Finally, monodispersed cadmium and lead selenides can be synthesized by heating solutions containing the corresponding metal salts in the presence of selenourea (Gobet and Matijević, 1984).

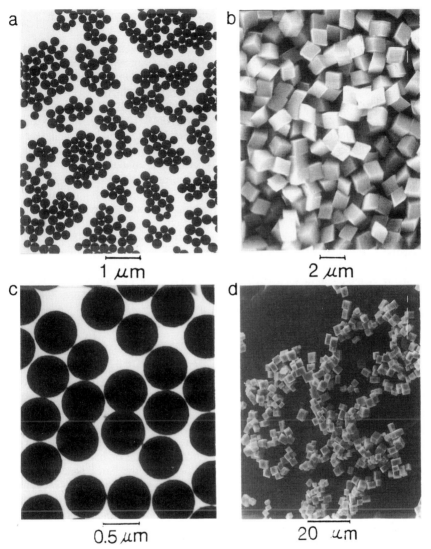

Figure 2.6 (a) *Transmission electron micrograph (TEM) of basic gadolinium carbonate particles obtained by ageing at 90°C for 1 h a solution 5.0 × 10⁻³ mol dm⁻³ in GdCl₃ and 0.5 mol dm⁻³ in urea (Matijević and Hsu, 1987). (b) Scanning electron micrograph (SEM) of CdCO₃ particles obtained by mixing 40 cm³ of a 10 mol dm⁻³ urea solution (preheated before addition at 80°C for 24 h) and 40 cm³ of a 2 × 10⁻³ mol dm⁻³ solution of CdCl₂ at room temperature (Janeković and Matijević, 1985). (c) TEM of ZnS particles obtained by ageing a 'seed' sol at 60°C for 1.5 h in the presence of 0.062 mol dm⁻³ HNO₃ and 0.11 mol dm⁻³ thioacetamide (TAA) (Murphy-Wilhelmy and Matijević, 1984). (d) SEM of PbS particles obtained by ageing at 26°C for 20 min a 'seed' sol in the presence of 1.25 × 10⁻³ mol dm⁻³ TAA (Murphy-Wilhelmy and Matijević, 1985)*

2.2.3 Precipitation by mixing reactants

In some, albeit a limited number, of cases it has been possible to produce monodispersed sols by direct mixing of reactant solutions with subsequent ageing, usually at somewhat elevated temperature. Specifically, different metal phosphates were prepared in the described manner, including spherical or ellipsoidal particles of $AlPO_4$ (Katsanis and Matijević, 1982). The X-ray analysis of these solids was consistent with variscite ($Al(OH)_2H_2PO_4$). In the case of ferric phosphate the nature of the products depended in a sensitive manner on the experimental parameters (pH, the phosphate salt used, concentration of reactants etc.) (Kratohvil *et al.*, 1984), although spherical amorphous particles of $FePO_4$ were obtained under a rather limited set of conditions (Wilhelmy and Matijević, 1987).

The precipitation of (basic) metal phosphates could be promoted by the simultaneous addition of urea, and the particle uniformity enhanced by the presence of surfactants or polyelectrolytes. Such combinations led to mono-dispersed colloidal cadmium(II), nickel(II), and manganese(II) (basic) phosphates (Springsteen and Matijević, 1989). In $CoSO_4$ solutions these combinations of reactants produced $Co_3(PO_4)_2$, $CoCO_3 \cdot H_2O$, or $Co(NH_4)PO_4 \cdot H_2O$ of different morphologies, by the change in the concentrations of cobalt(II) and phosphate salts (Ishikawa and Matijević, 1988).

2.3 Precipitation by the double-jet process

Well-defined monodisperse particles can be rapidly generated by the controlled double-jet precipitation (CDJP). This technique is employed in the photographic industry for making precisely defined silver halide crystals of narrow size distribution, as well as of well-developed habit, internal composition and epitaxy (Berry, 1977; Wey, 1981; Stávek *et al.*, 1989). The rate of production of such materials is about 1 mol AgX per litre per hour and the batch process can be scaled up to about 2000 l. Recently, it has been found that the CDJP technique may be employed to generate various other sparingly soluble salts, including metal oxalates and sulphates (Stávek *et al.*, 1990a, b).

In the CDJP technique the cation and anion reactant solutions are added simultaneously through separate input lines to a stirred solution of a protective polymer, which prevents coagulation and agglomeration of the growing crystals. While it is understood that in this process the nucleation or formation of primary particles and their growth occur simultaneously during the entire run, monodisperse microcrystals may be obtained, if it is ensured that the unstable nuclei disappear during the reaction.

The convenient crystallizer for the CDJP technique is designed to have two defined zones: a region of extreme supersaturation where the highly concentrated solutions of reactants are introduced into the system (primary zone), and the well-mixed bulk of the vessel (secondary zone). The actual supersaturations of reactant solutions near the jet exits are typically 10^5–10^8 times the solubility,

Figure 2.7   (a) *Scanning electron micrograph (SEM) of lead sulphate crystals prepared by the controlled double-jet precipitation (CDJP) technique under the following conditions: 1.0 mol dm^{-3} solutions of Pb(NO$_3$)$_2$ and Na$_2$SO$_4$ were added at a flow rate of 10 cm^3 min^{-1} over 10 min into 1000 cm^3 aqueous solution of 2% inert gelatin at pH = 2.0 and an excess of Pb^{2+} is 1 × 10^{-2} mol dm^{-3} (courtesy of J. Stávek). (b) SEM of strontium sulphate crystals prepared by the CDJP technique under the following conditions: 0.1 mol dm^{-3} solutions of Sr(NO$_3$)$_2$ and Na$_2$SO$_4$ were added at a flow rate of 10 cm^3 min^{-1} over 10 min into 1000 cm^3 aqueous solution of 2% inert gelatin at pH = 1.0 and an excess of SO$_4^{-2}$ is 1 × 10^{-2} mol dm^{-3} (courtesy of J. Stávek)*

which are then released by nucleation. If the bulk supersaturation is low enough, the unstable nuclei (continuously fed into the secondary mixing zone) are dissolved by Ostwald ripening and act as a source of additional matter for the growing crystals. By another mechanism of the precipitation of monodisperse solids by the CDJP technique, the primary particles formed near the jets are fed into the bulk where they become unstable and agglomerate with much larger secondary particles. Obviously, neither of the described mechanisms fits the LaMer model.

As examples, the electron micrographs in Figure 2.7 show lead sulphate and strontium sulphate crystals, obtained by the CDJP technique under the conditions given in the legend.

The advantage of this technique is in rapid precipitation using solutions of high concentrations. Furthermore, with mixed electrolytes, doped crystals of desired composition can be synthesized.

2.4 Precipitation in surfactant-rich media

In recent years surfactant solutions forming micelles, lyotropic mesophases and vesicles have been used as media for the precipitation of inorganic colloids (Fendler, 1987; Osseo-Asare and Arriagada, 1990). Specifically, emulsions, microemulsions and liquid crystals have been shown to exhibit, under certain conditions, a significant effect on particle size, shape and even chemical composition (Ward and Friberg, 1989).

Exceedingly small (30–50 Å) particles of noble metals (platinum, palladium, rubidium, indium and silver) were obtained by reduction of metal salts or organometallic compounds with hydrazine or hydrogen in microemulsions (Boutonnet *et al.*, 1982; Barnickel *et al.*, 1992). Using the same type of media, different compounds have also been prepared, such as silver bromide (Chew *et al.*, 1990) or yttrium (hydrous) oxide (Akinc and Celikkaya, 1987). In the latter case yttrium ions in the aqueous phase of a water in oil emulsion were reacted with triethanolamine (TEA), resulting in spheres of about 1 μm in modal diameter. Much smaller magnetite particles were generated in water/oil microemulsions using a mixture of $FeCl_3$, $FeCl_2$ and NH_3 (Gobe *et al.*, 1983). Carbonates of barium and calcium of different morphologies were produced by bubbling CO_2 through microemulsions of $Ba(OH)_2$ and $Ca(OH)_2$ (Kandori *et al.*, 1987).

Lyotropic liquid crystals represent a specially structured environment which can affect the nucleation and growth processes. Thus, ellipsoidal copper basic sulphate particles (Figure 2.8) have been obtained by precipitation in the aqueous part of the water/Tween 80 system containing $CuSO_4$ (Ward and Friberg, 1989).

Finally, surfactant vesicles can be used to synthesize finely dispersed inorganic particles, especially of metal sulphides or selenides, which have semiconductor properties (Chang *et al.*, 1990; Heywood *et al.*, 1990).

10 μm

Figure 2.8 *Scanning electron micrograph of basic copper sulphate particles precipitated from a copper sulphate solution as the aqueous part in the liquid crystals of the water/Tween 80 system (Ward and Friberg, 1989)*

2.5 Phase transformation

Many uniform particles have now been obtained by first preparing a precursor solid and then transforming it through chemical manipulations into a dispersion of particles that differ either in composition or shape, or both. This topic deals with many techniques, some of which are common in a given area of application, such as the sol–gel process in ceramics, or reduction of inorganic compounds to pure metals for catalysis.

The simplest approach in the phase transformation is to use a powder of a given compound and to alter the chemical composition, while maintaining the particle shape and uniformity. Such method is especially useful, if the powder needed cannot be obtained directly in the desired form, either chemical or morphological. It has been demonstrated that pure colloidal metals can be prepared from monodispersed precursors using appropriate reducing agents either in suspended state (such as hydrazine in aqueous sols) or hydrogen in the solid state (Matijević, 1992).

Figure 2.9a shows metallic cubic copper particles obtained by the reduction of a dispersion of copper(II) basic carbonate with $NH_2OH \cdot HCl$ (Hsu *et al.*, 1990), while Figure 2.9b shows the electron micrograph of fine iron particles prepared by treating an ellipsoidal haematite powder with hydrogen (Ishikawa and Matijević, 1988a). In the latter case the original powder was treated with silica to prevent irreversible agglomeration on heating during the reduction process. In neither of the cases described could solids of the illustrated shapes be obtained by direct synthesis.

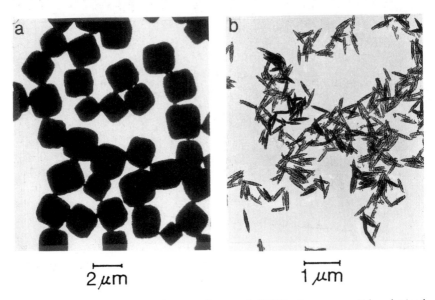

2 μm   1 μm

Figure 2.9   (a) *Transmission electron micrograph (TEM) of copper particles obtained by reduction with $NH_2OH \cdot HCl$ of copper(II) basic carbonate precursors (Hsu* et al.*, 1990). (b) TEM of iron particles obtained by reduction of silica coated ellipsoidal haematite particles by hydrogen at 450°C (Ishikawa and Matijević, 1988)*

It was mentioned above (Section 2.2.2) that metal (basic) carbonates are readily prepared by hydrolysis of electrolyte solutions in the presence of urea. In most instances powders generated in this way can be simply transformed into the corresponding oxides by calcination, usually at moderate temperatures, as was done with a number of spherical lanthanide compounds (Akinc and Sordelet, 1987; Matijević and Hsu, 1987). As another example, Figure 2.10 shows the γ-manganese(IV) oxide of two different morphologies obtained from the corresponding manganese(II) carbonate precursors (Hamada *et al.*, 1987).

Still another phase transformation procedure was proven useful, by forming first a gel-type precipitate in which crystalline particles are subsequently generated on ageing in the presence of certain reactants. Thus, spherical particles of magnetite were produced by keeping at 90°C for several hours a ferrous

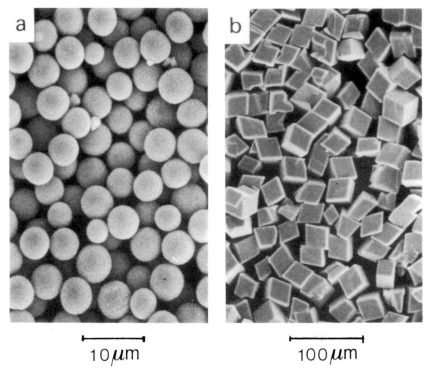

Figure 2.10 *Scanning electron micrograph of manganese(IV) oxide particles* (b) *obtained by heating the corresponding manganese(II) carbonate particles at* (a) *400°C under oxygen atmosphere at 50% relative humidity (Hamada et al., 1987)*

hydroxide gel, in which the added nitrate ion acts as a mild oxidizing agent (Sugimoto and Matijević, 1980). By partial substitution of iron(II) with Ni^{2+} or Co^{2+} (or both) spherical ferrite particles of different compositions could be obtained (Regazzoni and Matijević, 1982, 1983; Tamura and Matijević, 1982). Recently, monodispersed haematite particles of different shapes were obtained in large quantities from condensed ferric hydroxide gels (Sugimoto *et al.*, 1993a,b,c).

In some instances the phase transformation takes place through a dissolution/ crystallization mechanism. Under certain conditions, which may involve additives or changing solvents, the precursor particles dissolve and a new phase of different morphology, structure, or even chemical composition, precipitates. Two examples of this type of process are offered here. In the first case, a reducing and a complexing agent (TEA and N_2H_4) are mixed with an alkaline aqueous haematite dispersion consisting of spherical particles, which on ageing at elevated temperatures yielded magnetite (Figure 2.11a). After a somewhat longer reaction time, all of the haematite is converted to magnetite (Sapieszko and Matijević, 1980a). The prismatic particles shown in Figure 2.11b are of

a   b

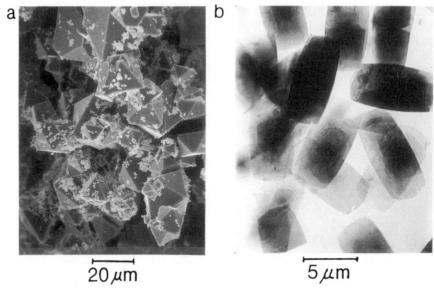

$20\,\mu m$ $5\,\mu m$

Figure 2.11   (a) *Scanning electron micrograph of magnetite particles obtained by ageing for 3 h at 250°C a dispersion of spherical haematite particles in the presence of 0.2 mol dm⁻³ TEA, 0.85 mol dm⁻³ hydrazine and 1.2 mol dm⁻³ NaOH (Sapieszko and Matijević, 1980a). (b) Transmission electron micrograph of manganese(II) phosphate particles obtained by redispersing in distilled water perfectly spherical hydrated manganese(II) phosphate particles, after equilibration for 48 h at room temperature (Springsteen and Matijević, 1989)*

manganese phosphate which crystallized after the mother liquor, in which perfectly spherical particles precipitated, was replaced by pure water (Springsteen and Matijević, 1989).

Phase transformations often take place even though they may not be recognized. There is much evidence available that haematite dispersions produced from $FeCl_3$ solutions are actually formed from initially precipitated β-FeOOH precursor particles (Bailey *et al.*, 1993).

2.6 Concluding remarks

In view of the rapid progress in the science of monodispersed fine particles, it was not possible to describe in the available space all methods presently used to produce such colloids. Thus, this chapter essentially deals with various precipitation techniques, which are most common and practical. Even so, only limited aspects of the procedures have been discussed and a small selection of available systems exemplified. For the same reason, the literature citations had to be restricted. No effort has been made to describe more complex systems, such as consisting of internally composite, coated or hollow particles. Furthermore, the writing was restricted to the preparation procedures of uniform particles. The mechanisms of their formation (chemical and physical), as well as characterizations of such dispersions in terms of different properties (optical, magnetic, electric, adsorptive etc.) as a function of the composition, size and shape could not be dealt with. To gain information on these topics, the reader is referred to original papers describing well-defined fine particles.

References

Aiken, B., Hsu, W. P. and Matijević, E. (1988) Preparation and properties of monodispersed colloidal particles of lanthanide compounds. III. Y(III) and mixed Y(III)/Ce(III) systems. *J. Am. Ceram. Soc.*, **71**, 845–853

Aiken, B., Hsu, W. P. and Matijević, E. (1990) Preparation and properties of uniform mixed and coated colloidal particles. V. Zirconium compounds. *J. Mater. Sci.*, **25**, 1886–1894

Akinc, M. and Celikkaya, A. (1987) Preparation of yttria powders by emulsion precipitation. *Adv. Ceram.*, **21**, 57–67.

Akinc, M. and Sordelet, D. J. (1987) Preparation of yttrium, lanthanium, cerium, and neodymium basic carbonate particles by homogeneous precipitation. *Adv. Ceram. Mater.*, **2**, 232–238

Akinc, M., Sordelet, D. J. and Munson, M. (1988) Formation, structure, and decomposition of lanthanide basic carbonates. *Adv. Ceram. Mater.*, **3**, 211–216

Andreasen, A. H. M. (1943) Einige Untersuchungen über die Herstellung von monodispersen Stoffen. *Kolloid-Z.*, **104**, 181–189

Bailey, J. K., Brinker, C. J. and Macartney, M. L. (1993) Growth mechanisms of iron oxide particles of different morphologies from the forced hydrolysis of ferric chloride solutions. *J. Colloid Interface Sci.*, **157**, 1–13

Barnickel, P., Wokaun, A., Sager, W. and Eicke, H.-F. (1992) Size tailoring of silver colloids by reduction in W/O emulsions. *J. Colloid Interface Sci.*, **148**, 80–90

Barringer, E. A. and Bowen, H. K. (1982) Formation, packing and sintering of monodisperse TiO_2 powders. *J. Am. Ceram. Soc.*, **65**, C199–C201

Berry, C. R. (1977) Precipitation and growth of silver halide emulsion grains. In *The Theory of the Photographic Process*, 4th edn (ed. James, T. H.), Macmillan Publishing Company, New York, 88–104

Blakemore, R. (1975) Magnetotactic bacteria. *Science*, **190**, 377–379

Blendell, J. E., Bowen, H. K. and Coble, R. L. (1984) High purity alumina by controlled precipitation from aluminum sulfate solutions. *Ceram. Bull.*, **63**, 797–802

Bogush, G. H., Tracy, M. A. and Zukoski, C. F. (1988) Preparation of monodisperse silica particles: control of size and mass fraction. *J. Non-Cryst. Solids*, **104**, 95–106

Boutonnet, M., Kizling, J., Stenius, P. and Maire, G., (1982) The preparation of monodispersed colloidal metal particles from microemulsions. *Colloids Surf.*, **5**, 209–225

Brace, R. and Matijević, E. (1973) Aluminium hydrous oxide sols. I. Spherical particles of narrow size distribution. *J. Inorg. Nucl. Chem.*, **35**, 3691–3705

Castellano, M. and Matijević, E. (1974) Uniform colloidal zinc compounds of various morphologies. *Chem. Materials*, **1**, 78–82

Catone, D. L. and Matijević, E. (1974) Aluminum hydrous oxide sols. II. Preparation of uniform spherical particles by hydrolysis of Al sec-butoxide. *J. Colloid Interface Sci.*, **48**, 291–301

Celikkaya, A. and Akinc, M. (1990). Preparation and mechanism of formation of spherical submicrometer zinc sulfide powders. *J. Am. Ceram. Soc.*, **73**, 2360–2365

Chang, A. C., Pfeiffer, W. F., Guillaume, B., Baral, S. and Fendler, J.H. (1990). Preparation and characterization of selenide conductor particles in surfactant vesicles. *J. Phys. Chem.*, **94**, 4284–4289

Chew, C. H., Gan, L. M. and Shah, D. O. (1990) The effect of alkanes on the formation of ultrafine silver bromide particles in ionic W/O emulsions. *J. Dispersion Sci. Technol.*, **11**, 593–609

Chiu, G. (1977) The preparation of monodispersed copper sulfide sols. *J. Colloid Interface Sci.*, **62**, 193–194

Chiu, G. (1981) The preparation of monodispersed zinc sulfide sols. *J. Colloid Interface Sci.*, **83**, 309–310

Demchak, R. and Matijević, E. (1969) Preparation and particle size analysis of chromium hydroxide sols of narrow size distributions. *J. Colloid Interface Sci.*, **31**, 257–262

Faraday, M. (1857) Experimental relations of gold (and other metals) to light. *Philos. Trans. Roy. Soc., London*, **147**, 145

Fegley, Jr, B., Barringer, E. A. and Bowen, H. K. (1984). Synthesis and characterization of monosized doped TiO$_2$ powders. *Commun. Am. Ceram. Soc.*, C113–C116

Fendler, J. H. (1987) Atomic and molecular clusters in membrane mimetic chemistry. *Chem. Rev.*, **87**, 877–899

Gobe, M., Kon-No, K., Kandori, K. and Kitahara, A. (1983). Preparation and characterization of monodisperse magnetite sols in W/O microemulsions. *J. Colloid Interface Sci.*, **93**, 293–295

Gobet, J. and Matijević, E. (1984) Preparation of monodispersed colloidal cadmium and lead selenides. *J. Colloid Interface Sci.*, **100**, 555–560

Hamada, S. and Matijević, E. (1981) Ferric hydrous oxide sols. IV. Preparation of uniform cubic hematite particles by hydrolysis of ferric chloride in alcohol water solutions. *J. Colloid Interface Sci.*, **84**, 274–277

Hamada, S., Bando, K. and Kudo, Y. (1986) The formation process of hydrous gallium(III) oxide particles obtained by hydrolysis at elevated temperatures. *Bull. Chem. Soc. Jpn.*, **59**, 2063–2069

Hamada, S., Kudo, Y., Okada, J. and Kano, H. (1987) Preparation of monodispersed managenese(IV) oxide particles from managanese(II) carbonate. *J. Colloid Interface Sci.*, **118**, 356–365

Hamada, S., Kudo, Y. and Minagawa, K. (1990) The formation of monodisperse indium(III) hydroxide particles by forced hydrolysis and elevated temperatures. *Bull. Chem. Soc. Jpn.*, **63**, 102–107

Haruta, M. and Delmon, B. (1986) Preparation of monodisperse solids. *J. Chim. Phys.*, **83**, 859–868

Her, Y.-S., Matijević, E. and Wilcox W. R. (1990) Continuous precipitation of monodispersed yttrium basic carbonate powders. *Powder Technol.*, **61**, 173–177

Herak, M. J., Kratohvil, J., Herak, M. M. and Wrischer, M. (1958) A light scattering and electron microscope investigation of monodispersed metal iodate hydrosols. *Croat. Chem. Acta*, **30**, 221–230

Hesleitner, P., Babić, D., Kallay, N. and Matijević, E. (1987) Adsorption at solid/solution interfaces. III. Surface charge and potential of colloidal hematite. *Langmuir*, **3**, 815–820

Heywood, B., Fendler, J. H. and Mann, S. (1990) *In situ* imaging of cadmium sulfide and zinc sulfide semiconductor particles in surfactant vesicles. *J. Colloid Interface Sci.*, **138**, 295–298

Hsu, W. P., Yu, R. and Matijević, E. (1993) Paper whitener. I. Titania coated silica. *J. Colloid Interface Sci.*, **156**, 56–65

Hsu, W. P., Rönnquist, L. and Matijević, E. (1988) Preparation and properties of monodispersed colloidal particles of lanthanide compounds. II. Cerium(IV). *Langmuir*, **4**, 31–37

Hsu, W. P., Yu, R. and Matijević, E. (1990) Preparation and characterization of uniform particles of metallic copper and coated metallic copper with inorganic compounds and polymers. *Powder Technol.*, **63**, 265–275

Ishikawa, T. and Matijević, E. (1988a) Formation of monodispersed spindle-type pure and coated iron particles. *Langmuir*, **4**, 26–31

Ishikawa, T. and Matijević, E. (1988b) Preparation and properties of uniform colloidal metal phosphates. III. Cobalt(II) phosphate. *J. Colloid Interface Sci.*, **123**, 122–128

Janeković, A. and Matijević, E. (1985). Preparation of monodispersed cadmium compounds. *J. Colloid Interface Sci.*, **103**, 436–447

Jean, J. H. and Ring, T. A. (1986) Nucleation and growth of monodispersed TiO_2 powders from alcohol solutions. *Langmuir*, **2**, 251–255

Kandori, K., Kon-No, K. and Kitahara, A. (1987) Dispersion stability of nonaqueous calcium carbonate dispersions prepared in water core of W/O emulsions. *J. Colloid Interface Sci.*, **115**, 579–582

Katsanis, E. P. and Matijević, E. (1982) Preparation and properties of uniform colloidal aluminum phosphate particles. *Colloids Surf.*, **5**, 45–53

Kratohvil, S. and Matijević, E. (1991) Preparation of copper compounds of different morphologies. *J. Mater. Res.*, **6**, 766–777

Kratohvil, S., Matijević, E. and Ozaki, M. (1984) Precipitation phenomena in $FeCl_3–NaH_2PO_2$ solutions at 245°C. *Colloid Polymer Sci.*, **262**, 804–810

LaMer, V. K. (1952) Nucleation in phase transitions. *Ind. Eng. Chem.*, **44**, 1270–1277

LaMer, V. K. and Barnes, M. D. (1946) Monodispersed hydrophobic colloidal dispersions and light scattering properties. I. Preparation and light scattering properties of monodispersed colloids. *J. Colloid Sci.*, **1**, 71–77

Livage, J. and Henry, M. (1988) A predictive model for inorganic polymerization reactions. In *Ultrastructure Processing of Advanced Ceramics* (ed. Mackenzie, J. D. and Ulrich, D. R.) Wiley-Interscience, New York, 183–185

Livage, J., Henry, M. and Sanchez, C. (1989) Sol–gel processing of transition metal oxides. *Prog. Solid State Chem.*, **18**, 259–341

Livage, J., Henry, M., Jolivet, J. P. and Sanchez, C. (1990) Chemical synthesis of fine powders. *MRS Bull.*, **15**, 18–25

Look, J.-L., Bogush, G. and Zukoski, C. F. (1990) Colloidal interactions during the precipitation of uniform micrometre particles. *Faraday Discuss. Chem. Soc.*, **90**, 345–357.

Matijević, E. (1985) Production of monodispersed colloidal particles. *Ann. Rev. Mater. Sci.*, **15**, 483–516

Matijević, E. (1992) Preparation and properties of well defined finely dispersed metals. *Discuss. Faraday Soc.*, **92**, 229–239

Matijević, E. (1993) Preparation and properties of uniform size colloids. *Chem. Materials*, **5**, 412–426

Matijević, E. and Hsu, W. P. (1987) Preparation and properties of monodispersed colloidal particles of lanthanide compounds. I. Gadolinium, europium, terbium, samarium, and cerium(III). *J. Colloid Interface Sci.*, **118**, 506–523

Matijević, E. and Murphy-Wilhelmy, D. (1982) Preparation and properties of monodispersed spherical colloidal particles of cadmium sulfide. *J. Colloid Interface Sci.*, **86**, 476–484

Matijević, E. and Scheiner, P. (1978) Ferric hydrous oxide sols. III. Preparation of uniform particles by hydrolysis of Fe(III) chloride, nitrate, and perchlorate solutions. *J. Colloid Interface Sci.*, **63**, 509–524

Matijević, E., Lindsay, A. D., Kratohvil, S., Jones, M. E., Larson, R. I. and Cayey, N. W. (1971) Characterization and stability of chromium hydroxide sols of narrow size distribution. *J. Colloid Interface Sci.*, **36**, 273–281

Matijević, E., Sapieszko, R. S. and Melville, J. B. (1975) Ferric hydrous oxide sols. I. Monodispersed basic iron(III) sulfate particles. *J. Colloid Interface Sci.,*, **50**, 567–581

Matijević, E., Budnik, M. and Meites, L. (1977) Preparation and mechanism of formation of titanium dioxide hydrosols of narrow size distribution. *J. Colloid Interface Sci.*, **61**, 302–311

Milić, N. B. and Matijević, E. (1982) Formation of spherical colloidal thorium basic sulfate particles. *J. Colloid Interface Sci.*, **85**, 306–315

Miura, M., Nagakane, T. and Masaki, S. (1956) Uniform and spherical crystals of barium sulfate. *J. Sci. Hiroshima Univ.*, **A19**, 510–514

Murphy-Wilhelmy, D. and Matijević, E. (1984) Preparation and properties of monodispersed spherical colloidal particles of zinc sulphide. *J. Chem. Soc., Faraday Trans. 1*, **80**, 563–570

Murphy-Wilhelmy, D. and Matijević, E. (1985) Preparation of uniform colloidal particles of lead sulfide and of mixed sulfides of cadmium and zinc and cadmium and lead. *Colloids Surf.*, **16**, 1–8

O'Brien, N. R. and Slatt, R. (ed.) (1990) *Argillaceous Rock Atlas*, Springer-Verlag, New York

Ocaña, M., Hoffman, D. and Matijević, E. (1991) Preparation of uniform colloidal particles of hafnium compounds. *J. Mater. Chem.*, **1**, 87–90

Osseo-Asare, K. and Arrigada, F. J. (1990) Synthesis of nanosize particles in reverse microemulsions. *Ceram. Trans. (Ceram. Powder Sci. 3)*, **12**, 3–16

Ozaki, M., Kratohvil, S. and Matijević, E. (1984) Formation of monodispersed spindle-type hematite particles. *J. Colloid Interface Sci.*, **102**, 146–151

Petres, J., Deželić, Gj. and Težak, B. (1966) Monodisperse sols of barium sulfate: I. Preparation of stable sols. *Croat. Chem. Acta*, **38**, 277–282

Regazzoni, A. E. and Matijević, E. (1982) Formation of spherical colloidal nickel ferrite particles as model corrosion products. *Corrosion*, **38**, 212–218

Regazzoni, A. E. and Matijević, E. (1983) Formation of uniform colloidal mixed cobalt–nickel ferrite particles. *Colloids Surf.*, **6**, 189–201

Sapieszko, R. S. and Matijević, E. (1980a). Preparation of well defined colloidal particles by thermal decomposition of metal chelates. Iron oxides. *J. Colloid Interface Sci.*, **74**, 405–422

Sapieszko R. S. and Matijević, E. (1980b) Preparation of well defined colloidal particles by thermal decomposition of metal chelates. II. Cobalt and nickel. *Corrosion*, **36**, 522–530

Scott, W. B. and Matijević, E. (1978) Aluminum hydrous oxide sols. III. Preparation of uniform particles by hydrolysis of aluminum chloride and aluminum perchlorate. *J. Colloid Interface Sci.*, **66**, 447–454

Springsteen, L. and Matijević, E. (1989) Preparation and properties of uniform colloidal metal phosphates. IV. Cadmium, nickel, and manganese(II) phosphates. *Colloid Polymer Sci.*, **267**, 1007–1015

Stávek, J., Šípek, M. and Nývlt, J. (1989) Controlled double-jet precipitation of silver halides. In *Proceedings of International Symposium on Preparation of Functional Materials and Industrial Crystallization* 18–19 August 1989, Osaka, Bapan), 17–23

Stávek, J., Hamslík, T. and Zapletal, V. (1990a) Preparation of high-temperature superconductors by the controlled double-jet precipitation. *Mater. Lett.*, **9**, 90–95

Stávek, J., Muraoka, K. and Toyokura, K. (1990) Controlled double-jet precipitation as an approach to high-added value materials. In *Proceedings of Bremen International Workshop for Industrial Crystallization* (12–13 September 1990, Bremen, Germany), 82–89

Stöber, W., Fink, A. and Bohm, E. (1968) Controlled growth of monodisperse silica spheres in the micron size range. *J. Colloid Interface Sci.*, **26**, 62–69

Sugimoto, T. (1987) Preparation of monodispersed colloidal particles. *Adv. Colloid Interface Sci.*, **28**, 65–108

Sugimoto, T. and Matijević, E. (1980) Formation of uniform spherical magnetite particles by crystallization from ferrous hydroxide gels. *J. Colloid Interface Sci.*, **74**, 227–243

Sugimoto, T., Sakata, K. and Muramatsu, A. (1993a) Formation mechanism of monodispersed pseudocubic α-Fe_2O_3 particles from condensed ferric hydroxide gel. *J. Colloid Interface Sci.*, **159**, 372–382

Sugimoto, T., Muramatsu, A., Sakata, K. and Shindo, D. (1993b) Characterization of hematite particles of different shapes. *J. Colloid Interface Sci.*, **158**, 420–428

Sugimoto, T., Khan, M. M. and Muramatsu, A. (1993c) Preparation of monodispersed peanut-type α-Fe_2O_3 particles from condensed ferric hydroxide gel. *Colloids Surf.*, **70**, 167–169

Tamura, H. and Matijević, E. (1982) Precipitation of cobalt ferrites. *J. Colloid Interface Sci.*, **90**, 100–109

Towe, K. M. and Moench, T. T. (1981) Electron–optical characterization of bacterial magnetite. *Earth Planetary Sci. Lett.*, **52**, 213–220

Ward, J. I. and Friberg, S. E. (1989) Preparing narrow size distribution particles from amphiphilic association structures. *MRS Bull.*, **14**, 41–46

Watson, J. H. L., Heller, W. and Wojtowicz, W. (1948) Morphological changes of tactoid-forming particles. *J. Chem. Phys.*, **16**, 997–998

Watson, J. H. L., Heller, W. and Schuster, T. (1960) A study of monodisperse tactoid-forming crystals of β-FeOOH. In *Proceedings of the European Reg. Conference on Electron Microscopy* (Delft), Vol. 1, 229–234

Watillon, A., Grunderbeeck, F. and Hautecler, M. (1958) Preparation et purification d'hydrosols de selenium stables et homeodisperses. *Bull. Soc. Chem. Belg.*, **67**, 5–21

Wey, J. S. (1981) Basic crystallization processes in silver halide precipitation. In *Preparation and Properties of Solid State Materials*, Vol. 6 (ed. Wilcox, W. R.) Marcel Dekker, New York, 67–117

Wilhelmy, R. B. and Matijević, E. (1987) Preparation and growth kinetics of monodispersed ferric phosphate hydrosols. *Colloids Surf.*, **22**, 111–131

Yura, K., Fredrikson, K. C. and Matijević, E. (1990) Preparation and properties of uniform colloidal indium compounds of different morphologies. *Colloids Surf.*, **50**, 281–293

Zettlemoyer, A. C., Siddiq, M. and Micale, F. J. (1978) Surface properties of heat-treated chromia of narrow particle size distribution. *J. Colloid Interface Sci.*, **66**, 173–182

Chapter 3

Particle formation during agglomerative precipitation processes

A. G. Jones

3.1 Introduction

3.1.1 Precipitation and agglomeration

Particulate precipitates are widespread throughout the chemical and allied industries, whether as final products or as intermediates during processing. For example, fine chemicals and catalysts, agrochemicals and pharmaceuticals often comprise substantial low tonnage, high added value, precipitates in crystalline form, while processes for new materials for the electronics industry and ceramics often employ a precipitation step, and protein isolation forms an example in biotechnology.

Besides its chemical composition, the physical form of a crystalline precipitate is important for its effect on both product quality (e.g. surface 'activity') and downstream processing interactions (e.g. solid–liquid separation) (Jones, 1985; Rossiter and Douglas, 1986). The two most important particle characteristics are normally size and shape. Frequently, precipitates are required to be in the form of fine powders, while in other applications larger forms are desired. Precipitated particles, however, are particularly prone to aggregation during their formation, frequently comprising numerous primary crystals seemingly stuck together to form agglomerates.

Such crystal agglomeration can arise via several distinct mechanisms and give rise to a wide variety of physical forms of differing particle strength. These range from raspberry- to star-like assemblages, with properties which may be an advantage or a disadvantage depending on the circumstances. For example, such particles can deform, may be porous and retain liquor giving rise to filtration and purity problems, may break and form dust, or may simply be unacceptable in appearance. On the other hand, the formation of agglomerates can enhance the solid–liquid separation characteristics of fine precipitates and decrease the packing density of the product.

Thus in addition to size, specific surface area, degree of agglomeration, structure and voidage are all important particle characteristics. All these quantities are determined by the kinetic processes of nucleation, growth and agglomeration and each of these processes is dependent on supersaturation – the essential driving force for crystallization – and can also be influenced by particle-fluid dynamics, thus providing an interaction with processing conditions.

Precipitation and crystallization from solution can also be markedly affected by the presence of trace impurities or second solvents, changing the product characteristics beyond recognition; indeed, this can be used to advantage as a method for control of particle form.

The prediction and control of crystal size distribution in terms of crystal growth and nucleation kinetics and flow regimes has occupied many researchers in the field of industrial crystallization for several decades. Progress has been reviewed in several books, review articles and symposium proceedings (e.g. Walton, 1967; Nývlt *et al.*, 1985; Garside, 1985; Tavare, 1986; Mersmann and Kind, 1988; Randolph and Larson, 1988; Garside *et al.*, 1991; Söhnel *et al.*, 1991; Mullin, 1993).

The additional process of crystal agglomeration gives rise to substantial analytical complications, and has thus received relatively less attention. It has become a topic of increasing interest in recent years, however, coincident with increased interest in precipitation processes in general. Agglomeration itself is of considerable scientific and technological importance and has been the subject of several international conferences (Capes, 1961, 1977, 1981, 1985; Wyn-Jones and Gormally, 1983; Davis *et al.*, 1984; Family and Landau, 1984; Gregory, 1984a), selective reviews (Bell *et al.*, 1983; Gregory, 1984b, 1989; Lyklema, 1985) and a vast number of articles in journals of diverse disciplines which are related but have often developed separately.

3.1.2 Precipitation modes

In order to create the solid phase, industrial precipitators generally utilize some method for generating supersaturation by the addition of a third component including:

(a) inducing a chemical reaction to produce the solute at a concentration in excess of its solubility
(b) lowering solubility, e.g. by drowning out aqueous inorganic solutions with organic solvents
(c) watering-out from organic solutions
(d) salting-out (usually via the common-ion effect)
(e) reduction in pressure (e.g. using supercritical fluid solvents).

Such precipitation modes generally (although not always) create supersaturation at much higher levels than by simple cooling or evaporative crystallization. In consequence, a common characteristic of these systems is the relatively rapid formation of the solid phase – hence the term 'precipitate'. In this context, therefore, the term 'precipitation' is often meant to imply 'fast crystallization'. Crystal faces, of course, reflect a regular internal array of atoms or ions. Some precipitates, however, are amorphous and many comprise relatively small agglomerated particles rather than large discrete crystals, as referred to above. However, the underlying driving force (i.e. chemical potential difference) is the same in both cases. It is thus largely the particle *structure* and formation *kinetics* that differ and which determine the particle characteristics.

This chapter selectively reviews and analyses methods for characterizing, predicting and controlling the physical form of crystals precipitated from solution in solid–liquid suspensions from a process engineering perspective. Firstly, solubility and supersaturation are defined followed by a discussion of crystallizer modelling including the use of population balance theory to predict performance. The kinetics of nucleation, crystal growth and agglomeration are subsequently considered and, finally, the influences of precipitation conditions on precipitate particle formation are illustrated by some recent examples.

3.2 Solubility and supersaturation

Solubility and supersaturation are essential prerequisites for the prediction of precipitate form since they determine both the overall yield during a precipitation process and the kinetics of particle formation.

3.2.1 Equilibrium saturation concentration

The solubility of a substance in a solvent is simply the maximum concentration that can exist at equilibrium under a given set of conditions and usually increases (but sometimes decreases) with solution temperature (see Figure 3.1). However,

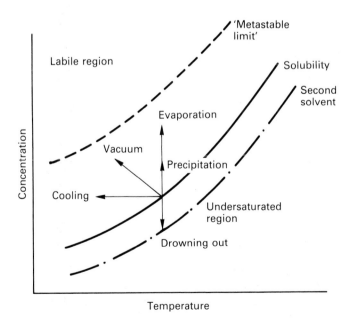

Figure 3.1 *Solubility–supersolubility diagram.*

theoretical prediction of the solublity in non-ideal systems (which are the most common) is not yet reliable, and thus resort is commonly made to empirical correlations, e.g. using expressions of the form

$$c^* = s_0 + s_1 \theta + s_2 \theta^2 \tag{3.1}$$

where c^* is the equilibrium saturation concentration at temperature θ, and s_0, s_1 and s_3 are coefficients of the components. The solubility of a substance can often be affected by the presence of a third component. For example, addition of a miscible second solvent, sometimes called a 'co-solvent' or 'diluent', often reduces solubility (see Figure 3.1). Again this effect can be correlated by an empirical expression, e.g. of the form

$$\ln c^* = a_0 + a_1 x + a_2 x^2 \tag{3.2}$$

where x is the concentration of the second solvent (for definition of the symbols used here and in the rest of this chapter see the Appendix, page 93). The second solvent can be thought of as an impurity and, indeed, some impurities can affect solubility even at the parts-per-million level.

Solubility is also affected by particle size, with small crystals (<1 μm, say) exhibiting a greater solubility than large ones. This relationship is quantified in the Gibbs–Thomson, Ostwald–Freundlich equation (Mullin, 1993)

$$c(L) = c^* \exp (4M\gamma/RT\rho L) \tag{3.3}$$

where $c(L)$ is the solubility of a crystal of diameter L and molecular weight M. Thus at a given solution concentration some small crystals dissolve while larger ones continue to grow at their expense. This process is known as 'ripening'.

3.2.2 Solubility product

In the precipitation of salt systems, the crystallizing species normally exist as free ions in solution and in this case the solubility of each ion has to be taken into account. Thus for a sparingly soluble substance A_pB_q, the solubility product is given by

$$k_{sp} = [A]^p[B]^q \tag{3.4}$$

In the simplest case $[A] = [B] = c^*$ and thus $k_{sp} = c^{*2}$. For higher concentrations the concepts of chemical activity are employed (Mullin, 1993).

3.2.3 Driving force for precipitation

As mentioned above, the essential prerequisite for crystallization is the presence of supersaturation in the liquor, i.e. concentration in excess of the solubility limit, whatever the system. This concentration difference tends to drive the system to equilibrium via the formation of the solid phase (see Figure 3.1). Thermodynamically, the driving force is given by the change in chemical potential between standing and equilibrium states. This quantity is not easy to measure, however, and is more conveniently expressed in terms of solution concentration via the following approximation:

$$\frac{\Delta\mu}{RT} \cong \ln\,(c/c^*) \cong \frac{c}{c^*} - 1 = \frac{\Delta c}{c^*} = S - 1 = \sigma \qquad (3.5)$$

where $\Delta\mu$ is the change in chemical potential, c is the standing concentration and c^* is the equilibrium saturation concentration. It is worth noting that, although commonly used, eqn (3.5) is strictly only valid for $c \approx c^*$, but many precipitations employ $c \gg c^*$. For crystallization processes, supersaturation is commonly expressed as Δc ($= c - c^*$) and is sometimes called the *concentration driving force*, while S is the *supersaturation ratio* (c/c^*) and σ is the *relative supersaturation* ($S - 1$).

For ionic systems, however, the definition of the appropriate driving force becomes more complex since the ionic concentrations are not necessarily in stoichiometric ratio and the solubility product generally applies. Thus the supersaturation ratio of sparingly soluble systems can be described by

$$\sigma = \sqrt[r]{[A]^p[B]^q/k_{sp}} - 1 \qquad (3.6)$$

where $r = p + q$. Again, more complex relationships strictly apply at higher concentrations. It can be seen from these definitions of driving force that sparingly soluble substances can easily exhibit high levels of relative supersaturation. Hence the crystallization process can be very fast and the precipitation difficult to control.

3.3 Modelling precipitation processes

3.3.1 Crystal yield

The simplest models for predicting crystal precipitator performance start with the assumption of a well-mixed vessel in which concentration, supersaturation, agitation and temperature are uniform. The first step is the calculation of crystal yield. This can easily be estimated from knowledge of solution concentration and equilibrium conditions, permitting calculation of the overall mass balance (Mullin, 1993)

$$Y = \frac{WH[c_1 - c_2\,(1 - F)]}{1 - c_2(R - 1)} \qquad (3.7)$$

where c_1 and c_2 are the initial and final concentrations (mass of anhydrous solute per mass of solvent), W is the initial mass of solvent, F is the fraction of solvent evaporated and H is the ratio of molecular weights of hydrate and anhydrous species. The quantity c_2 is generally unknown but if it is assumed that $c_2 = c_2^*$ then this calculation gives the maximum mass that can be deposited. In practice, of course, it will be somewhat less due to residual supersaturation which can be calculated starting from a mass (supersaturation) balance between the liquid and solid phases in which crystallization *kinetics* are introduced, as follows.

3.3.2 Mass (supersaturation) balance

During a precipitation, solute is depleted by crystallization, offsetting the increase in supersaturation as the precipitant reacts or the solubility product decreases. The solute mass balance between solution and solid phases is thus given in terms of supersaturation by

$$-\frac{d\Delta c}{dt} = \frac{dc^*}{d\theta}\frac{d\theta}{dt} + R_g A \tag{3.8}$$

where $dc^*/d\theta$ is the temperature dependence of solubility, R_g is the mass growth rate, and A is the total surface area of the crystals present per unit mass of solvent, which depends, of course, on the crystal size distribution of the precipitate mass.

3.3.3 Crystal-size distribution

Whilst knowlege of crystal yield is important, it is of only limited utility in predicting the *characteristics* of a crystalline product, in particular its mean size and size distribution. In the simplest case, if a mass W_{so} of seed crystals of size L_{so} are added to initiate crystallization and nucleation can be assumed to be totally suppressed, then the maximum final crystal size attainable can be calculated knowing the crystal yield. For example, for monosized seeds

$$L_{max} = L_{so}[(Y + W_{so})/W_{so}]^{1/3} \tag{3.9}$$

In the general case, of course, vast numbers of new crystal nuclei form which subsequently grow, thereby competing for solute and giving rise to a particle-size distribution of lower mean size.

3.3.4 Population density

The fundamental statistical quantity describing a size distribution is the population density n, and immediately a choice arises of which basis to use for its definition. For precipitating systems exhibiting breakage and agglomeration, particle volume is normally assumed to be conserved and it is often mathematically preferable similarly to define population density on a volume basis. However, since most experimental studies characterize particles by their overall linear dimension, the linear population density is adopted here and the equivalent equations derived (while still conserving crystal volume).

Thus, population density $n(L)$, is defined by

$$n(L) = \lim_{\Delta L \to 0} \frac{\Delta N}{\Delta L} \tag{3.10}$$

where ΔN is the number of crystals in the size range ΔL in unit suspension volume.

3.3.5 Population balance

Prediction of the crystal-size distribution when nucleation and growth proceed simultaneously requires knowledge of crystallization kinetics and slurry residence time, which together determine the total crystal number and the size of the crystals. These quantities are embodied within a mathematical framework known as the 'population balance' (Hulburt and Katz, 1964). When coupled with a mass balance on the liquid and solid phases, the crystal-size distribution can be predicted in a manner akin to prediction of a reactor product spectrum in the field of chemical reaction engineering. This approach has been developed extensively, particularly by Randolph and Larson (1988), for application to crystallization systems.

The general conservation equation is given by

$$\text{Input} - \text{Output} = \text{Accumulation} - \text{Net generation} \tag{3.11}$$

The population balance accounts for the number of crystals at each differential size interval in a continuous distribution. The quantity conserved is thus the number (population) density and may be thought of as an extension of the perhaps more familiar mass balance.

For an idealized continuous mixed-suspension, mixed-product removal (MSMPR) crystallizer (the analogue of the continuous stirred tank reactor (CSTR) in chemical reaction engineering, see Figure 3.2) the population balance is given by

$$\frac{\partial n}{\partial t} + \frac{\partial}{\partial L}(nG) + \frac{n - n}{\tau}0 = B - D \tag{3.12}$$

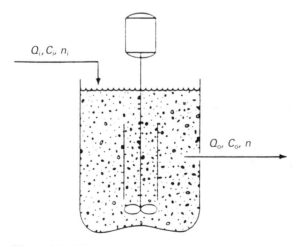

Figure 3.2   *Mixed-suspension, mixed-product-removal agitated vessel (after Randolph and Larson (1988))*

where $n(L,t)$ is the population density defined on a crystal size basis, G $(=dL/dt)$ is the linear growth rate (i.e. the rate of change of some overall characteristic particle dimension), and τ is the mean residence time within the vessel calculated as the ratio of the active volume of the crystallizer and the volumetric flow rate of suspension. The parameters B $(=B_a - D_a)$ and D $(=B_d - D_d)$ represent the net formation of particles of overall size L by aggregation and disruption, respectively. Equation (3.12) is solved subject to a boundary condition. Thus, if it is asssumed that the nuclei are so small that they are formed (for all practical pupose) at 'zero size' of number density n^0, then

$$n^0 = \left.\frac{dN}{dL}\right|_{L=0} = \left.\frac{dN}{dt}\middle/\frac{dL}{dt}\right|_{L=0} = B^0/G \tag{3.13}$$

where B^0 is the 'birth' rate of nuclei of zero size. Although it looks complicated, the population balance simply accounts for all the mechanisms by which particles can enter and leave a particular size interval – by nucleation, growth, inflow or outflow to the vessel, breakage and aggregation. From the population balance all the important characteristics of the crystal-size distribution including moments and other statistics can be estimated (for an excellent exposition, see Randolph and Larson, 1988).

The total solids holdup M_T is related to the crystal-size distribution by third moment of the population density (see later) by the integral equation

$$M_T = \alpha\rho_c \int_0^\infty n(L)L^3 dL \tag{3.14}$$

where α is the crystal volume shape factor. This integral value of M_T can thus be compared with that determined by mass balance from the change in solution concentration between inlet and outlet, i.e.

$$M_T = (c_i - c_o) \tag{3.15}$$

The supersaturation balance (eqn (3.8)) becomes

$$-\frac{d\Delta c}{dt} = \frac{dc^*}{dt} + \frac{1}{2}G\beta\rho_c \int_0^\infty n(L)L^2 dL \tag{3.16}$$

where β is the crystal surface area shape factor.

This interaction between solution supersaturation, crystallization kinetics, crystal-size distribution and solids holdup therefore presents a complex feedback loop – increasing supersaturation leads to increased crystal growth, but also increased nucleation rates, which generates more surface area and thus lowers supersaturation, and so on.

3.3.6 Performance prediction

Given expressions for the crystallization and agglomeration kinetics and solubility of the system, eqns (3.12) to (3.16) can be solved simultaneously to predict the performance of both batch and of continuous crystallizers, at either steady or unsteady state. As is evident, however, the population balance equations are complex and thus numerical methods are required for their general solution. Several different approaches have been proposed, for example by the finite-element method and the collocation method (e.g. Gelbard and Seinfeld, 1978), by moment methods (e.g. Liao and Hulbert, 1976) and discretization (e.g. Hounslow *et al.*, 1988; Marchal *et al.*, 1988).

3.3.7 Analytical approximations

Although the general solution of the population balance requires numerical methods, some useful analytic solutions are nevertheless available for particular cases. For example, if agglomeration and breakage can be neglected and crystal growth occurs at a constant rate, then an analytic solution of eqn (3.12) for a continuous MSMPR crystallizer at steady state results:

$$n(L) = n^0 \exp(-L/G\tau) \tag{3.17}$$

which is linear in a semilogarithmic plot of $\ln n(L)$ versus L

$$\ln n(L) = \ln \frac{B^0}{G} - \frac{L}{G\tau} \tag{3.18}$$

where τ ($=V/Q$) is the mean residence time within the crystallizer and G ($=dL/dt$) is the linear crystal growth rate of a crystal of size L. Equation (3.17) therefore indicates an exponential decay in population density with increasing crystal size. Thus, a plot of $\log n$ versus L (Figure 3.3) has a slope inversely proportional to

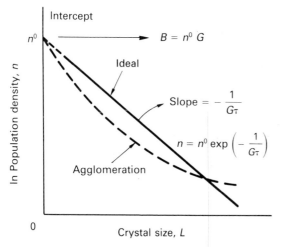

Figure 3.3 *Idealized mixed-suspension mixed-product removal population density distributions and effect of crystal agglomeration*

the linear crystal growth rate G and an intercept n^0, proportional to the nucleation rate B^0 ($=n^0 G$). This approach has been widely used to determine simultaneous crystal growth and nucleation kinetics from the analysis of MSMPR crystal-size distributions.

A corresponding analytical solution is also available for steady-state nucleation and agglomeration in the absence of crystal growth, but is of more complex form (see page 74).

3.4 Particle formation kinetics

3.4.1 Nucleation

Two modes of nucleation can be distinguished: primary and secondary.

Primary nucleation
This is the 'classical' mode of nucleation and the rate of homogeneous nucleation may be expressed theoretically from thermodynamic arguments by a relationship of the form

$$\frac{dN}{dt} = B_p^0 = k_n \exp\left(- k_m/\log^2 S\right) \tag{3.19}$$

This mode occurs mainly at high levels of supersaturation and is thus most prevalent during precipitations. Equation (3.19) predicts a rapid increase in nucleation with increasing level of supersaturation, once a certain threshold value is exceeded (the 'metastable limit', see Figure 3.1). This fact is made use of in batch operation in which an induction period is observed. The time taken before the appearance of the new phase may thus be related to the nucleation and growth kinetics (Söhnel and Mullin, 1988). The solubility relationship given by eqn (3.3) (above) can also be used to estimate the critical size of nuclei above which they continue to grow, below which they dissolve. In practice, liquors may contain impurities and foreign particles which enhance the rate of heterogeneous nucleation, but the form of eqn (3.19) is still often considered to apply empirically.

Secondary nucleation
In industrial crystallizers the additional presence of existing crystals also enhances the nucleation rate. It is generally observed that the order of secondary nucleation with respect to supersaturation is less than that of primary nucleation. Experimental data are thus normally correlated by an empirical expression of the form

$$\frac{dN}{dt} = B_s^0 = k_b M_T^j \sigma^b \tag{3.20}$$

where σ is the relative supersaturation $(S-1)$. The total solids holdup term in eqn. (3.20) accounts for nucleation originating from existing crystals. Theoretically, for crystal/crystallizer collisions $j = 1$, for crystal/crystal collisions $j = 2$ and for fluid shear j = 1 to 2. Secondary nuclei are known to be formed over a range of crystal sizes but are often characterized for practical purposes by an 'effective' size, e.g. for use in crystallizer design. Although secondary nucleation is frequently observed in crystallizing systems, it is commonly assumed that j = 0 during precipitation and that primary nucleation predominates.

3.4.2 Crystal growth

Crystal growth may also occur by a variety of mechanisms for which several theoretical models have been proposed. The overall linear crystal growth rate G (= dL/dt) is normally given by an expression of the form

$$\frac{dL}{dt} = G = k_g \sigma^g \qquad (3.21)$$

For diffusion controlled growth g = 1, for screw dislocations g = 1 to 2 and for polynuclear growth $g > 2$. Again, however, crystal growth rate data from industrial crystallizers are normally correlated empirically using eqn (3.21). Although it is often assumed for design purposes that crystals grow invariantly (i.e. all crystals grow at the same rate), in fact growth rate anomalies are often observed; crystals can exhibit size-dependent growth (i.e. the growth rate of a given crystal class changing with size) and/or growth rate dispersion (i.e. the growth rate varying within a crystal size class).

Variations occur in the shape of crystals when individual faces grow at different rates, the overall crystal habit being determined by the slowest growing face (Mullin, 1993). It is worth pointing out that the presence of impurities (endemic in industrial liquors) can affect crystallization kinetics markedly leading to variations in both crystal size and habit (shape), often spectacularly.

Thus for precipitating systems as considered above, the nucleation and linear crystal growth rates can be expressed empirically by

$$B = k_b \left[\sqrt[r]{[A]^p [B]^q / k_{sp}} - 1 \right]^b \qquad (3.22)$$

and

$$G = k_g \left[\sqrt[r]{[A]^p [B]^q / k_{sp}} - 1 \right]^g \qquad (3.23)$$

where b and g are the orders of nucleation and growth, respectively, k_{sp} is the solubility product (as before), and the coefficients k_n and k_g depend on other environmental conditions (temperature, presence of existing crystals, impurities etc.). More general forms of these equations together with their use in precipitator modelling are considered by Bhatia (1989).

Two important factors for precipitating systems emerge from the above discussion. Firstly, since primary nucleation predominates in most precipitations, the order of nucleation is generally much higher than that of crystal growth, certainly more so than in simple crystallizing systems exhibiting secondary nucleation. Secondly, for sparingly soluble substances in particular, the levels of relative supersaturation exhibited are much higher and precipitation is generally much more rapid than, say, cooling crystallization. The consequences of these features are again two-fold. Firstly, the primary crystal particle size is generally relatively small (of the order of micrometres or less). This can be desirable in terms of product quality, but certainly has undesirable consequences for solid–liquid separation. Secondly, precipitation processes are much more sensitive to local mixing conditions; high peaks in supersaturation with their consequences for product particle characteristics are generally to be avoided (see later).

3.4.3 Agglomeration

As mentioned above, it is often observed that crystals do not exist as discrete entities but comprise agglomerated particles. The situation is made yet more complicated by the differing forms of agglomerates that may exist. Thus again, broadly speaking$ two modes of crystal agglomeration can be distinguished, i.e. primary (as a result of malgrowth of crystals), and secondary (as a consequence of crystal–crystal aggregation). Both modes can arise by several mechanisms but it is often difficult to identify by visual observation alone whether agglomerates are indeed primary or secondary.

Primary agglomeration

Agglomerated forms of crystals are sometimes called 'composites', 'poly-crystals', 'dendrites', 'twins' etc. Primary agglomerate growth has been postulated to arise either as a consequence of impurity action or by diffusion field limitations especially at high growth rates (Walton, 1967) or surface nucleation, but there have been relatively few detailed studies of primary agglomerate solution growth kinetics. The crystallographic analysis of primary agglomerates, however, has long been known (McKie and McKie, 1986; Mullin, 1993). Twinned crystals in particular appear to be composed of two intergrown individuals, similar in form, joined symmetrically about a compatible axis (a twin axis) or plane (a twin plane).

For example, the effect of polycrystal growth on agglomeration of hexa-methylene tetramine (HMT) has been studied by Davey and Rutti (1976). Regular six-membered clusters were observed (Figure 3.4), the continued growth of which also resulted in occlusions. The creation of clusters was considered due to initial dendritic growth and a crystal packing model was proposed consistent with these observations. Mantovani *et al.* (1983) studied the twin growth of sucrose crystals. Aoki and Nakamoto (1984) studied the effect of supersaturation and lead ions on the twinning of potassium chloride crystals. These studies, however, were mainly concerned with the *occurrence* and *form* of primary agglomerates rather than their formation kinetics.

Aggregation and secondary agglomeration

In agitated particulate systems the motion of the crystals gives rise to collisions. Particles in contact may therefore stay aggregated, depending on their surface charge, as in the field of colloid science (Shaw, 1980). In the case of precipitation

Brownian motion

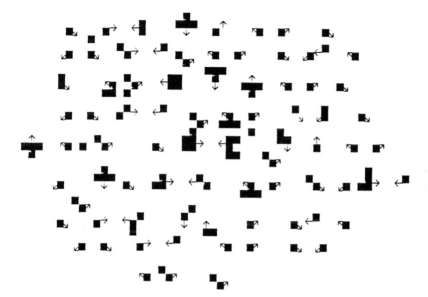

Figure 3.4   *Schematic representation of perikinetic aggregation*

from solution, however, the further process of crystal intergrowth may occur leading to the formation of relatively strong secondary agglomerates. These processes can be enhanced by agitation, thereby providing a second way by which mixing conditions are important in precipitation. At high agitation levels, however, disruption of large particles can also occur, tending to disperse the particles and limit the maximum attainable particle size. Thus the final form of secondary agglomerates is a result of the balance between these opposing rate processes. The population balance provides the mathematical framework, incorporating kinetic expressions for the various crystal formation, aggregation and disruption mechanisms (see below).

In the MSMPR crystallizer at steady state, assuming binary collisions, the increase in particle number density brought about by particle agglomeration at a particular particle size is compensated for by a corresponding reduction at two smaller sizes. The birth and death terms on the right-hand side of the population balance (eqn (3.12)) can then be written as

$$B = \frac{L^2}{2} \int_0^L \frac{K(L',L'')n(L')n(L'')\mathrm{d}L'}{L''^2} - n(L) \int_0^\infty K(L',L'')n(L')\mathrm{d}L' \qquad (3.24)$$

(note $L'^3 + L''^3 = L^3$) the factor 1/2 appearing in order not to count collisions twice. Formulation of the aggregation kernel, or collision frequency factor K, is chosen to correspond to the particular mechanism of aggregation. It accounts for the physicochemical forces that affect the aggregation process and is observed to depend on crystal size and solution supersaturation, amongst other factors (see page 80 et seq.). The effect of crystal aggregation on the crystal-size distribution is shown in Figure 3.3. Small particle numbers are reduced at the expense of larger ones, resulting in concave curvature and making kinetic analysis more difficult.

In general, both the aggregation kernel crystal and crystal growth rate are size dependent, but for simplicity both are often assumed to be independent of the particle size; conversely, the values of these parameters evaluated from experimental data represent certain averages over the size range involved.

A further analytic solution of the population balance (eqn (3.12)) is then possible if it is assumed that crystal growth is negligible compared with a purely (size-independent) agglomeration process (Hostomský and Jones, 1993).

$$\ln n(L) = \ln n(L_0) - (5/2) \ln (L/L_0) - [(L/L_0)^3 - 1] \ln [\kappa/(\kappa + 1/2)] \quad (3.25)$$

where L_0 is the primary particle size and

$$\ln n(L_0) = \ln \left[\frac{3}{2} \left(\frac{1 + 2\kappa}{\pi} \right)^{1/2} \frac{1}{K\tau} \right] + \ln \frac{\kappa}{\kappa + 1/2} - \ln L_0 \quad (3.26)$$

and the dimensionless parameter κ is defined as $\kappa = K B^0 \tau^2$

In the purely agglomerative process with a typical value of the parameter of $\kappa \gg 1$, the number density distribution $n(L)$ according to eqn (3.25) is linear when plotted in logarithmic coordinates, i.e. $\ln n(L)$ versus $\log L$, and exhibits an initial slope equal to $-5/2$ viz:

$$n(L) = n^0 (L/L_0)^{-5/2} \quad (3.27)$$

Several alternative, non-linear, aggregation kernels are summarized in Hartel *et al.*, (1976), Drake (1972) and Tavare (1987). Selection of the most appropriate one gives rise to the main difficulty of using the population balance approach. The two simplest models are as follows.

3.4.4 Perikinetic aggregation

Perikinetic aggregation results from Brownian motion of particles caused by random thermal fluctuations in a quiescent liquid (see Figure 3.4). By substituting the Stokes–Einstein equation for diffusivity into Fick's second law, Smoluchowski (1916) showed that the collision frequency factor takes the form (Gregory, 1984a, b, 1989)

$$K(L'',L'') = \frac{2kT}{3\mu} \frac{(L' + L')^2}{L'L''} \quad (3.28)$$

For monodisperse particles the aggregation kernel becomes independent of particle volume and analytical solutions for the total number of particles (in the absence of growth and nucleation) may be given as

$$N_T(t) = N_0/(1 + \frac{4kT}{3\mu} N_0 t) \qquad (3.29)$$

Thus a plot of $1/N_T$ against t is a straight line (Figure 3.5) and is independent of the agitation rate. Perikinetic motion is generally thought to apply to particles of less than about 1 μm depending on the particle-fluid motion and fluid viscosity.

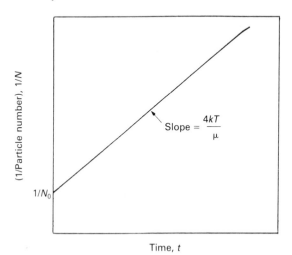

Figure 3.5   *Transient particle number (perikinetic aggregation)*

3.4.5 Orthokinetic aggregation

It has long been observed that gentle stirring or agitation promote aggregation (see Figure 3.6). As mentioned above, this is due to the velocity gradients that are induced in the liquid, causing relative motion of the particles which are present and, consequently, increasing the rate of collision. This mechanism is generally considered to apply to particles greater than about 1 μm. Smoluchowski (1917) also presented a simple theory of aggregation kinetics (assuming collisions of perfect collection efficiency) for predicting spherical particle-size distributions in a uniform liquid shear field of constant velocity gradient. The aggregation kernel is then expressed as

$$K(L',L'') = 4/3 G'(L' + L'')^3 \qquad (3.30)$$

For agitated vessels, the following equation is frequently used to estimate the average shear rate (Camp and Stein, 1943):

$$\overline{G'} = \sqrt{P/(\mu V_L)} \qquad (3.31)$$

where P is the power input to a liquid volume V_L of viscosity μ and is a function of the agitation rate (Harnby *et al.*, 1992).

Fluid shear

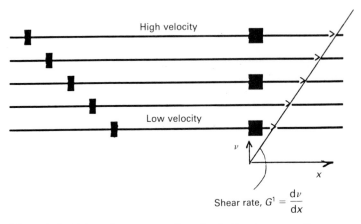

Figure 3.6   *Schematic representation of orthokinetic aggregation*

If the particles are initially monodisperse and the total solids volume fraction $(f = \pi L^3 N/6)$ remains constant (in the absence of crystal growth and nucleation), then

$$N_T(t) = N_0 \exp(-4fG't/\pi) \tag{3.32}$$

A plot of $\ln N_T$, or alternatively $\ln L$ (ignoring agglomerate porosity), against t thus yields a straight line (Figure 3.7) the slope of which depends on agitation rate. Thus, under ideal circumstances, the modes of aggregation (i.e. perikinetic or orthokinetic) can be discriminated between. Deviations below the expected slopes are usually attributed to collision inefficiency leading to imperfect aggregation. Of course deviations may also occur in a crystallization or precipitation process due to growth and nucleation, unless these factors are properly accounted for when using the population balance (see above).

Turbulence effects
Turbulence induces collisions between neighbouring particles by increasing their random motion, and so may enhance aggregation and disruption rates. Camp and Stein (1943) developed the Smoluchowski (1917) theory of orthokinetic aggregation to include such turbulent aggregation processes, while Saffman and Turner (1956) considered the effect of spatial variations of turbulent motion.

In turbulent flow, two different collision mechanisms may operate: collisions within eddies and inertia effects. By assuming that the size of the particles is much less than the scale of turbulence Levich (1962) concluded that for colloids in turbulent fluid flow aggregation within the eddies is the dominant mechanism.

These theories indicate that aggregation rates depend on the square root of dispersed power input. Note that the mean dispersed power input $\epsilon = P/V$ is frequently proposed as a scale-up criterion for agitated vessels.

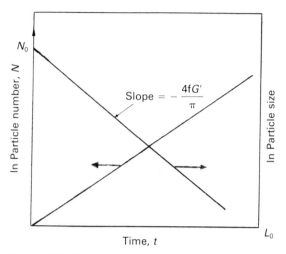

Figure 3.7   *Transient particle size and number (orthokinetic aggregation)*

Collection efficiency

In the Smoluchowski model, orthokinetic aggregation is predicted to increase with increasing particle volume and fluid shear rate, but in practice the collection efficiency is relatively low and decreases with increasing particle-size (Adler, 1981; Higashitani *et al.*, 1983), thus imposing particle size limitations. This reduction is generally thought to be due to the shear field tending to break up the aggregate on overcoming its particle strength (see later), although short range van der Waals attractive and electrostatic repulsive forces have also been invoked (Brakalov, 1987).

3.4.6 Particle dispersion

Within an agitated suspension the process of aggregation is opposed by that of dispersion. Thus new, small particles may form from existing ones by turbulent break up and dispersion of flocs or aggregates, or by the erosion, attrition or breakage of crystals, these processes being related to secondary nucleation, a form of comminution tending to reduce particle size. Two floc break-up mechanisms have been suggested in the literature: surface erosion of primary particles, and particle splitting. Furthermore, these processes can occur either by fluid–, impeller–, crystallizer– or particle–particle interactions.

With regard to solving the population balance equation, there are many forms of the particle disruption terms B_d and D_d (Remillard *et al.*, 1978, 1980; Petanate and Glatz, 1983; Randolph and Larson, 1988), but a particularly simple form which requires no integration of a fragment distribution is the two-body equal-volume breakage function. It is assumed that each particle breaks into two smaller pieces, each of which comprises half the original volume. Thus

$$B_d(v) = 2D_d(2v) \tag{3.33}$$

where, for convenience, the particle volume v is used as the independent variable in place of the crystal size L in defining population density. Randolph (1969) proposed a form for D_d that was proportional to the volume of the rupturing particle and the population density at that volume. Petanate and Glatz (1983) arrived at the same form through empirical considerations, assuming that local fluid shear forces were responsible for that disruption. The disruption functions then become

$$D_d = K_d n(v)v \tag{3.34}$$

where K_d is a disruption parameter, and

$$B_d = 4K_d vn(2v) \tag{3.35}$$

At the other end of the spectrum, however, breakage via microattrition will result in many fine crystals being generated and a relatively unchanged parent particle (see Secondary Nucleation, above) with a correspondingly more complex distribution function (Remillard *et al.*, 1978).

Hydrodynamic particle or aggregate break-up mechanisms depend on the turbulent hydrodynamic regime for which a special application of the energy spectrum approach (Kolmogoroff's universal equilibrium range theory developed by Shinnar and Church (1960) and Levich (1962)), has proven useful (Parker *et al.*, 1972). Thus, Argaman and Kaufman (1970) suggested that the principal mode of surface erosion of flocs is by turbulent drag, while Thomas (1964) proposed that floc break up is principally the result of dynamic pressure differences on opposite sides of the floc, giving rise to deformation and ultimately splitting when the particle shear strength is exceeded. From these various analyses, the rate of disruption of particle sizes occurring in the viscous subrange is predicted to be linearly dependent on the dispersed power input, while those in the inertial subrange are second order.

3.4.7 Maximum aggregate size

Further studies assuming that aggregative particle growth is limited by disruptive hydrodynamic forces have been made by several workers (e.g. Healy and LaMer, 1964; Bagster and Tomi, 1974; Tomi and Bagster, 1978; Tambo and Hozumi, 1979; and Miyanami *et al.*, 1982). In addition to collision energies increasing with increasing particle size and thus changing from viscous force to inertial force dominance, the yield stress of an aggregate has also been found to decrease with increased particle size (Tambo and Hozumi, 1979) making larger particles easier to break. In practice, the maximum aggregate size is normally correlated to shear rate by an empirical expression of the form

$$L_{max} = bG'^{-c} \tag{3.36}$$

where c is in the range 0.3–1.0.

3.5 Some precipitation process studies

A vast number of studies of precipitation processes have been reported in the literature. The examples cited below are simply intended to illustrate in particular studies of particle formation kinetics and their dependence on processing conditions, from which some particle control strategies emerge.

3.5.1 Particle formation kinetics

The crystallization of alumina has been investigated by several authors (Misra and White, 1971; Low and White, 1975; Halfon and Kaliaguine, 1976; Remillard *et al.*, 1978, 1980). Misra and White found that agglomeration was the dominant particle formation process, but crystal intergrowth also occurred and gave rise to impurity inclusions. Remillard *et al.* determined that particle abrasion also occurs and proposed correlations for the birth and death functions, which include the effects of hydrodynamic intensity.

Grabenbauer and Glatz (1981), Petanate and Glatz (1983), Glatz *et al.* (1986) and Brown and Glatz (1987) studied the precipitation of soy protein crystals, as referred to in part above. In the latter study, however, using previously precipitated agglomerates, aggregate breakage arising from collisions was found to be the dominant phenomenon and the data were modelled using an empirical breakage function in the population balance; empirical models of the form of eqn (3.36) were found wanting in describing the process.

Hartel *et al.* (1986) carried out an elegant study of the precipitation and subsequent aggregation of calcium oxalate dihydrate (CaOx) crystals in an MSMPR mininucleater/Couette aggregator sequence. Aggregation increased strongly with increasing CaOx supersaturation while disruption rates decreased, indicating that a change in ionic conditions at the crystal surface enhanced the probability and strength of particle attachment on collision. Aggregation was only weakly dependent on rotation speed, while rupture of the aggregates increased and the latter was presumed to be due to increasing turbulence. A general form of the population balance including aggregation and rupture terms was solved numerically to model the experimental particle-size distributions. While excellent agreement was obtained using semiempirical two-particle aggregation and disruption models (Figure 3.8), crystal-size distribution predictions of theoretical models based on laminar and turbulent flow considerations deviated from the data, indicating that more work is required to deduce the precise aggregation and disruption mechanisms.

In an alternative approach, i.e. using batch operation to determine crystalliza-tion and agglomeration kinetics in sequence, Söhnel *et al.* (1988) studied the precipitation of strontium molybdate crystals. Analysis of induction times implied that crystals were formed by primary nucleation followed by diffusion-controlled growth. Soon after the induction period, however, the small individual crystals aggregate, which is consistent with an orthokinetic mechanism (Figure 3.9), and form agglomerates (the size of which depends on both the stirring intensity and initial supersaturation) and eventually stabilizes. Again, a supersaturation dependence of attachment efficiency was detected.

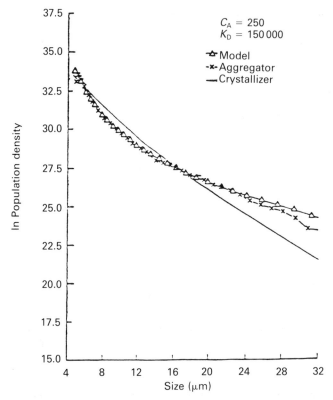

Figure 3.8   *Predicted and experimental calcium oxalate agglomerate size distributions (Hartel* et al., *1986)*

Frank *et al.* (1988) studied the batch precipitation kinetics of salicylic acid and extended it to continuous operation, while David *et al.* (1991a) reported a kinetic study of the semibatch precipitation of adipic acid using the method of classes due to Marchall *et al.* (1988) extended to include hydrodynamic effects (David *et al.*, 1991b) to model the agglomerate particle-size distributions.

3.5.2 Effect of supersaturation

In order to agglomerate, aggregated crystals must form a firm attachment. There are, however, remarkably few reports of the supersaturation dependence of agglomeration. Beckman and Farmer (1987), for example, modelled the precipitation of barium sulphate in an MSMPR crystallizer using the population balance, and concluded empirically that agglomeration efficiency is lowered by high precipitation rates which are related to the level of supersaturation. It is generally observed however, that agglomeration rates increase with super-

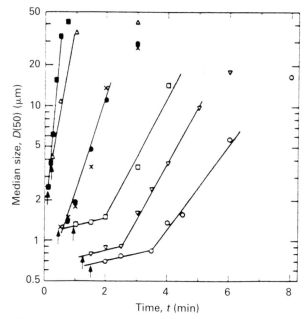

Figure 3.9 *Transient mean particle size of strontium molybdate crystals for initial concentrations (mol m^{-3} of (○) 15;, (△) 17; (□) 20; (●) 30; (×) 40; (A) 50; (■) 70 SrMoO$_4$ (Söhnel et al., 1988)*

saturation as though the crystals become more 'sticky' (Mullin, 1992). Similarly, crystal intergrowth is also observed, and this would also be expected to depend on supersaturation. Thus the agglomeration kernel can be related empirically to supersaturation by an expression of the form

$$K = k_a \sigma^a \tag{3.37}$$

In their studies of alumina precipitation, Remillard *et al.* (1978, 1980) inferred second-order dependence of crystal growth rate and fourth-order dependence of crystal agglomeration.

In a study of CaOx precipitation, Hounslow *et al.* (1988) report a linear dependence of both growth and agglomeration rates on supersaturation, while David *et al.* (1991a, b) assumed that agglomeration depends directly on growth and reported a linear dependence for adipic acid precipitation. Thus in these studies the agglomeration order $a = 1$.

3.5.3 Effect of pH

Agglomeration is similarly dependent on solution pH due to its effect on surface charge and hence the propensity of aggregated crystals to remain together. Maruscak *et al.* (1971) studied agglomeration during the continuous precipitation of calcium carbonate, and subsequently Baker and Bergougnou (1974) provided a model of agglomeration (i.e. aggregation) controlled growth which qualitatively predicted their findings. Large precipitates were formed under conditions favouring high agglomeration rates; namely, high concentrations, temperatures and mixing speeds. Agglomeration rate was also affected by solution pH and it was inferred that this was due to its effect on the surface charge, electrical double layer and, therefore, particle repulsion forces. In a subsequent study, Jones *et al.* (1992a) determined that calcium carbonate crystal agglomeration occurs when the surface charge (zeta-potential) is at a minimum corresponding to low pH and calcium ion concentration. Particle size increases rapidly thereafter, which is consistent with an orthokinetic aggregation mechanism.

3.5.4 Effects of impurities

As mentioned above, agglomeration can also be affected by the presence of impurities. Leeden and Rosmalen (1984) concluded that 1-hydroxy-1,1-biphosphonic acid (HEDP) enhances the agglomeration of barium sulphate crystals. Since this result was contrary to the expectations of zeta-potential determina-

Figure 3.10   *Precipitated calcium carbonate primary crystals at* (a) *100 rpm and* (b) *450 rpm.* (c) *Agglomerates (Wachi and Jones, 1991b)*

tions, it was concluded that some other factor was responsible for the observed agglomeration. In another study, Budz *et al.* (1986) observed bimodal distributions of calcium sulphate (gypsum) in an MSMPR crystallizer using laser light scattering. The presence of agglomerates was confirmed by microscopy. Aluminium and maleate ions both enhanced the degree of agglomeration, but reduced overall growth and nucleation rates.

3.5.5 Effect of non-ideal mixing

Finally, it should be noted that the degree of mixing within a crystallization processing vessel is important in determining performance due to the effect of spatial non-uniformities of supersaturation. Plainly, molecules or ions have to be intimately mixed at the molecular level for supersaturation to be generated and for both nucleation and subsequent crystal growth to proceed. As mentioned above, this factor becomes increasingly important for fast precipitation processes. Danckwerts' (1958) pioneering work characterized mixing at the molecular level and noted its importance in determining particle size during precipitation. Becker and Larson (1969) considered the two extremes of micromixing during continuous MSMPR crystallization, while Pohorecki and Baldyga (1983, 1988) proposed a model of turbulent mixing for precipitation with reaction illustrated by barium sulphate and demonstrated the importance of the manner of feed-stream input. Garside and Tavare (1985) further analysed the limits of micromixing and their sensitivity to reaction and crystallization kinetics.

The effects of liquid mixing on particle size can have profound effects on subsequent downstream processing of the crystal slurry. Frequently, crystals from a precipitation are small and difficult to separate from their mother liquor, and this is made worse if poor mixing results in high local levels of supersaturation. In an effort to control this behaviour, however, Jones *et al.* (1986) and Jones and Mydlarz (1989) demonstrated the effect of solvent dilution (i.e. premixing) in improving the crystal size of potassium sulphate crystals obtained by drowning out with aqueous organic solvents and their subsequent filtrability.

In a recent study of the gas–liquid precipitation of calcium carbonate crystals (Wachi and Jones, 1991b), primary crystal size was observed by means of microscopy to be dependent on agitator speed (Figure 3.10) and the presence of agglomerates gave rise to a consequent bimodal crystal-size distribution (Figure 3.11). A film model of mass transfer with chemical reaction and precipitation has been presented (Wachi and Jones, 1991a; Jones *et al.* 1992b) which predicts an observed increase in primary crystal size with increasing agitator speed (Figure 3.12) due to the effect of liquid mixing rate on gas–liquid mass transfer into the bulk.

It was also observed that, subsequent to primary crystal growth, particle size again increased rapidly (which is consistent with an orthokinetic mechanism) and that the ultimate particle size decreased with increased estimated fluid shear rate (Figure 3.13) (Jones *et al.*, 1992a).

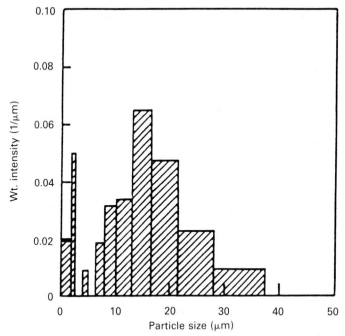

Figure 3.11 *Particle-size distribution of precipitated calcium carbonate crystals: 200 rpm for 300 s (Wachi and Jones, 1991b)*

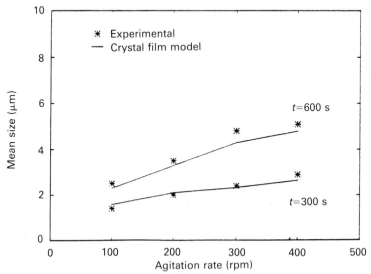

Figure 3.12 *Effect of liquid mixing rate on primary particle size of calcium carbonate crystals. Initial [Ca(OH)$_2$] = 5.2 mol m^{-3} (adapted from Jones et al., 1992b)*

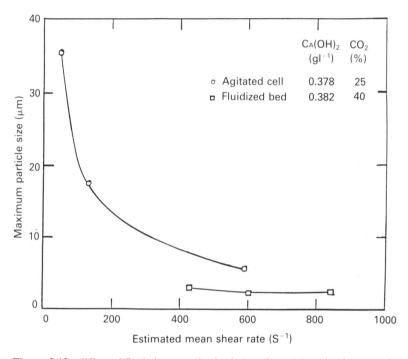

Figure 3.13   *Effect of fluid shear on the final size of precipitated calcium carbonate crystals (Jones et al., 1992a)*

Finally, all of the studies described above involved the prediction and analysis of the *overall* particle size, and thus information about the size of the constituent primary particles was lost. In order to predict both overall and primary particle size during precipitation with agglomeration, a dual particle characteristic (i.e. overall agglomerate size and the number of primary particles therein) population balance model incorporating simultaneous particle aggregation and disruption by removal of primary particles together with nucleation and growth has been proposed recently by Wachi and Jones (1992). Bimodal distributions comprising both large agglomerates and small primary particles of high specific surface area predicted in this way (Figure 3.14) are qualitatively consistent with experimental observations (Figure 3.11).

3.6 Control of particle size

It is plain even from the above selected review of the literature that precipitation can be a very complicated phenomenon in which contrary behaviour is often observed, or deduced. Despite the large amount of work done, the theoretical basis has yet to be fully developed, particular problems tending to require

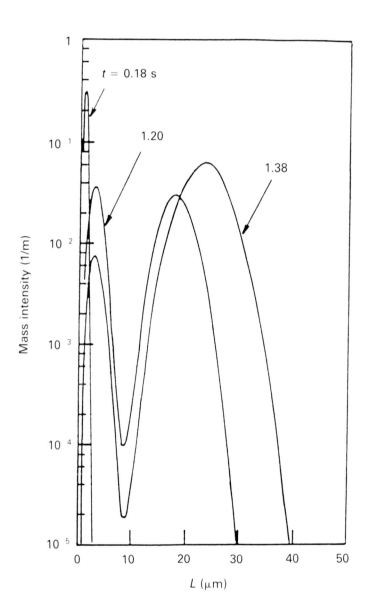

Figure 3.14 *Predicted transient particle-size distributions. Initial [Ca(OH)₂] = 5.18 mol m⁻³; 200 rpm for 3600 s (Wachi and Jones, 1992)*

specific solutions. The main factors that appear to control particle size in many systems, however, are:

(a) ionic strength
(b) level of supersaturation
(c) suspension density
(d) presence of impurities/solvents/additives
(e) mixing conditions.

In order to generate fine precipitates, very high supersaturations and fast mixing rates should generally be employed. One particular process which is often dominant in determining product-size distributions, however, is crystal agglomeration, the rate of which tends to increase with the level of supersaturation, suspension density, agitation and particle size (each of which will, of course, be related and the effects may exhibit maxima). Thus agglomeration may often be reduced by operation at low levels of supersaturation, e.g. by controlled operation of a batch crystallization or precipitation and the prudent use of solvent dilution and seeding, but at the probable expense of larger primary particles being formed.

In some cases measurement of the zeta potential is useful as an indicator of stability, i.e. agglomeration of small particles. The occurrence of either perikinetic or orthokinetic agglomeration can be deduced by the effect of agitation – only the latter is affected and tends to occur to particles larger than $1–10\,\mu m$. Agitation rates should be decreased with caution, however, consistent with achieving adequate particle suspension and homogeneous liquid mixing to avoid local peaks in supersaturation levels. On the other hand, very high shear rates of course tend to break up and diminish agglomerates, despite increasing collision rates.

To reduce or avoid agglomeration, removal of certain impurities (once identified) can be effective, while to enhance agglomeration certain additives can be used, e.g. those commonly employed as flocculating agents to enhance solid–liquid separation in the water industry, or those used to stabilize fine particle dispersions. These effects must be determined empirically with care, however, since they can be pH and supersaturation dependent.

3.7 Summary and conclusion

Crystal precipitation is an increasingly widespread industrial process, and the importance of processing conditions on product quality and downstream processing are becoming more generally recognized. Crystal agglomeration is particularly significant in precipitation processes and this aspect is also receiving increasing attention. Most process engineering studies of precipitation have been largely empirical and, while the population balance provides a mathematical framework, models of particle formation kinetics and their interaction with suspension hydrodynamics still lack a sound theoretical basis for use in design; scale-up data remain virtually non-existent.

Precipitate particle size and degree of crystal agglomeration can be affected by several variables, including level of supersaturation, suspension density, particle size, hydrodynamic intensity, pH and temperature. Few general rules apparently apply, since particle formation rates can pass through extrema and interact with each of these parameters. The main controlling factor, however, is often the level of supersaturation. For example, to produce fine crystals high levels of supersaturation should probably be employed, but to avoid excessive agglomeration low levels of supersaturation, suspension density and agitation should probably be used, consistent with other process requirements. For some systems, precipitation from supercritical fluids provides a promising alternative ultrafine particle formation technique.

It is known, however, that precipitation can also be affected by the presence of impurities (ions, surfactants, polymers etc.) but the magnitude and even the direction of the effect at a given composition is, as yet, largely unpredictable in general. Crystallography provides important information in understanding primary agglomerate growth processes, but useful kinetic models are virtually absent.

The field of colloid and surface science is highly relevant to aggregation processes and, although mostly based on idealized systems and models, should provide more at least qualitative insight into the secondary agglomeration process in the future. Similarly, advances in understanding the hydrodynamics of dispersed phases in mixing equipment and its effect on particle collision and disruption should be better incorporated into future studies of crystal precipitation and agglomeration from solution.

Finally, it is noted that the characterization of precipitates is most frequently made at the secondary level of determining agglomerate size, with little information being available on the primary distribution and porosity within the particles. Recent studies indicate that both of these characteristics should be determined if accurate crystallization kinetics, solid–liquid separation behaviour and product quality are to be accurately determined, predicted and ultimately controlled.

References

Adler, P. M. (1981) Heterocoagulation in shear flow. *J. Colloid Interface Sci.*, **83**, 106–115

Aoki, Y. and Nakamuto, Y. (1984) Penetration twins of potassium chloride. *J. Crystal Growth*, **67**, 579–586

Argaman, Y. and Kaufman, W. J. (1970) Turbulence and flocculation. *J. San. Eng. Div. ASCE*, **96**, 223–241

Bagster, D. F. and Tomi, D. (1974) The stresses within a sphere in simple shear flow flow fields. *Chem. Eng. Sci.*, **29**, 1773–1783

Baker, C. G. J. and Bergougnou, M. A. (1974) Precipitation of sparingly soluble salts: a model of agglomerate controlled growth. *Can. J. Chem. Eng.*, **52**, 246–250

Bhatia, S. K. (1989) Dynamics of continuous precipitation under non-stoichiometric conditions. *Chem. Eng. Sci.*, **44**, 751–762

Becker, G. W. and Larson, M. A. (1969) Mixing effects in continuous crystallization. *Chem. Eng. Prog. Symp. Ser.*, **95**(65), 14–23

Beckman, J. R. and Farmer, R. W. (1987) Bimodal CSD barite due to agglomeration in an MSMPR crystallizer. In *Fundamental Aspects of Crystallization and Precipitation Processes* (American Institution of Chemical Engineers Symposium Series No. 253). Vol. 83, American Institution of Chemical Engineers, 85–94

Brakalov, L. B. (1987) A connection between the orthokinetic coagulation capture efficiency of aggregates and their maximum size. *Chem. Eng. Sci.*, **42**, 2373–2383

Brown, D. L. and Glatz, C. E. (1987) Aggregate breakage in protein precipitation. *Chem. Eng. Sci.*, **42**, 1831–1839

Budz, J., Jones, A. G. and Mullin, J. W. (1986) Effect of selected impurities on the continuous precipitation of calcium sulphate (gypsum). *J. Chem. Technol. Biotechnol.*, **36**, 153–161

Camp, T. R. and Stein, P. C. (1943) Velocity gradients and internal work in fluid motion. *J. Boston Soc. Civil Eng.*, **30**, 219

Capes, C. E. (ed.) (1961, 1977, 1981, 1985) *International Symposia on Agglomeration*, Iron and Steel Society Inc. (Canada) Bookcrafters, Chelsea, MI

Danckwerts, P. V. (1958) The effect of incomplete mixing on homogeneous reactions. *Chem. Eng. Sci.*, **8**, 93–102

Davey, R. J. and Rutti, A. (1976) Agglomeration in the crystallization of hexamethylene tetramine from aqueous solution. *J. Crystal. Growth*, **32**, 221–226

David, R., Marchal, P., Klein, J.-P., and Villermaux, J. (1991a) Crystallization and precipitation engineering – III. A discrete formulation of the agglomeration of crystals in a crystallization process. *Chem. Eng. Sci.*, **46**, 205–213

David, R., Villermaux, J., Marchal, P. and Klein, J.-P. (1991b) Crystallization and precipitation engineering – IV. Kinetic model of adipic acid crystallization. *Chem. Eng. Sci.*, **46**, 1129–1136

Davis, R. F., Palmour, H. and Porter, R. L. (1984) *Emergent Process Methods for High-temperature Ceramics*, Plenum Press, New York

Drake, R. L. (1972) A general mathematical survey of the coagulation equation. In *Topics in Current Aerosol Research*, Part 2 (ed. Tidy, G. M. and Brocks, J. R.) Pergamon Press, New York

Family, F. and Landau, D. P. (eds) (1984) *Kinetics of Aggregation and Gelation*, Elsevier, Amsterdam

Frank, R., David, R., Villermaux, J. and Klein, J.-P. (1988) Crystallization and precipitation engineering – II. A chemical reaction engineering approach to salicylic acid precipitation: modelling of batch kinetics and application to continuous operation. *Chem. Eng. Sci.*, **43**, 69–77

Garside, J. (1985) Industrial crystallization from solution. *Chem. Eng. Sci.*, **40**, 3–26

Garside, J. and Tavare, N. S. (1985) Mixing, reaction and precipitation: limits of micromixing in an MSMPR crystallizer. *Chem. Eng. Sci.*, **40**, 1485–1493

Garside, J., Davey, R. J. and Jones, A. G. (1991) *Advances in Industrial Crystallization*, Butterworth-Heinemann, Oxford

Gelbard, F. and Seinfeld, J. W. (1978) Numerical solution of the dynamic equation for particulate systems. *J. Comput. Phys.*, **28**, 357

Glatz, C. E., Hoare, M. and Landa-Vertiz, J. (1986) The formation and growth of protein precipitates. *AIChE J.*, **32**, 1196–1204

Grabenbauer, G. C. and Glatz, C. E. (1981) Protein precipitation-analysis of particle size distribution and kinetics. *Chem. Eng. Commun.*, **12**, 203–219

Gregory, J. (ed.) (1984a) *Solid–liquid Separation*, Society of Chemical Industry/Ellis Horwood, Chichester

Gregory, J. (1984b) Flocculation and filtration of colloidal particles. In *Emergent Process Methods for High Technology Ceramics* (ed. Davis, R. F. *et al.*) Plenum Press, London, 59–70

Gregory, J. (1989) *CRC Crit. Rev. Environ. Control*, **19**, 185–230

Halfon, A. and Kaliaguine, S. (1976) Aluminiatrihydrate crystallization. 2: A model of agglomeration. *Can. J. Chem. Eng.*, **54**, 168–172

Harnby, N., Edwards, M. F. and Nienow, A. W (1992) Mixing in the Process Industries, 2nd edn, Butterworth-Heinemann, Oxford 137 ff.

Hartel, R. W., Gottung, B. E., Randolph, A. D. and Drach, G. (1986) Mechanisms and kinetic modeling of calcium oxalate crystal aggregation in urine-like liquor. Part I. Mechanisms. *AIChE J.*, **32**, 1176–1185; Part II. Kinetic modelling. *AIChE J.*, **32**, 1186–1195

Healy, T. W. and LaMer, V. K. (1964) The energetics of flocculation and redispersion of polymers. *J. Colloid Sci.*, **19**, 323–336

Higashitani, K., Yamauchi, K., Matsuno, Y. and Hosokawa, G. (1983) Turbulent coagulation of particles in a viscous fluid. *J. Chem. Eng. Jpn.*, **16**, 299–304

Hostomský, J. and Jones, A. G. (1991) Calcium carbonate crystallization, agglomeration and form during continuous precipitation from solution. *J. Phys. D: Appl. Phys.*, **24**, 165–170

Hostomský, J. and Jones, A. G. (1993) Modelling and analysis of agglomeration during precipitation from solution. In *Industrial Crystallization '93* (ed. Rojkowski, Z. H.) University of Warsaw, Poland, 2.–037–2.–42. Idem, Crystallization and agglomeration kinetics of calcium carbonate and barium sulphate in the MSMPR crystallizer. Ibid, 2.–049–2.–054

Hounslow, M. J., Ryall, R. L. and Marshall, V. R. (1988) Modelling the formation of urinary stones. In *CHEMECA 88* (Sydney, Australia), 1097

Hulbert, H. M. and Katz, S. (1964) Some problems in particle technology. A statistical mechanical formulation. *Chem. Eng. Sci.*, **19**, 555–574

Jones, A. G. (1985) Crystallization and downstream processing interactions. In *POWTECH 85* (*Institution of Chemical Engineers Symposium Series No. 91*) Institution of Chemical Engineers, Rugby, 1–11

Jones, A. G. and Mydlarz, J. (1989) Continuous crystallization and subsequent solid–liquid separation of potassium sulphate. Part I: MSMPR kinetics. *Chem. Eng. Res. Des.*, **67**, 283–293; Part II: Slurry filtrability, *Chem. Eng. Res. Dev.*, 294–300

Jones, A. G., Budz, J. and Mullin, J. W. (1987) Batch crystallization and solid–liquid separation of potassium sulphate. *Chem. Eng. Sci.*, **42**, 619–629

Jones, A. G., Wachi, S. and Delannoy, C.-C. (1992a) Precipitation of calcium carbonate in a fluidised bed reactor. In *Fluidization VII* (ed. Potter, O. E. and Nicklin, D. J.) Engineering Foundation, New York, 407–414

Jones, A. G., Hostomský, J. and Li, Zhou (1992b) Effect of liquid mixing rate on primary crystal size during the gas–liquid precipitation of calcium carbonate. *Chem. Eng. Sci.*, **47**, 3817–3824

van der Leeden, M. C. and van Rosmalen, G. M. (1984) The role of additives in the agglomeration of barium sulphate. In *Industrial Crystallization 84* (ed. Jančic, S. J. and de Jong, E. J.) Elsevier, Amsterdam, 325–330

Levich, V. G. (1962) *Physicochemical Hydrodynamics*, Prentice Hall, Englewood Cliffs, NJ

Liao, P. F. and Hulbert, H. M. (1976) Crystallization and agglomeration kinetics of potassium sulphate. Paper presented at the AIChE Annual Meeting (Chicago) G1–G14

Low, G. C. and White, E. T. (1975) Agglomeration effects in alumina precipitation. In *Proceedings of the Symposium on Extractive Metallurgy* (Melbourne) III.5.1–III.5.10

Lyklema, J. (1985) The colloidal background of agglomeration. In *4th International Symposium on Agglomeration* (ed. Capes, C. E.) Iron and Steel Society Inc. (Canada) AIME/Bookcrafters, Chelsea, MI

Mantovani, G., Vaccari, G., Accorsi, C. A., Aquilano, D. and Rubbo, M. (1983) Twin growth of sucrose crystals. *J. Crystal Growth*, **62**, 595–602

Marchal, P., David, R., Klein, J.-P. and Villemaux, J. (1988) Crystallization and precipitation engineering – I. An efficient method for solving population balance in crystallization with agglomeration. *Chem. Eng. Sci.*, **43**, 59–67

Maruscak, A., Baker, C. G. J. and Bergougnou, M. A. (1971) Calcium carbonate precipitation in a continuous stirred tank reactor. *Can. J. Chem. Eng.*, **49**, 819–824

McKie, D. and McKie, C. (1986) *Essentials of Crystallography*, Blackwell, Oxford

Mersmann, A. and Kind, M. (1988) Chemical engineering aspects of precipitation from solution. *Chem. Eng. Technol.*, **11**, 264–276

Misra, C. and White, E. T. (1971) Crystallization of Bayer aluminium hydroxide, *J. Crystal Growth*, **8**, 172–178

Miyanami, K., Tojo, K., Yokota, M., Fujiwara, Y. and Aratini, T. (1982) Effect of mixing on flocculation. *Ind. Eng. Chem. Fundam.*, **21**, 132–135

Mullin, J. W. (1993) *Crystallization*, 3rd edn, Butterworth-Heinemann, Oxford

Nývlt, J., Söhnel, O., Matuchova, M. and Broul, M. (1985) *The Kinetics of Industrial Crystallization*, Elsevier, Amsterdam

Parker, D. S., Kaufman, W. J. and Jenkins, D. (1972) Floc breakup in turbulent flocculation processes. *J. San. Eng. Div., ASCE*, **SA1**, 79–99

Petanate, A. M. and Glatz, C. E. (1983) Isoelectric precipitation of soy protein. I: factors affecting particle size distribution. *Biotechnol. Bioeng.*, **25**, 3049–3058; II: Kinetics of protein aggregate growth and breakage. *Biotechnol. Bioeng.*, **25**, 3059–3078

Pohorecki, R. and Baldyga, J. (1983) The use of a new model of micromixing for determination of crystal size in precipitation. *Chem. Eng. Sci.*, **38**, 79–83

Pohorecki, R. and Baldyga, J. (1988) The effects of micromixing and the manner of reactor feeding on precipitation in stirred tank reactors. *Chem. Eng. Sci.*, **43**, 1949–1954

Randolph, A. D. (1969) Effect of crystal breakage on crystal size distribution from a mixed suspension crystallizer. *Ind. Eng. Chem. Fundam.*, **8**, 58

Randolph, A. D. and Larson, M. A. (1988) *Theory of Particulate Processes*, 2nd edn, Academic Press, New York

Remillard, M., Cloutier, L. and Methot, J. C. (1978) Crystallisation du trihydrate d'allumine: un model pour l'attrition. *Can J. Chem. Eng.*, **56**, 230–235

Remillard, M., Cloutier, L. and Methot, J. C. (1980) Crystallisation du trihydrate d'allumine: effects des conditions d'agitation, *Can. J. Chem. Eng.*, **58**, 348–356

Rossiter, A. P. and Douglas, J. M. (1986) Design and optimisation of solids processes. *Chem. Eng. Res. Des.*, **64**, 175–196

Saffman, P. G. and Turner, J. S. (1956) On the collision of drops in turbulent clouds. *J. Fluid. Mech.*, **1**, 16

Shaw, D. J. (1980) *Introduction to Colloid and Surface Chemistry*, 3rd edn, Butterworth-Heinemann, Oxford

Shinnar, R. and Church, J. M (1960) Predicting particle size in agitated dispersions. *Ind. Eng. Chem.*, **52**, 253–256

Smoluchowski, M. V. (1916) Three lectures on diffusion, Brownian movement and coagulation of colloidal systems. *Phys. Z.*, **17**, 557

Smoluchowski, M. V. (1917) Mathematical theory of the kinetics of coagulation of colloidal systems. *Z. Phys. Chem.*, **92**, 129–168

Söhnel, O. and Mullin, J. W. (1988) Interpretation of crystallization induction periods. *J. Colloid Interface Sci.*, **123**, 43–50

Söhnel, O., Mullin, J. W. and Jones, A. G. (1988) Nucleation, growth and agglomeration during the batch precipitation of strontium molybdate. *Ind. Eng. Chem. Res.*, **27**, 1721–1728

Söhnel, O., Chianese, A. and Jones, A. G. (1991) Theoretical approaches to the design of precipitation systems: the present state of the art. *Chem. Eng. Commun.*, **106**, 151–175

Tambo, N. and Hozumi, H. (1979) Physical characterisation of flocs. II: Strength of floc. *Water Res.*, **13**, 421–427

Tavare, N. S. (1986) Mixing in continuous crystallizers. *AIChE J.*, **32**, 705–732

Thomas, D. G. (1964) Turbulent disruption of flocs in small particle suspensions. *AIChE J.*, **10**, 517–523

Tomi, D. T. and Bagster, D. F. (1978) The behaviour of aggregates in stirred vessels. I: Theoretical considerations of the effects of agitation. *Trans. Inst. Chem. Eng.*, **56**, 1–8

Wachi, S. and Jones, A. G. (1991a) Mass transfer with chemical reaction and precipitation. *Chem. Eng. Sci.*, **46**, 1027–1033

Wachi, S. and Jones, A. G. (1991b) Effect of gas–liquid mass transfer on crystal size distribution during the batch precipitation of calcium carbonate. *Chem. Eng. Sci.*, **46**, 3289–3293

Wachi, S. and Jones, A. G. (1992) Dynamic modelling of particle size distribution and degree of agglomeration during precipitation. *Chem. Eng. Sci.*, **47**, 3145–3148

Walton, A. G. (1967) *The Formation and Properties of Precipitates*, Wiley Interscience, New York

Wyn-Jones, E. and Gormally, J. (eds) (1983) Aggregation processes in solution. In *Studies in Physical and Theoretical Chemistry*, Vol. 26, Elsevier Amsterdam

Appendix: List of symbols

a_i	coefficients, $i = 0, 1, 2$ (eqn (3.2))	L	characteristic particle dimension (m)
A	total crystal surface area ($m^2 m^{-3}$)	L_{so}	characteristic seed crystal size (m)
B^0	nucleation rate (number s^{-1} m^{-3})	M	molecular weight
		M_T	suspension density (kg m^{-3})
B_a	birth rate of particles due to aggregation (number s^{-1} m^{-3})	$n(L,t)$	population density distribution function (number m^{-4})
B_d	birth rate of particles due to disruption (number s^{-1} m^{-3})	n_o	outlet population density distribution function (number m^{-4})
c	concentration in solution (kg m^{-3})		
D_a	death rate of particles due to aggregation (number s^{-1} m^{-3})	N	number of particles (number m^{-3})
D_d	death rate of particles due to disruption (number s^{-1} m^{-3})	N_0	initial total number of particles (number m^{-3})
D_I	impeller diameter (m)	N_T	total number of particles (number m^{-3})
f	solids fraction	P	power input (W m^{-3})
F	fraction of solvent evaporated	R	universal gas constant
G'	mean velocity gradient (shear rate) (s^{-1})	R_g	mass crystal growth rate (kg m^{-2} s^{-1})
G	linear crystal growth rate (m s^{-1})	s_i	coefficients, $i = 0, 1, 2$ (eqn (3.1))
H	ratio of molecular weights of hydrate and anhydrous species	S	supersaturation ratio
		t	time (s)
k	Boltzman constant	T	temperature (K)
k_a	coefficient (eqn (3.37))	u	particle volume (m^3)
k_b	coefficient (eqn (3.20))	v	particle volume (m^3)
k_g	coefficient (eqn (3.21))	V_L	vessel liquid volume (m^3)
k_{sp}	solubility product	W	mass of solvent (kg)
$K(L)$	aggregation kernel (m^3 s^{-1})	W_{so}	seed crystal mass (kg kg^{-1})
K_d	disruption parameter ((m^3 $s)^{-1}$)	x	concentration ratio of second solvent (kg kg^{-1})
		Y	crystal yield (kg kg^{-1})

Exponents

b supersaturation order of nucleation rate

c shear rate dependence of maximum agglomerate size

g supersaturation order of crystal growth

Greek symbols

α volume shape factor

β surface shape factor

γ surface energy (J m^{-2})

$\Delta\mu$ chemical potential difference

ϵ mean dispersed energy input (W m^{-3})

κ dimensionless aggregation kernel (eqn (3.27))

μ fluid viscosity (Pa s)

υ kinematic viscosity (m^2 s^{-1})

ρ_c crystal density (kg m^{-3})

ρ_L fluid density (kg m^{-3})

σ relative supersaturation

θ temperature (°C)

τ mean residence time in vessel (s)

Chapter 4

Computational studies of the morphology of molecular crystals through solid-state intermolecular force calculations using the atom–atom method

G. Clydesdale, K. J. Roberts and R. Docherty

4.1 Introduction

Crystallization represents an important separation, purification and preparation technique for the chemical industry. For the processing of organic chemical particulates, consideration of product stability often precludes the use of other separation techniques such as distillation, and the technique confers the additional advantage of being a more energy-efficient process.

Traditionally the process engineering community has utilized mass and energy balance considerations to predict the size distribution of particulates prepared in industrial crystallizers (e.g. Davey, 1987). For continuous crystallizers, Randolph and Larson's mixed-suspension mixed-product removal (MSMPR) (see e.g. Randolph and Larson, 1988; Larson, 1991) has enabled the prediction of particle-size distributions on a routine basis. This approach provides an appropriate and useful modelling technique for optimizing the crystallization of bulk inorganic particulates such as those relevant to the agrochemical and industrial-chemical sectors. However, its use is not always appropriate when dealing with speciality particulates produced by batch crystallization processes, such as those related to the production of fine chemicals, e.g. dyes and pigments, pharmaceuticals and toiletries, effect chemicals and additives. Over the past decade this limitation has become increasingly pronounced as the European chemical industry has moved away from labour- and capital-intensive bulk chemical production to batch-produced higher value-added products.

It is now clear that in industrial crystallization the control of macroscopic aspects such as the size distribution alone is not sufficient and that structural factors such as particle shape and size, morphological stability and purity, structural polymorphism and surface structure are becoming increasingly more important in process control and optimization. This in turn demands a better understanding of the underlying molecular processes that occur during crystallization. As such structural factors tend to be system specific, the use of empirical routes for their optimization is both economically demanding and ultimately self-defeating as it offers very limited scope for process development and enhancement. As a result, significant effort has been devoted to evolving and developing structurally based predictive routes for modelling batch crystallization processes and in this chapter we overview recent work in this area which has been directed towards the computational modelling of crystal morphology and the assessment of the structural role played by crystal habit modifiers.

4.2 Crystal growth mechanisms

Crystallization from the liquid phase can be effected experimentally by a number of approaches which can produce the essential driving force of chemical potential (i.e. supersaturation) to initiate the growth process, these include:

(a) physical techniques such as solvent evaporation, solubility change by temperature variation, electrolytic solvent decomposition, drowning out by a miscible second solvent
(b) chemical techniques such as a reactive double decomposition induced by, for example, the mixing of two salts.

The overall supersaturation-driven crystal growth process can be subdivided into two distinct phases: nucleation and growth.

In the nucleation stage supersaturation causes the formation and agglomeration of molecular solute clusters in solution. These clusters are maintained in a dynamic equilibrium which is dictated by the balance between the volume and surface contributions to their excess free energy. At sufficient cluster sizes the bulk term dominates over that due to the surface and a viable particle is formed. This nucleation stage is a genuinely three-dimensional process in contrast to the growth process that follows in which each stable cluster or grain develops by a cooperative series of growth events taking place on a number of structurally

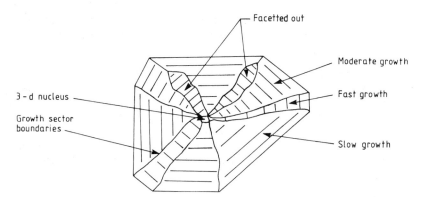

Figure 4.1 *Schematic representation of a section cut through a particulate grain demonstrating that the three-dimensional nature of such a grain is built up from a cooperative series of two-dimensional surfaces processes. Note that the 'as-grown' crystal morphology is dominated by the slow growing crystal growth surfaces* (hkl) *and that faster growing faces will 'grow out' and not be represented in the final crystal habit*

distinct crystal growth interfaces (as shown in Figure 4.1), our three-dimensional grain is in reality just the addition of a number of two-dimensional events which occur on individual crystal surfaces which are usually aligned on low-index Miller planes (*hkl*).

Not only are each of the crystal faces likely to possess a different surface chemistry, but they can grow by one of a number of different interface kinetic mechanisms which depend, in turn, on additional factors including super-saturation, solvent and additives. From this we can see that the traditional treatment by process engineers of a crystalline grain on a macroscopic rather than at the molecular level is likely to lead to problems in trying to understand and predict the size and shape of particulates.

Over the past 40 years our understanding of growth interface kinetics has developed significantly. The seminal work of Burton, Cabrerra and Frank (BCF) in 1951 (Burton *et al.*, 1951) laid the foundation for research in this area and has been the basis for the development of crystal growth theory in the years since. In terms of particulate technology the main advances in crystal growth theory have been in our understanding and definition of the three mechanistic regimes of crystal growth (see Figure 4.2) which define the relationship between the surface growth rate (R) and the supersaturation (S) of the crystallization system:

(a) at low supersaturations the BCF mechanism requires screw dislocations to provide crystal surface attachment sites to enable growth at the crystal–liquid interface to proceed; a parabolic rate dependence, i.e. $R \propto S^2$, takes place here

(b) at moderate supersaturations surface integration can take place by two-dimensional surface nucleation via a birth and spread mechanism (see e.g. Gilmer and Jackson, 1977); here the rate law becomes $R \propto S^{5/6}$

(c) at high supersaturations we have a rough interface growth (RIG) model; interface (kinetic) roughening (see e.g. Jackson, 1958) takes place and growth can proceed without the need to form any well-defined surface layers at the interface with a linear rate dependence, i.e. $R \propto S$. Under linear growth rate conditions the growth interface is inherently unstable.

Due to the different surface chemistry of the various crystal faces each crystallographically unique face may grow by any one of the three interface kinetic processes. This variation in surface chemistry, and hence surface binding, with crystal orientation means that the particulate growth and development is anisotropic. In addition, depending on the relative growth rates of the crystal faces in three dimensions the particle morphology can be highly variable, e.g. prismatic, tabular, needle-like etc. Thus the size and shape of a single particulate grain is governed by:

(a) the number of crystallographically independent faces
(b) which of the three mechanisms the face grows by.

The implications of morphological control in particulate technology can mean that, although the economics of a process may dictate fast precipitation rates and hence high supersaturation, the surface processes require the particle morphology to be maintained such that none of the crystal surfaces should grow under a RIG

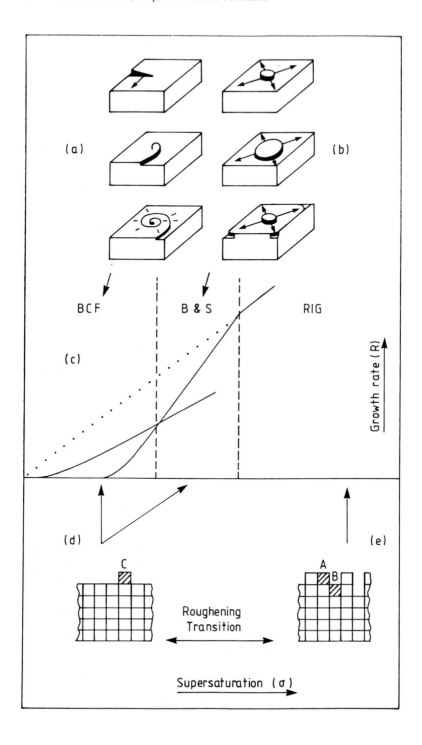

mechanism if the resulting particle is to avoid becoming excessively contaminated with impurities. Additionally, in the post-crystallization stage a poorly defined or an undesirable particle morphology (i.e. in terms of size and shape) is likely to:

(a) be difficult to filter and hence cause reduced separation efficiency
(b) result in the creation of hazardous dusts due to attrition and fragmentation during particle flow
(c) lead to caking due to a poorly defined product packing density
(d) produce a variable dissolution rate due to variable surface area.

Our understanding of these processes can be obtained from careful calculations of the intermolecular forces involved in the growth process which take place on each of the individual crystal growth surfaces (*hkl*). From these it is possible to predict the growth rate of crystal surfaces, and of the surfaces in contact with solvent and habit-modifying additives during the growth or dissolution processes. From such simulations we can predict the overall particle morphology. Advances in the computational power of dedicated computer workstations mean that such calculations can be provided from advanced molecular modelling techniques in a fairly routine manner.

4.3 The crystal chemistry of molecular materials

The structures and crystal chemistry of molecular materials are often classified into different categories according to the type of intermolecular forces present. The more important interatomic interactions contributing to the intermolecular forces include:

(a) simple van der Waals attractive and repulsive interactions
(b) classical hydrogen bonding
(c) electrostatic interactions between specific polar regions in the molecule
(d) non-classical hydrogen bonding of the type $X-H\cdots O$, where $X = sp$, sp^2, sp^3 or aromatic carbon
(e) short directional van der Waals contacts epitomized by Cl–Cl, S–Cl and S–S contacts.

Molecules can essentially be regarded as impenetrable systems whose shape and volume characteristics are governed by intramolecular interactions which result in molecular conformation and the atomic radii of the constituent atoms. The

---

Figure 4.2 (Opposite)   *Schematic illustrating the interrelationship between kinetic mechanisms and supersaturation: (a) the dislocation-controlled (BCF) mechanism; (b) two-dimensional surface nucleation by a birth and spread mechanism; (c) the relation between mechanisms and supersaturation; (d), (e) rough interface growth (RIG) showing the formation of kink (A) and step (B) sites after transformation from a singular (C) to atomically rough interface*

atomic radii are essentially exclusion zones around atoms that no other atom may enter except under special circumstances (e.g. bonding).

The awkward van der Waals shapes of organic molecules result only in a limited number of low symmetry crystal systems and space groups being adopted by organic molecules (Kitaigorodskii, 1973). The general shape of molecules (see the example of azobenzene given in Figure 4.3a) tends to result in unequal

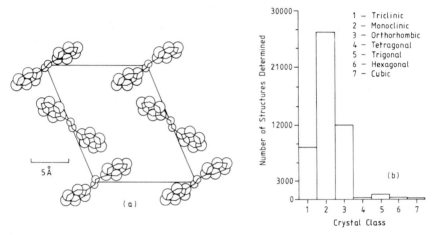

Figure 4.3   *The crystal chemistry of molecular crystals:* (a) *molecular packing for azobenzene showing the tendency of the structure to close pack whilst the anisotropic molecular shape gives rise to unequal crystal lattice parameters and hence low crystal symmetry;* (b) *histogram based on data taken from the CSSR Database showing the propensity of molecular crystals for low symmetry crystal systems*

unit cell parameters. A further consequence of their unusual shape is that organic molecules prefer to adopt space groups (such as $P2_1/c$ and $P2_1$) which possess a translational symmetry element as this allows the most efficient spatial packing of protrusions of one molecule into gaps left by the arrangement of others (Desiraju, 1989) (see Figure 4.3a). These tendencies are reflected in Figure 4.3b which is an analysis of the CSSR Database (CSE Applications Group, 1989) a version of the Cambridge Crystallographic Database (Taylor and Kennard, 1982; Allen *et al.*, 1983). The vast majority of the structures solved prefer the triclinic, monoclinic and orthorhombic crystal systems with monoclinic as the favourite.

One of the 'simplest' models of a molecular crystal is to consider the intermolecular forces to consist of only van der Waals attractive and repulsive forces. Molecules should pack to give the best lattice (cohesive) energy through minimizing the repulsive forces and maximizing the attractive interactions. A useful parameter for judging the efficiency of a molecule for using space in a given solid-state arrangement is the packing coefficient C (Kitaigorodskii, 1973):

$$C = ZV_m/V_{cell} \qquad\qquad (4.1)$$

where V_m is the molecular volume, V_{cell} is the unit cell volume and Z is the number of molecules in the unit cell. Table 4.1 shows the coefficients for a range of organic molecules including hydrocarbons, hydrogen-bonded systems, planar and twisted molecules. The molecular structures are given in Figure 4.4.

An interesting feature of Table 4.1 is the rough correlation between higher packing coefficients and increasing aromatic size (Wright, 1987). The perylene-based pigment (Figure 4.4j) has a coefficient of 0.78, only slightly lower than perylene itself ($C = 0.8$) due to the slight disruptive effect of the methyl substituents. The large, flat, planar aromatic molecules tend to stack face-to-face with slight lateral displacements to minimize atom–atom repulsions, as demonstrated by the typical structure shown in Figure 4.3a. Variations in crystal packing such as sandwich and herring-bone have been studied as a function of aromatic surface area (Gavezotti and Desiraju, 1988, 1989) and attempts to generate crystal structures from such information have been recently investigated (Gavezotti, 1991).

Table 4.1 *The packing coefficients of some organic molecules*

Structure[a]	Name	C
(a)	Benzene	0.68
(b)	Naphthalene	0.70
(c)	Anthracene	0.72
(d)	Perylene	0.80
(e)	Benzophenone	0.64
(f)	Benzoic acid	0.62
(g)	Urea	0.65
(h)	Indigo	0.68
(i)	β-Phthalocyanine	0.73
(j)	N,N-dimethylperylene-3,4:9,10-bis(dicarboxamide)	0.78
(k)	6'-Acetamido-6-bromo-2-cyanodiethylamino-4-nitroazobenzene	0.67
(l)	1,4-Diketo-3,6-diphenylpyrrolo(3,4-c)pyrrolo	0.66

[a] See Figure 4.4.

The planar aromatic systems in Table 4.1 seem to have the highest packing coefficients. Once even slight deviations from planarity are introduced the packing coefficient falls. The azo dyestuff (Figure 4.4l) is planar except for the end-group. This results in the packing coefficient falling to 0.67. For benzophenone the phenyl rings are tilted by 54° with respect to each other and the coefficient is 0.64.

A second distinctive class of molecular species is the hydrogen-bonded system. Analysis of Table 4.1 shows that hydrogen-bonded materials such as urea

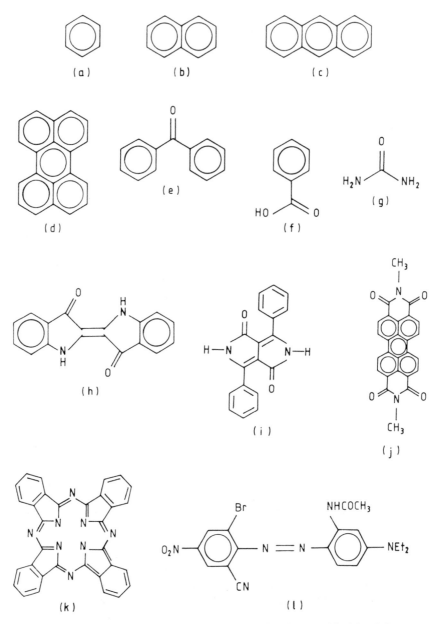

Figure 4.4   *Molecular structures for the molecules detailed in Table 4.1:* (a) *benzene;*
(b) *naphthalene;* (c) *anthracene;* (d) *perylene;* (e) *benzophenone;* (f) *benzoic acid;*
(g) *urea;* (h) *indigo;* (i) *β-phthalocyanine;* (j) *N,N dimethylperylene-3,4:9,10-*
bis(dicarboxamide); (k) *6′-acetamido-6-bromo-2-cyanodiethylamino-4-nitroazobenzene;*
(l) *1,4-diketo-3,6-diphenylpyrrolo(3,4-c)pyrrolo*

(0.65) and benzoic acid (0.62) have packing coefficients less than benzene (0.68). This surprising feature is due to the rather open nature of hydrogen-bonded structures, which is a result of attempts to maximize the hydrogen-bonding network. Benzoic acid (see Figure 4.4f) has a dimer structure of the type shown below:

$$
\begin{array}{ccc}
& O ::::: H - O & \\
// & & \backslash \\
Ar - C & & C - Ar \\
\backslash & & // \\
& O - H ::::: O &
\end{array}
$$

Benzamide (designated Y in Figure 4.15) has an extended two-dimensional hydrogen-bonding network of N–H:::OC interactions. Urea (Figure 4.4g) has a three-dimensional arrangement of hydrogen bonds where each urea molecule is hydrogen bonded to six surrounding urea molecules. This cluster of urea molecules accounts for over 90% of the lattice energy (Docherty, 1989). The desire to achieve these complicated networks of hydrogen bonds is not always consistent with achieving the most efficient packing. Hydrogen-bonding interaction continues to be an area of intense research as it is a key element in molecular solid-state chemistry (Etter, 1991), in the design of three-dimensional molecular aggregates (Simard *et al.*, 1991) and in the understanding and construction of molecular recognition complexes for biologically interesting substrates (Chang and Hamilton, 1988).

The role of special hydrogen bonds such as C–H:::O interactions remains a subject of debate. The crystallographic evidence for their role in determining packing was considered in detail (Desiraju *et al.*, 1987) through an analysis of the Cambridge Structural Database (Taylor and Kennard, 1982; Allen *et al.*, 1983). These 'special' hydrogen bonds are generally weaker than traditional hydrogen bonds and, although spectroscopic, crystallographic and theoretical investigations continue to probe their magnitude and geometry, their overall role in determining structural patterns, in competition with other interactions, remains questionable (Desiraju *et al.*, 1987).

A similar doubt exists about the role of polar contacts and directional van der Waals contacts. Debate continues over whether these weak interactions have a primary or secondary role in determining the structure arrangement. Short Cl–Cl contacts have been identified in a number of structures including 1,4-dichloro-benzene (Desiraju, 1989). It is clear that a network of Cl–Cl contacts can be observed holding interactions together. In *p*-chlorobenzoic acid, Cl–Cl interactions can be seen holding the layers of benzoic acid molecules together.

In general, it is probably safe to suggest that, in the majority of cases, the desire to pack effectively in the solid state is the biggest single driving force towards selected structural arrangements. The notable exceptions will be cases in which the need to form complex hydrogen-bonded networks overrides the desire to pack in the most efficient manner. Weaker interactions such as polar interactions, 'special' hydrogen bonds and directional van der Waals contacts are probably not solely primary structure-determining features, but will tend to be optimized within a given efficient arrangement. Ultimately, the structural arrangement for a molecule will depend on molecular structure, its constituent atoms and the delicate balance of subtle intermolecular interactions achievable through certain packing arrangements. All these structural factors play a significant role in the crystallization of organic particulates and are considered in more detail in the later sections of this chapter.

4.4 Predicting the morphology of molecular crystals from the bulk crystal structure

4.4.1 Model assumptions

The growth rate of a flat crystal surface is related to the enthalphy of dissolution at the growth interface. If we consider the dissolution of a crystallizing unit fixed at a kink site on a 'flat' growth surface of a simple Kossel crystal, then we can derive (see e.g. Bennema and van der Eerden, 1987) the molar enthalpy associated with bond (Φ) formation of

$$\Delta H_{\text{form}} = 6\Phi^{\sigma f} + 3\Phi^{sf} + 3\Phi^{ff} \tag{4.2}$$

and with bond breakage of

$$\Delta H_{\text{break}} = 6\Phi^{ff} + 3\Phi^{sf} + 3\Phi^{\delta\delta} \tag{4.3}$$

This gives

$$\Delta H(T)_{\text{diss}} = 6\Phi^{\sigma f} - 3\Phi^{ff} - 3\Phi^{\delta\delta} \tag{4.4}$$

or, more generally for any crystal structure, we can derive the molar enthalpy by summing over all atoms (i) to get

$$\Delta H(T)_{\text{diss}} = N \sum_{i=1}^{n} (\Phi_i^{\sigma f} - 0.5(\Phi_i^{ff} + \Phi_i^{\delta\delta})) \tag{4.5}$$

where the superscripts $\delta\delta$ and sf refer to solid–solid and solid–fluid bonds at the growth interface and σf and ff refer to solid–fluid and fluid–fluid bonds in the bulk fluid. Due to the lack of structural data on solid–fluid interactions at the crystal growth interface, and in order to use this theory on practical systems, three significant assumptions must be introduced:

(a) equivalent wetting – we assume that $\Phi_i^{\sigma f} = \Phi_i^{sf}$, i.e. that solid–fluid intermolecular bonds formed at the interface are the same as they would be if formed in the bulk

(b) surface/bulk structure equivalence – we assume that $\Phi_i^{\delta\delta} = \Phi_i^{ss}$ where the superscript ss refers to solid–solid interaction in the bulk crystal structure, i.e. we assume surface structure equals bulk structure and that no molecular rearrangement takes place at the surface

(c) the proportionality relationship – we assume that the three bonds formed during crystallization Φ_i^{ss}, Φ_i^{sf} and Φ_i^{ff} are in in the same ratio for any crystal surface (*hkl*).

Using these assumptions we can describe the growth process in terms of only Φ_i^{ss} interactions and can derive

$$\alpha_{hkl} = (E_{slice}^{ss}/E_{cr}^{ss})\,(\Delta H(T)_{diss}/RT) \tag{4.6}$$

where the factor α, which was first introduced by Jackson (1958) to assess the morphological stability of the growth interface, is now described only in terms of the solid-state interactions. In this expression E_{slice}^{ss} and E_{cr}^{ss} are the slice and lattice energies which are related in terms of the surface attachment energy E_{att}^{ss} (first defined by Hartman and Perdock, 1955) by

$$E_{att}^{ss} = E_{cr}^{ss} - E_{slice}^{ss} \tag{4.7}$$

For crystals growing at low supersaturations by either the BCF or birth-and-spread mechanisms, Hartman and Bennema (1980) have shown that:

$$R \propto E_{att}^{ss} \tag{4.8}$$

Thus from the calculations of the intermolecular bonds (Φ_i^{ss}) in relation to a given crystal surface (*hkl*), we are able to model the relative growth rates of the crystal faces prominent in the external polyhedral morphology.

4.4.2 The selection of likely growth forms

Simple rules for the selection of likely growth forms were first proposed by Bravais (1866), Friedel (1907) and later improved by Donnay and Harker (1937). This approach, hereafter referred to as the BFDH method (see also Phillips, 1963), suggests that the importance of a crystallographic form {*hkl*} is directly dependent on the corresponding interplanar spacing d_{hkl} after allowance has been made for space group symmetry. That is to say, the greater the interplanar spacing the greater the morphological importance of a form.

In the BFDH method allowance needs to be made for translational symmetry elements in the crystal unit cell which have the effect of effectively reducing the fundamental crystallization unit which may be repeated more than once in the period d_{hkl}. For example, if the surface structure is repeated twice then the effective interplanar spacing is halved. Such morphological rules based on symmetry reduction can be predicted directly from the diffraction extinction conditions given for the 230 space groups in *International Tables for X-ray Crystallography* (1983). Thus the overall BFDH rule can be summarized as:

taking into account submultiples of the interplanar spacing d_{hkl} due to space group symmetry the most important crystallographic forms will have the greatest interplanar spacings.

The overall approach for the BFDH method is given in more detail elsewhere (Docherty and Roberts, 1988; Docherty, 1989) and nowadays the assessment of these rules on likely morphological forms can be aided through the use of computer programs such as MORANG (Docherty *et al.*, 1988).

4.4.3 Calculating the attachment energy

Hartman and Perdok (1955) were amongst the first to attempt to quantify crystal morphology in terms of the interaction energy between crystallizing units. They used the assumption by Born (1923) that surface energy is directly related to the chemical bond energies and identified uninterrupted chains of 'strong bonds' called 'periodic bond chains' (PBC). A detailed consideration of the PBC approach is beyond the scope of this chapter, but the founding work of Hartman's group over the following 20 years (see also Hartman, 1973) led to the acceptance of the importance of calculating the slice and attachment energies. E_{slice}^{ss} is formally defined as the energy released on the formation of a growth slice of thickness d_{hkl} whilst the attachment energy (E_{att}^{ss}) is defined as the fraction of the total lattice energy (E_{cr}^{ss}) released on the attachment of this slice to a growing crystal surface (Hartman, 1973; Hartman and Bennema, 1980; Berkovitch-Yellin, 1985).

For molecular materials the overall strength of intermolecular interactions can be calculated by summing the interactions between a central molecule and each of the surrounding molecules within the crystal. Each intermolecular interaction can be considered to be the sum of the constituent atom–atom interactions, 'the atom–atom method' (Williams, 1966). If there are n atoms in the central molecule and n' atoms in each of the N surrounding molecules, then the lattice energy (E_{cr}^{ss}) can be calculated the using summation

$$E_{cr}^{ss} = 1/2 \sum_{k=1}^{N} \sum_{i=1}^{n} \sum_{j=1}^{n'} V_{kij} \qquad (4.9)$$

where V_{kij} is the interaction energy between atom i in the central molecule and atom j in the kth surrounding molecule. The interaction V_{ij} between two non-bonded atoms i and j can be described by Lennard–Jones (Jones, 1923) and Buckingham-type (Buckingham and Corner, 1947) potentials. Equation (4.10), for example, shows a simple Lennard–Jones potential combined with an additional Coulombic term to describe electrostatic interactions:

$$V_{ij} = -A/r_{ij}^6 + B/r_{ij}^{12} + q_i q_j/D r_{ij} \qquad (4.10)$$

where A and B are parameters for describing a particular atom–atom interaction, q_i and q_j are the fractional charges on atoms i and j separated by distance r and D is the dielectric constant. In some cases additional hydrogen-bonding parameters may have to be added to complete the description of the interactions (see e.g. Momany *et al.*, 1974).

Calculation of all the interactions between a central molecule and the other molecules within a crystal can be a large computational task. On inspection of the potential function shown in eqn (4.10) it is clear that since A, B, q_i, q_j and D are constants for a particular atom–atom interaction then the magnitude of an atom–atom interaction depends on the distance r between the atoms. The van der Waals interactions act over a small range as they depend on large inverse powers of r. The electrostatic interactions act over a slightly longer range depending only on $1/r$. However, at large separation distances the interaction between two molecules (sum of the atom–atom interactions) becomes negligible. This is shown in Figure 4.5 where the trend for the lattice energy to reach a plateau value is illustrated for some organic systems.

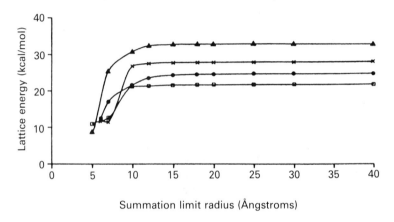

Figure 4.5   *Calculated lattice energy as a function of the summation limit for a number of organic compounds:* (□) *urea;* (●) *anthracene;* (×) β-*succinic acid;* (Δ) α-*glycine*

Using the atom–atom method to calculate the factors E_{att}^{ss} and E_{slice}^{ss}, which are important in predicting crystal growth rate, simply requires the partition of the intermolecular bonds calculated through the lattice energy summation between the slice and attachment energies (see Figure 4.6) by deciding which intermolecular interactions contribute to each. The slice energy is calculated by summing the interactions between a central molecule and all the molecules within a slice of thickness d_{hkl}. The attachment energy is calculated similarly by summing all the interactions between a central molecule and all the molecules outwith the slice. The centre of the slice can be defined as the centre of gravity of a molecule or an atom in the molecule or indeed at a lattice point of each of the symmetrically independent sites in the unit cell. The slice and attachment energies are then averaged over all these sites. This model of predicting crystal morphology is referred to as the 'AE (attachment energy) model'. The computer program HABIT (Clydesdale *et al.*, 1991a) has been specifically written for such calculations, and results using it are presented later in this chapter.

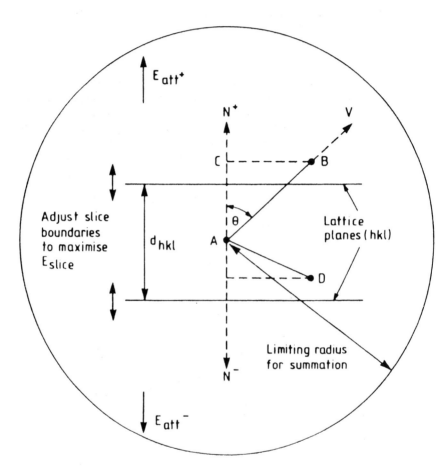

Figure 4.6 *Basic approach for calculation of intermolecular interactions using atom–atom method showing how the lattice energy is partitioned between the slice and attachment energies within a limiting sphere. A is the central molecule, B is a molecule outside the slice, D is a molecule inside the slice, N^+ is the growth normal to the planes (hkl), $N-$ is the growth normal to the planes ($\overline{hkl}$), AB and AD are bonding vectors, d_{hkl} is the interplanar spacing, θ is the angle between the growth normal and the bonding vector and AC is the component of the vector AB parallel to N the growth normal. Note that the slice boundaries defined by d_{hkl} may be shifted along the growth normal to obtain the energetically most stable slice*

4.4.4 Plotting the three-dimensional crystal shape

Gibbs (1875) proposed that the shape of a crystal will be one in which the total free energy is at a minimum, i.e.

$$\sum_i \gamma_i A_i = \text{Minimum} \tag{4.11}$$

where γ_i and A_i are the surface energies and areas of the ith face, respectively. Wulff (1901) extended this theory with a theorem which stated that the equilibrium form of a crystal should be bounded by faces whose distance from an origin should be proportional to the specific surface energies of the faces and by analogy to the relative rates at which the faces grow away from the nucleation centre of the crystal. The resulting 'Wulff plot' (see Figure 4.7) is simply a three-

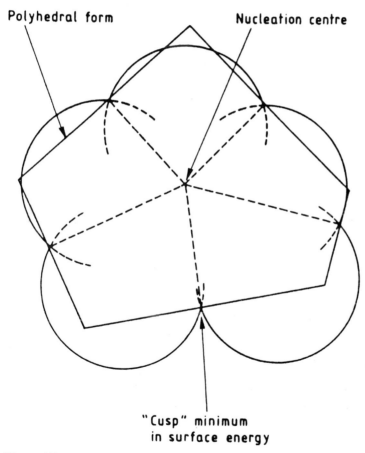

Polyhedral form   **Nucleation centre**

**"Cusp" minimum
in surface energy**

Figure 4.7 *Schematic representation of the classical Wulff plot for the derivation of equilibrium form*

dimensional polar plot of the predicted growth rate, which in our case we take to be equal to E_{att}^{ss}, as a function of the crystal growth plane (*hkl*) normal. It follows from Gibbs' theorem that a construction of planes tangent to the cusp minimum of each growth vector will give the equilibrium form and that this will be a polygon. Wulff plots can be prepared using the gnomonic projection (e.g. Phillips, 1963) which has been greatly aided with the availability of dedicated computer programs such as SHAPE (Dowty, 1980).

4.5 Examples of morphological predictions

A number of morphological predictions are presented in this section to demonstrate the suitability of this method to predict the morphology of a variety of organic molecular solid particulates.

4.5.1 Suitability of intermolecular potentials

Figure 4.5 shows the profiles for the lattice energy for a number of representative organic materials plotted as a function of the intermolecular interaction distance limit. These data show typical atom–atom summation distances and demonstrate that materials exhibiting a range of different types of intermolecular bonding (including hydrogen bonding) and different degrees of electrostatic interactions show similar summation-limit profiles.

The suitability of the intermolecular potentials used is also demonstrated by the comparison of the calculated lattice energy with the 'experimental lattice energy' (V_{exp}) shown in Table 4.2. This latter quantity is obtained from the experimentally determined sublimation enthalpy (ΔH_{sub}) via

$$V_{exp} = -\Delta H_{sub} - 2RT \qquad (4.12)$$

where $2RT$ represents a correction factor for the difference between the gas phase enthalpy and the vibrational contribution to the crystal enthalpy (Momany *et al.*, 1974). The sublimation enthalpies used in Table 4.2 were taken from Cox and Pilcher (1970). The values in this table show that it is possible to carry out calculations on a range of materials with some confidence.

As the crystal morphology is related to relative rather than absolute intermolecular bond strengths, changing the intermolecular potential does not necessarily have a significant effect of the predicted morphology. This was demonstrated in previous work (Docherty and Roberts, 1988; Docherty, 1989) for the case of the dicarboxylic acid succinic acid ($HOOC–CH_2–CH_2–COOH$). Succinic acid crystallizes in the monoclinic space group $P2_1/c$ in a unit cell of dimensions $a = 5.519$, $b = 8.862$ and $c = 5.101$ Å with $\beta = 91.59°$ (Leviel *et al.*, 1981) with two molecules in the unit cell. The lattice energy for succinic acid was calculated using three intermolecular parameter sets:

Table 4.2 *Calculated lattice energies and 'experimental' lattice energies for a range of organic materials*

Material	Lattice energy (kcal mol⁻¹)		Reference
	Calculated	'Experimental'	
Biphenyl	−21.6	−20.7[a]	Kitaigorodskii *et al.* (1968)
Naphthalene	−19.4	−18.6[a]	Williams (1966)
Anthracene	−24.9	−26.2[a]	Williams (1966)
n-Octadecane	−35.2	−37.8[a]	Williams (1966)
Benzophenone	−24.5	−23.9[a]	Momany *et al.* (1974)
Trinitrotoluene	−25.1	−24.4[a]	Momany *et al.* (1974)
α-Glycine	−33.0	−33.8[a]	Momany *et al.* (1974)
L-Alanine	−33.3	−34.2[a]	Momany *et al.* (1974)
Benzoic acid	−20.4	−23.0[a]	Lifson *et al.* (1979)
Benzamide	−20.9	—	Lifson *et al.* (1979)
Urea	−22.7	−22.2[a]	Lifson *et al.* (1979)
β-Succinic acid	−30.8	−30.1[b]	Lifson *et al.* (1979)
β-Succinic acid	−28.2	−30.1[b]	Momany *et al.* (1974)
β-Succinic acid	−31.9	−30.1[b]	Derissen and Smit (1978)

[a] Based on sublimation enthalpies from Cox and Pilcher (1970).
[b] Based on sublimation enthalpies from Momany *et al.* (1974).

(a) set I – this set of parameters derived by Momany *et al.* (1974) and Némethy *et al.* (1983) were applicable to both hydrocarbons and carboxylic acids. The authors used the 6–12 potential to describe the interactions between atoms except for the case of a hydrogen bond where a 10–12 potential (similar in structure to the 6–12 function) was used
(b) set II – this set of parameters from Lifson *et al.* (1979) were derived to deal specifically with carboxylic acids and use the 'classic' 6–12 potential function. Unlike set I or set III, the authors found that it was not necessary to introduce a special hydrogen-bonding function
(c) set III – like set II this set of parameters derived by Smit and Derissen (1978) is particular to carboxylic acids. These workers used the 6-exp function to describe the interactions between atoms except in the case of hydrogen bonds of the form OH:::O where the interactions (V_{hb}) are described by a Lippincott–Schroeder potential (Lippincott and Schroeder, 1955).

For sets I and III the charge distributions were provided by Momany *et al.* (1974), while those for set III were provided by Lifson *et al.* (1979). The calculated lattice energies (see Table 4.2) are in very good agreement with the 'experimental' lattice energy of −30.1 kcal mol⁻¹ as derived from the sublimation energy of 28.9 kcal mol⁻¹ (Momany *et al.*, 1974).

The morphology of succinic acid crystals grown by sublimation (Berkovitch-Yellin, 1985) reveals a crystal habit dominated by the {020}, {100}, {011} and {111} forms, with smaller {110} forms (see Figure 4.8a). The morphologies predicted from the attachment energy calculations using potential sets I, II and III are shown in Figures 4.8b to 4.8d and are in excellent agreement with the observed morphology, although all these models overestimate the importance of the {111} face and underestimate the importance of the {011} face. Between the simulated morphologies there are some slight variations in the relative importance of some of the growth forms; sets I and III give a slightly better fit to the observed morphology than set II. This reflects the fact that the former

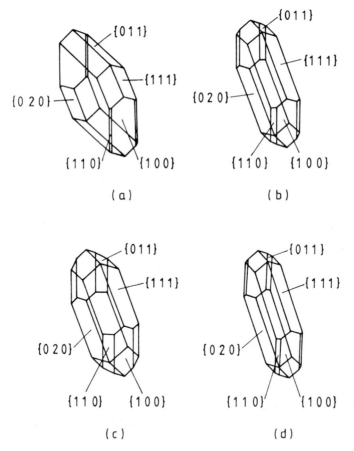

Figure 4.8 *Morphologies for succinic acid:* (a) *observed morphology (Berkovitch-Yellin, 1985);* (b) *attachment energy model (parameter set I);* (c) *attachment energy model (parameter set II);* (d) *attachment energy model (parameter set III)*

potential sets provide a better description of hydrogen bonding than the latter which is also reflected in their better fit to the experimental sublimation enthalpy shown in Table 4.2.

4.5.2 Molecular shape and crystal morphology

The shape of an organic molecule plays an important role in the kind of intermolecular forces present within the solid state (see Section 4.3) and naturally, to some extent, dictates the overall crystal shape during crystal growth. To illustrate this point we consider the role played by intermolecular forces in producing equant, needle-like and plate-like crystal morphologies.

An equant morphology – hexamine
Hexamine (hexamethylenetetramine, urotropine, $(CH_2)_6N_4$, see Figure 4.9a) crystallizes unusually for an organic material in the highly symmetrical cubic

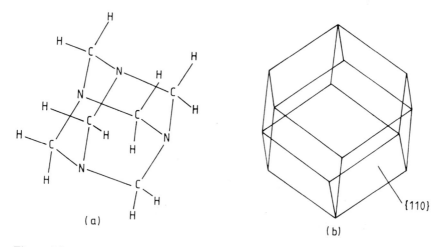

Figure 4.9 *Hexamine:* (a) *molecular structure;* (b) *morphology exhibited by observed and predicted crystals*

space group $I\bar{4}3m$ with two molecules in a body-centred unit cell of side 7.021 Å (Roscoe *et al.*, 1923). The results of the intermolecular force calculations together with the BFDH and attachment energy calculations are given in Tables 4.3 and 4.4. The latter were calculated using parameters from Momany *et al.* (1974) with the partial charges calculated using the program MNDO (Dewar and Theil, 1979). The high multiplicity (number of symmetry-equivalent faces) of the predicted forms is due to the high symmetry of the system under study and indicates that a small number of forms dominate the crystal habit. The observed

Table 4.3 *Interplanar spacings and attachment energies for hexamine*

Form {hkl}	d_{hkl} (Å)	E_{att} (kcal mol^{-1})	Multiplicity
{1 1 0}[a]	4.96	−8.36	12
{2 0 0}	3.51	−11.86	6
{2 1 1}	2.87	−11.96	24

[a] Observed forms.

Table 4.4 *Intermolecular bonding analysis for hexamine: most important bond types; bonds are between the central molecule 1(000) and Z(UVW) where Z is the symmetry position in the unit cell and U, V and W are multiples of the unit cell áxes a, b and c, respectively*

U	V	W	Z	Type	Bond energy (kcal mol^{-1})
0	−1	−1	2	a	−1.17
−1	0	−1	2	a	−1.17
−1	−1	0	2	a	−1.17
−1	0	0	2	a	−1.17
0	−1	0	2	a	−1.17
0	0	−1	2	a	−1.17
0	0	0	2	a	−1.17
−1	−1	−1	2	a	−1.17
0	0	−1	1	b	−0.60
0	0	1	1	b	−0.60
1	0	0	1	b	−0.60
0	1	0	1	b	−0.60
0	−1	0	1	b	−0.60
−1	0	0	1	b	−0.60
0	−1	−1	1	c	−0.05
−1	0	−1	1	c	−0.05
0	1	1	1	c	−0.05
1	0	1	1	c	−0.05
0	−1	1	1	c	−0.05
0	1	−1	1	c	−0.05
−1	0	1	1	c	−0.05
1	0	−1	1	c	−0.05
−1	−1	0	1	c	−0.05
1	1	0	1	c	−0.05
−1	1	0	1	c	−0.05
1	−1	0	1	c	−0.05

morphology is a rhombic dodecahedron dominated by {110} forms (Roscoe *et al.*, 1923) as shown in Figure 4.9b. Both the BFDH and AE models also predict this type of morphology.

A needle-like morphology – L-alanine

L-Alanine ($CH_3CH(NH_2)COOH$) is a naturally occurring α-amino acid which crystallizes in the orthorhombic space group $P2_12_12_1$ with four molecules in a unit cell of dimensions $a = 6.025$, $b = 12.324$ and $c = 5.783$ Å (Lehman *et al.*, 1972). The four molecules in each unit cell form a network of hydrogen bonds which continues throughout the L-alanine crystal. The intermolecular forces, detailed in Table 4.5, were calculated using parameters from Momany *et al.*

Table 4.5 *Intermolecular bonding analysis for L-alanine: most important bond types; bonds are between the central molecule 1(000) and Z(UVW) where Z is the symmetry position in the unit cell and U, V and W are multiples of the unit cell axes a, b and c, respectively*

U	V	W	Z	Type	Bond energy (kcal mol^{-1})
0	0	−1	3	a	−7.21
−1	0	−1	3	a	−7.21
0	0	−1	1	b	−5.54
0	0	1	1	b	−5.54
−1	0	−1	2	c	−1.90
−1	0	0	2	c	−1.90
−1	0	−2	4	d	−1.13
−1	−1	−2	4	d	−1.13
−1	0	−2	2	e	−0.84
−1	0	1	2	e	−0.84
0	0	−2	4	f	−0.66
0	−1	−2	4	f	−0.66
0	0	−2	1	g	−0.61
0	0	2	1	g	−0.61
1	0	−1	3	h	−0.54
−2	0	−1	3	h	−0.54

(1974) which include a hydrogen-bond potential. The lattice energy of −33.3 kcal mol^{-1} agrees well with the 'experimental' lattice energy of −34.2 kcal mol^{-1} derived from a reported sublimation enthalpy of 33.0 kcal mol^{-1} (Cox and Pilcher, 1970). The attachment energy results are given in Table 4.6.

The morphology of L-alanine crystals is generally described as prismatic dominated by {120}, {011}, {020} and {110} forms (Lehmann *et al.*, 1972). This is shown in Figure 4.10a. The AE model of the morphology (Figure 4.10b) agrees well with the observed morphology, although the predicted crystal is thicker along the *c* direction.

Table 4.6   *Calculated attachment energy values for L-alanine*

Form {hkl}	E_{att} (kcal mol^{-1})
{0 2 0}[a]	−4.5
{1 1 0}[a]	−5.2
{0 1 1}[a]	−7.4
{1 2 0}[a]	−5.3
{0 2 1}	−12.4
{1 0 1}	−10.2
{1 1 −1}	−10.3
{1 2 −1}	−12.8
{1 3 0}	−5.4

[a] Observed forms.

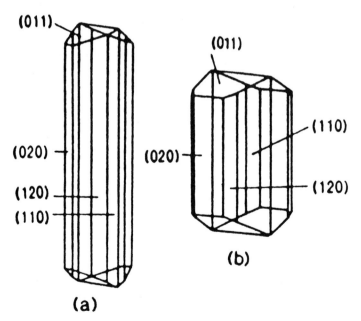

Figure 4.10   *The morphology of L-alanine: (a) observed, (b) attachment energy model*

A plate-like morphology – triclinic normal alkanes

The morphologies of a series of even-numbered normal alkanes in the range $C_{18}H_{38}$ to $C_{28}H_{58}$ were predicted (Clydesdale and Roberts, 1991; Clydesdale, 1991d; Clydesdale *et al.*, 1991b). A number of polymorphs are known in this range (see e.g. Nyburg and Lüth, 1972); for the purpose of this chapter the triclinic polymorph (space group P$\bar{1}$) is presented. The unit-cell dimensions were taken from the predicted values given by Nyburg and Potworowski (1973). The *n*-alkane homologues were created using the molecular graphics package CHEM-X (Chemical Design Ltd, 1988) by polymerizing ethane. The molecular structures were optimized using molecular mechanics procedures incorporated in the molecular graphics package INTERCHEM (Bladon and Breckinridge, 1988), while crystal packing was optimized using the program PCK83 (Williams, 1983).

For *n*-octadecane ($C_{18}H_{38}$) there is a known crystal structure (Nyburg and Lüth, 1972) which (using parameters from Williams, 1966) gave a lattice energy of -35.2 kcal mol^{-1}. For the modelled structure the calculated lattice energy was -34.4 kcal mol^{-1}. Both of these agree reasonably well with the 'experimental' lattice energy of -37.8 kcal mol^{-1} (Cox and Pilcher, 1970). The strengths of the intermolecular forces and attachment energies of the dominant crystal faces are summarized in Tables 4.6 and 4.7, respectively.

Table 4.7 *Intermolecular bonding analysis for triclinic n-alkanes: most important bond types; bonds are between the central molecule 1(000) and Z(UVW) where Z is the symmetry position in the unit cell and* U, V *and* W *are multiples of the unit cell axes* a, b *and* c, *respectively*

U	V	W	Z	Type	Bond energy (kcal mol^{-1})
1	0	0	1	a	-6.17
-1	0	0	1	a	-6.17
0	1	0	1	b	-4.80
0	-1	0	1	b	-4.80
1	-1	0	1	c	-2.35
-1	1	0	1	c	-2.35
1	1	0	1	d	-0.57
-1	-1	0	1	d	-0.57
1	0	-1	1	e	-0.52
-1	0	1	1	e	-0.52
0	-1	1	1	f	-0.36
0	1	-1	1	f	-0.36
0	0	1	1	g	-0.21
0	0	-1	1	g	-0.21

The observed morphologies of triclinic *n*-alkane crystals show a plate-like habit dominated by the {001} form. The predicted morphologies are all similar to Figures 4.11a and 4.11b for $C_{18}H_{38}$ and $C_{28}H_{58}$, respectively, and agree well

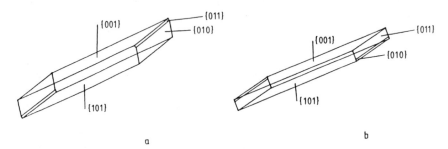

Figure 4.11 *Examples of the morphological simulation of the triclinic* n-*alkanes:* (a) $C_{18}H_{38}$; (b) $C_{28}H_{58}$

with the observed morphology (Boistelle, 1980; K. Lewtas, 1991, unpublished results). In the predicted model the relative importance of the minor forms {011} and {010} changes with increasing chain length: for $C_{18}H_{38}$ (Figure 4.10a) {010} is the more important; by $C_{22}H_{46}$ both are of equal morphological importance; by $C_{28}H_{58}$ (Figure 4.11b) {011} is the more important. Table 4.9 details the aspect ratios as a function of homologue which can be seen to increase with the carbon chain length (compare Figures 4.11a and 4.11b).

Table 4.8 *Calculated attachment energy values for* n-*octadecane*

Form {hkl}	E_{att} (kcal mol^{-1})
{001}[a]	−2.63
{010}[a]	−19.68
{01$\overline{1}$}	−21.13
{011}[a]	−20.06
{01$\overline{2}$}	−21.34
{012}	−21.26
{102}	−24.33
{103}	−24.56
{01$\overline{3}$}	−21.38
{101}[a]	−23.04

[a] Observed forms.

Table 4.9 *Predicted aspect ratios of triclinic* n-*alkanes in the range* $C_{18}H_{38}$ *to* $C_{28}H_{58}$

n-Alkane	Aspect ratio
$C_{18}H_{38}$	7.07
$C_{20}H_{42}$	7.62
$C_{22}H_{46}$	8.91
$C_{24}H_{50}$	9.33
$C_{26}H_{54}$	10.32
$C_{28}H_{58}$	11.47

4.5.3 Probing the effect of intermolecular interaction distance: example of benzoic acid

Benzoic acid ($C_6H_5CO_2H$, see Figure 4.4f) is an aromatic carboxylic acid which crystallizes in the monoclinic space group $P2_1/c$ with four molecules in a unit cell of dimensions $a = 5.510$, $b = 5.157$ and $c = 21.973$ Å with $\beta = 97.41°$ (Bruno and Randaccio, 1980). The molecules pack in the unit cell as hydrogen-bonded dimers as shown in Figure 4.15.

To assess the effects of individual intermolecular bonds on the crystal morphology, in particular the dimer formations, the AE model of the crystal was first predicted. The lattice energy was calculated using parameters from Lifson *et al.* (1979) which includes partial charges. The value of -20.4 kcal mol^{-1} is in good agreement with the experimental lattice energy of -23.0 kcal mol^{-1} (based on an experimental sublimation enthalpy of 21.8 kcal mol^{-1} (Cox and Pilcher, 1970)). The slice and attachment energies were then calculated. These are listed in Table 4.10.

Table 4.10 *Interplanar spacings and attachment energies with the AE model (E_{att} 1), with the dimer bond omitted (E_{att} 2) and with a lower limit of -0.9 kcal mol^{-1} (E_{att} 3) for benzoic acid*

Form {hkl}	d_{hkl} (Å)	E_{att} 1 (kcal mol^{-1})	E_{att} 2 (kcal mol^{-1})	E_{att} 3 (kcal mol^{-1})
{002}[a]	10.89	−3.05	−3.41	−0.00
{100}[a]	5.46	−6.52	−6.99	−3.30
{10$\bar{2}$}[a]	5.16	−6.47	−6.98	−3.30
{011}[a]	5.02	−13.77	−7.20	−10.44
{012}	4.66	−13.82	−7.27	−10.44
{102}	4.65	−14.53	−7.90	−11.11
{013}	4.20	−13.08	−10.02	−9.68
{10$\bar{4}$}	4.13	−7.77	−8.27	−3.30
{$\bar{1}\bar{1}$1}[a]	3.75	−16.71	−10.12	−13.74
{110}	3.75	−16.72	−10.12	−13.74

[a] Observed forms.

The observed morphology for benzoic acid (Berkovitch-Yellin, 1985) given in Figure 4.12a shows a {001} main face with smaller {011}, {100} and {10$\bar{2}$} side faces. The AE model of the morphology (Figure 4.12b, is in close agreement with the observed form. The main differences are in the overestimation of the {100} form and the underestimation of the {10$\bar{2}$} form.

The effect on the morphology of certain individual intermolecular interactions was then considered via a 'bonding window' routine. This allows the user to

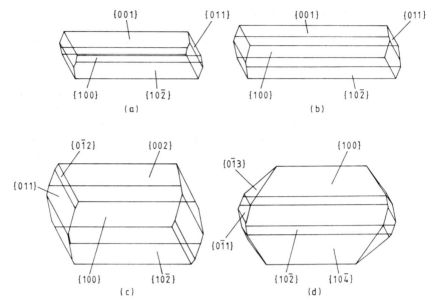

Figure 4.12 *Morphologies of benzoic acid showing the effect of intermolecular interaction distance on the calculation of* E_{att}*: (a) observed morphology; (b) simulated morphology with no restriction on intermolecular bonding distance; (c) simulated morphology with an upper limit of –2.0 kcal mol^{-1}, i.e. removing the dimer bond from the calculation; (d) simulated morphology with a lower limit of –0.9 kcal mol^{-1}*

specify upper and lower energy cut-off points: any intermolecular interactions stronger than the maximum value or weaker than the minimum value are not considered in the energy calculations. Table 4.11 gives a list of the most important intermolecular interactions with all bonds included and shows that the dimer bond holding together molecules 1 and 3 inside the unit cell is very much the strongest intermolecular interaction. (There is a similar bond between molecules 2 and 4.) Using the bonding window the contribution of these dimer bonds to the slice and attachment energies were omitted and the morphology recomputed. The morphology is shown in Figure 4.12c, there is a substantial change in the thickness of the crystal although the same forms are present compared to the AE model with no bonding limits (Figure 4.12b). The energy results (see Table 4.10) show this to be a result of large decreases in the attachment energies of the {011} and {012} forms. The {10$\bar{2}$} forms have a higher value, thus they decrease in importance and the crystal also becomes shortened along the crystallographic *b* direction.

A further morphology was computed using a lower limit of –0.9 kcal mol^{-1}. This is shown in Figure 4.12d. This model is substantially altered as very few bonds are contributing to the morphology here: the attachment energy of the {002} forms is zero, showing that the contributions to this are all included in the

Table 4.11   *Intermolecular bonding analysis for benzoic acid: most important bond types; bonds are between the central molecule 1(000) and Z(UVW) where Z is the symmetry position in the unit cell and U, V and W are multiples of the unit cell axes* a, b *and* c, *respectively*

Bond[a]	U	V	W	Z	Bond energy (kcal mol^{-1})
a	0	0	0	3	−7.037
b	0	1	0	3	−1.767
c	0	−1	0	1	−1.702
d	0	1	0	1	−1.702
e	1	0	0	1	−1.155
f	−1	0	0	1	−1.155
g	1	1	0	3	−0.989
h	1	−1	0	2	−0.604
i	1	0	0	2	−0.604
j	0	−1	0	2	−0.542
k	0	0	0	2	−0.542
l	−1	−1	0	1	−0.413
m	1	1	0	1	−0.413

[a] See Figure 4.15.

first coordination sphere. Apart from this model the greatest effect is observed when the dimer bond is omitted. This results in the prediction becoming much closer to the BFDH prediction (Docherty, 1989) which agrees closely with the observed morphology, i.e. the crystal has been reduced from a complicated hydrogen-bonded material to a less complex one where only simple van der Waals interactions control the growth in specific directions. This theoretical approach demonstrates that the crystallization process probably involves monomers not dimers.

4.6 Modelling the effect of additives on particle shape

4.6.1 Background

It is well known that the growth of crystalline materials from the liquid phase is strongly influenced by the nature of the crystallizing environment. Factors such as solvent, impurity content and temperature can often produce large changes to the growth rates of individual habit faces and hence modify the morphology of the crystallized material. It is often necessary to optimize growth conditions to produce a required crystal habit, hence a predictive capability both to model the basic crystal morphology as well as assessing the effects of habit-modifying additives is highly desirable.

In this section we describe the computer modelling techniques that have been developed for the simulation of the crystal morphology of molecular materials in the presence of 'tailor-made' additive molecules. These are 'designer' impurities which resemble the host system with differences which affect the crystal growth processes in a number of ways (Weissbuch *et al.*, 1985). Mechanistically it is thought that they have enough molecular compatibility with the host system to be able to incorporate onto the surface of the growing crystal and affect the growth rate of individual crystal faces by modifying properties of the crystallizing solution or by affecting the solid-state structure of the precipitated material. In the latter case additive molecules affect the growth rate of crystal surfaces either by blocking the movement of surface step/kink terraces (referred to as blocking modifiers) or by incorporating in the solid-state and disrupting the intermolecular bonding networks (disrupting modifiers).

In order to model the change of growth rate as a function of additive we need to modify Hartman and Perdok's classical definition of attachment energy and introduce some new energetic terms (these are defined in detail in Section 4.6.2). The overall scheme is given in Figure 4.13.

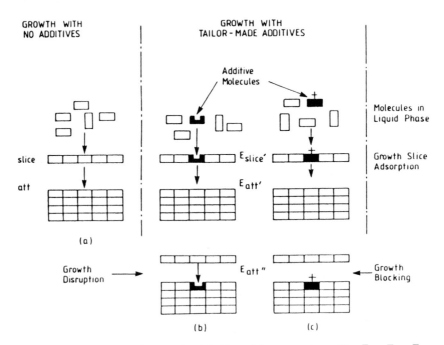

Figure 4.13 *Schematic showing the definition of the energy terms* E_{att}, $E_{att'}$, $E_{att''}$, E_{sl} *and* $E_{sl'}$ *used in morphological modelling for* (a) *pure systems and systems having* (b) *disruptive-type and* (c) *blocker-type 'tailor-made' additives*

Disruptive additives (Figure 4.13b) are generally smaller than the host system, but with a high degree of molecular similarity. Thus they can adsorb onto specific surface sites during crystal growth and this results in a failure to complete the proper bonding sequence normally adopted by the host material (Figure 4.13a). This in turn changes the value of the attachment energy associated with the adsorption of subsequent growth layers and affects the crystal habit.

The blocker type of molecular additive (Figure 4.13c) is usually larger than the host system. Again the additive is structurally similar to the host material but with an end-group which differs significantly. The section of the molecule which is structurally similar to the host system is accepted into certain specific sites on some faces as a host molecule. The differing end-group (the blocker) then prevents the oncoming molecules getting into their rightful positions at the surface.

4.6.2 Modification of attachment energy calculations to account for additive action on crystal surfaces

The nature of the intermolecular potential calculation for a crystal lattice without additive ensures that the growth slice is apolar and hence in Hartman and Perdok notation we define an E_{att} term which is really a summation of the two equal attachment energy contributions along the positive and negative growth directions, i.e.

$$E_{att} = E_{att(+)} + E_{att(-)} \qquad (4.13)$$

where

$$E_{att(+)} = E_{att(-)} \qquad (4.14)$$

as shown in Figure 4.4. However, the incorporation of an additive violates the local crystal symmetry and eqn (4.14) no longer holds and we have to calculate both $E_{att(+)}$ and $E_{att(-)}$ separately. Thus in order to assess fully the structural ramifications of the additive adsorption we need to calculate five additional terms in all: $E_{sl''}$ $E_{att(+)''}$ $E_{att(-)''}$ $E_{att(+)''}$ and $E_{att(-)''}$. These are illustrated in Figure 4.13. $E_{sl'}$ is the slice energy calculated with an additive at the centre of the slice (Weissbuch *et al.*, 1985a); $E_{att'}$ is the attachment energy of a growth slice containing an additive onto a pure surface (Weissbuch *et al.*, 1985a) and $E_{att''}$ is the attachment energy of a pure growth slice onto a surface containing an additive (Docherty, 1989). The computer program HABIT (Clydesdale *et al.*, 1991a) has been extended (G. Clydesdale, R. Docherty and K. F. Roberts, unpublished results) to allow calculation of these modified slice and attachment energy terms associated with the additive adsorption process.

In order to assess how easily an additive will adsorb on a given crystal surface (*hkl*) the parameters E_{sl} and E_{att} can be compared to $E_{sl'}$ and $E_{att'}$ and in doing so probe the molecular compatibility of the host and additive system. Weissbuch *et al.* (1985) found it useful to combine these terms by defining the relative 'binding or incorporation energy' (Δb) as

$$\Delta b = E_{sl'} + E_{att(+)'} + E_{att(-)'} - E_{cr} \tag{4.15}$$

Thus crystal faces where there is minimum change in the incorporation energy are where the additives are likely to incorporate. If Δb is strongly dependent upon crystal orientation then the incorporation will be specific to one crystal face, and vice versa.

In our approach we propose the use of the further parameter $E_{att''}$ which reflects the energy released on the addition of a pure growth slice onto a surface on which an additive has adsorbed. This additional parameter can be used as a direct measure of the growth rate of a crystal face 'poisoned' with a 'tailor-made' additive molecule, thus enabling calculation of the crystal morphology with an additive present. When there is a noticeable difference between $E_{att(+)''}$ and $E_{att(-)''}$ the former values should be used for faces (hkl) and the latter values used for faces $(\overline{hkl})$. In this way the polar effect of the additive may be modelled. In this paper the latter approach is referred to as 'model 2' with 'model 1' being the case where the average value ($E_{att''}$) is used for all faces.

The additive molecules are either obtained from the CSSR X-ray database or constructed using a conventional molecular graphics package (in our work we use either CHEM-X (Chemical Design Ltd, 1988) or INTERCHEM (Bladon and Breckinridge, 1988)). Available structural parameters are converted from fractional to Cartesian coordinates and the additive molecule is then fitted onto the host molecule system. The Cartesian coordinates for the fitted host–additive pair are then converted back to fractional coordinates using the unit cell parameters of the host using the computer program CRYSTLINK (Clydesdale *et al.*, 1992b). The attachment energy parameters as described above for predicting the new morphology with the presence of an additive can then be calculated.

4.7 Examples of simulations of crystal morphology in the presence of 'tailor-made' additives

4.7.1 Disrupting additives: example of benzamide crystallizing in the presence of benzoic acid

Benzamide ($C_6H_5CONH_2$, Figure 4.15) crystallizes in the monoclinic space group $P2_1/c$ with four molecules in a unit cell of dimensions $a = 5.607$, $b = 5.460$ and $c = 22.053$ Å with $\beta = 90.66°$ (Blake and Small, 1972). Benzoic acid (see Section 4.5.3) crystallizes similarly. The calculated lattice energy for benzamide of -21.0 kcal mol^{-1} (using the parameters of Momany *et al.* (1974)) is in agreement with previous calculations on benzoic acid (Docherty *et al.*, 1991). The calculated lattice energy for benzoic acid was -20.4 kcal mol^{-1}. This compares to a sublimation enthalpy of 22.4 kcal mol^{-1} (Cox and Pilcher, 1970). The computed morphology of benzamide is given in Figure 4.14a and reveals a plate-like morphology dominated by {002} forms and bounded by smaller {011}

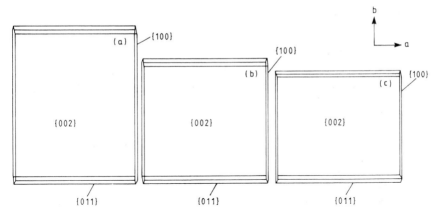

Figure 4.14 *Morphologies of benzamide:* (a) *attachment energy model;* (b) *with benzoic acid additive (model 1);* (c) *with benzoic acid additive (model 2)*

and {100} forms. The simulated and observed (Berkovitch-Yellin, 1985) morphologies show good agreement.

The effect of the benzoic acid additive can be seen from Table 4.12. The change in incorporation energy with a benzoic acid molecule in each of the crystallographic sites is given along with the host and additive attachment energies. The host attachment energy is the sum of $E_{att(+)}$ and $E_{att(-)}$ (which are of equal magnitude). From these results it seems that sites 3 and 4 for the (011) face are the most likely sites for the benzoic acid additive to adsorb since there is no loss in incorporation energy at these sites. Additive adsorption onto other sites results in a slight loss in incorporation energy.

Since the $E_{att''}$ parameter is a measure of the relative growth rate with an additive present in the surface site, two attachment energy models can be developed. In model 1 it can be assumed that the additive only adsorbs onto the {011} faces so that there will be a reduction in the growth rate of these faces relative to the other faces. The morphology computed for model 1 using $E_{att''}$ for the {011} faces and E_{att} for the other faces is shown Figure 4.14b.

Since the energy loss at the other faces is small it seems possible that the additive can adsorb onto the other faces to a lesser extent. In model 2, $E_{att''}$ was used for all faces, the computed morphology is shown in Figure 4.14c. For both models there is a reduction in the growth rates along the *b* direction, which is in agreement with experimental results (Weissbuch *et al.*, 1985).

Energetic considerations show that this effect is due to the additive preferentially adsorbing onto {011} faces which affect growth along the *b* direction. Table 4.13 demonstrates the effect of the additive from a consideration of the intermolecular bonding. There is little difference between the interactions with a host molecule at the centre of the slice and those with an additive at the centre except along the *b* direction. Here the interaction is much decreased by the

Table 4.12   *Changes in incorporation energy (Δb) and attachment energies for benzamide with benzoic acid additive (in kcal mol^{-1}), and the symmetry position in the unit cell (Z)*

Face (hkl)	Z	Δb	$E_{att(+)''}$	$E_{att(-)''}$	$E_{att''}$	E_{att}
(002)	1	1.73	−1.65	−1.65	−3.30	−3.31
	2	1.74				
	3	1.74				
	4	1.73				
(100)	1	2.56	−4.81	−4.81	−9.63	−8.64
	2	2.17				
	3	2.17				
	4	2.56				
(10$\bar{2}$)	1	2.55	−5.17	−5.17	−10.33	−9.35
	2	2.17				
	3	2.17				
	4	2.55				
(102)	1	2.57	−5.62	−5.62	−11.23	−10.24
	2	2.18				
	3	2.18				
	4	2.57				
(01$\bar{1}$)	1	1.42	−4.22	−4.22	−8.45	−10.89
	2	1.41				
	3	−0.25				
	4	−0.26				
(01$\bar{2}$)	1	1.39	−6.31	−6.31	−12.62	−15.14
	2	1.31				
	3	−0.28				
	4	−0.36				
(01$\bar{3}$)	1	0.86	−8.47	−8.47	−16.94	−19.73
	2	1.29				
	3	−0.81				
	4	−0.38				
(10$\bar{4}$)	1	1.86	−4.56	−4.56	−9.12	−8.83
	2	1.47				
	3	1.47				
	4	1.86				
(104)	1	2.74	−6.92	−6.92	−13.83	−12.66
	2	2.36				
	3	2.36				
	4	2.74				
(1$\bar{1}$0)	1	0.53	−6.80	−7.96	−14.76	−16.53
	2	0.34				
	3	2.19				
	4	2.00				

Table 4.13 *Intermolecular bonding analysis for benzamide/benzoic acid: most important bond types; bonds are between the central molecule 1(000) and Z(UVW) where Z is the symmetry position in the unit cell and* U, V *and* W *are multiples of the unit cell axes* a, b *and* c, *respectively*

Bond	U	V	W	Z	Bond energy (kcal mol^{-1})	
					Host	Additive
a	−2	0	0	3	−5.08	−5.18
b	0	1	0	1	−3.55	−1.44
c	0	−1	0	1	−3.55	−3.11
d	−1	0	0	3	−2.77	−2.69
e	−1	0	0	2	−0.72	−0.75
f	−1	−1	0	2	−0.72	−0.73
g	1	1	0	1	−0.72	−0.68
h	−1	−1	0	1	−0.72	−0.63

presence of the additive subsequently reducing growth along this direction. (Note also the polar effect of the additive – bonds *b* and *c* and bonds *e–h* in Table 4.13 are of subtly different magnitudes with the additive present.)

The molecular mechanism associated with this process can be clearly seen from the molecular packing diagram given in Figure 4.15. In this case it can be seen that due to the molecular similarity a benzoic acid molecule can get into the growing crystal at point (X) where it completes the normal bonding sequence. A further benzamide molecule trying to continue growth along the *b* direction at point (Y) cannot complete the proper bonding sequence as it encounters O:::O repulsion when expecting a hydrogen atom to complete an N–H:::O hydrogen bonding sequence.

4.7.2 Blocking additives: example of naphthalene crystallizing in the presence of biphenyl

Naphthalene ($C_{10}H_8$, Figure 4.4b) is an aromatic hydrocarbon consisting of two fused benzene rings. It crystallizes in the monoclinic space group P2$_1$/a in a bimolecular unit cell of dimensions $a = 8.098$, $b = 5.953$ and $c = 8.652$ Å with $\beta = 1\ 24.4°$ (Ponomarev *et al.*, 1976). The calculated lattice energy (using the intermolecular forces of Williams (1966)) of −19.4 kcal mol^{-1} agrees well with the 'experimental' lattice energy of −18.8 kcal mol^{-1} (based on a sublimation enthalpy of 17.4 kcal mol^{-1} (Cox and Pilcher, 1970)).

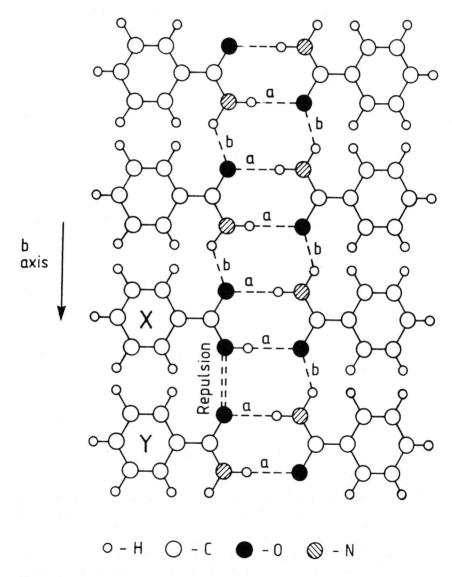

Figure 4.15 *Schematic illustration of the disruption of bonding between benzamide molecules (Y) along the crystal b direction by benzoic acid additive (X). a and b are intermolecular interactions defined in Table 4.11*

Attachment energies were calculated for the forms likely to dominate the crystal habit (identified via the geometrically based BFDH model). These are given in Table 4.14. The observed morphology (Pavlovska and Nenow, 1972)

Table 4.14 *Changes in incorporation energy (Δb) and attachment energies (in kcal mol^{-1}) for naphthalene with biphenyl additive, and the symmetry position in the unit cell (Z)*

Face (hkl)	Z	Δb	$E_{att(+)''}$	$E_{att(-)''}$	$E_{att''}$	E_{att}
(001)	1	4.11	−3.25	−2.02	−5.27	−5.97
	2	4.11				
(1$\bar{1}\bar{1}$)	1	3.12	−3.71	−5.34	−9.04	−12.24
	2	2.07				
(1$\bar{1}$0)	1	3.49	−4.26	−4.68	−8.94	−11.77
	2	2.43				
(20$\bar{1}$)	1	2.15	−4.27	−6.16	−10.43	−13.08
	2	2.15				
(20$\bar{2}$)	1	1.77	−4.93	−7.45	−12.39	−15.39
	2	1.77				
(11$\bar{2}$)	1	1.75	−4.35	−6.43	−10.77	−14.32
	2	2.74				
(200)	1	2.46	−5.84	−6.61	−12.45	−14.79
	2	2.46				
(1$\bar{1}$1)	1	2.84	−5.57	−5.19	−10.76	−14.25
	2	1.77				
(020)	1	2.03	−6.55	−6.55	−13.09	−17.46
	2	0.81				

(Figure 4.16a) is dominated by {001} flat main faces with {110} and {1$\bar{1}\bar{1}$} side faces and {20$\bar{1}$} and {101} front faces. The computed AE model (Figure 4.16b) agrees well with this.

The effect of 'blocker' additives is modelled by replacing unfavourable steric repulsions encountered as a result of the blocking nature of the additive with vacancies, thus lowering the attachment energies for faces which are 'blocked', simulating the slower growth allowing direct prediction of morphologies. The modified $E_{att''}$ parameter was used to obtain a morphological representation of the effects of a blocker additive which was not previously possible by workers such as Weissbuch *et al.* (1985) and Addadi *et al.* (1986) who presented only incorporation energy values and experimentally grown crystals to demonstrate the additive effects.

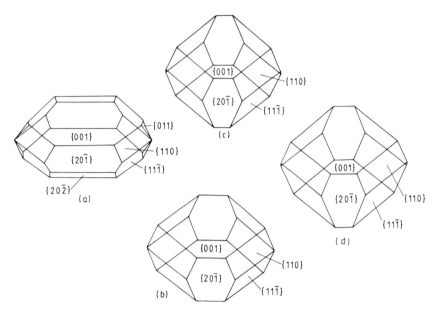

Figure 4.16   *The influence of biphenyl on the morphology of naphthalene:* (a)
naphthalene – observed morphology; (b) *naphthalene – attachment energy model;* (c)
effect of biphenyl additive on the morphology (model 1); (d) *effect of biphenyl additive
on the morphology (model 2)*

As a test of this method biphenyl was considered as an additive in the
naphthalene crystal. Biphenyl ($C_{10}H_{10}$) is an aromatic compound consisting of
two unfused benzene rings. A biphenyl molecule (using the crystal structure of
Trotter (1961)) was included into the naphthalene lattice by fitting one of the
rings to one of the naphthalene rings and obtaining coordinates in terms of this
lattice as outlined in Section 4.6.2.

The observed effect of biphenyl on the naphthalene morphology (Jetten, 1983)
is an increase in the growth rates of the {110} faces due to preferential additive
incorporation on these faces. This effect is predicted by the attachment energy
models given in Figures 4.16c and 4.16d for models 1 and 2, respectively. These
should be compared to the AE model morphology of naphathlene given in Figure
4.16b. The {110} forms are clearly more important with biphenyl present. This
effect is even clearer in the polar model 2 morphology which contains a very
small additional (20$\bar{2}$) face. The incorporation energy results (Table 4.14)
indicate that the additive can incorporate on all the faces, although site 2 for
(020) is the most favoured.

Table 4.15 gives the intermolecular bonding analysis. This shows a very large
repulsive interaction between the two molecules in the unit cell due to the effect
of the blocker part of biphenyl (one of the benzene rings). The other strong
interactions are not greatly affected.

Table 4.15 *Intermolecular bonding analysis for naphthalene/biphenyl: most important bond types; bonds are between the central molecule 1(000) and Z(UVW), where Z is the symmetry position in the unit cell and U, V and W are multiples of the unit cell axes* a, b *and* c, *respectively*[a]

Bond	U	V	W	Z	Bond energy (kcal mol^{-1})	
					Host	Additive
a	0	0	0	2	−2.36	303.61
b	−1	0	0	2	−2.36	−2.32
c	0	−1	0	2	−2.36	−2.14
d	−1	−1	0	2	−2.36	−1.98
e	0	1	0	1	−1.56	−4.19
f	0	−1	0	1	−1.56	−1.21
g	−1	−1	−1	2	−0.75	−0.40
h	0	0	1	2	−0.75	−0.86
i	−1	0	−1	2	−0.75	−0.64
j	0	−1	1	2	−0.75	−0.70
k	0	0	1	1	−0.53	−0.55
l	0	0	−1	1	−0.53	0.21

[a] Bonds [00$\bar{1}$1] (strength 0.208), [0002] (strength 303.61), [01$\bar{1}$1] (strength 0.158) and [0101] (strength 4.19 kcal mol^{-1}) were omitted and replaced by vacancies.

4.8 Conclusions

In this overview we have presented the development of computational modelling techniques for the prediction of crystal morphology as a function of crystal structure as well as the morphology when mediated by additive. The agreement between theory and experiment is in all cases excellent. However, the modelling approaches developed so far only allow the prediction of the morphology of molecular crystals growing by the birth-and-spread mechanism. Future developments are likely to be directed towards:

(a) the routine simulation of particulates crystallizing with faces growing under the RIG mechanism using Ising models (see e.g. van der Eerden *et al.*, 1978; Roberts *et al.*, 1985; Docherty *et al.*, 1991; Bennema *et al.*, 1992.

(b) simulating the growth rate of crystal surfaces mediated by the presence of steps and kinks provided by screw dislocations during growth by the BCF mechanism

(c) modelling the structural role played by additives and impurities such as solvent which do not mimic the crystal chemistry of the host lattice by optimizing the orientation and conformation of molecules with respect to the host structure

(d) extension of the types of molecules currently considered into those exhibiting a high degree of polarization such as ionic compounds through the development of new intermolecular potential functions and use of accelerated convergence techniques to cope with long-range dipolar interactions

(e) developing new structural models for considering the role played by lattice strain on crystal growth rate and growth rate dispersion so that mechanical effects such as attrition in crystallization processes can be considered.

The development and extension of the morphological modelling techniques described here can be expected to provide a routine and accessible approach to modelling the particulate properties of materials prepared by crystallization on a structural basis.

References

Addadi, L., Berkovitch-Yellin, Z., Weissbuch, I., Lahav, M. and Leiserowitz, L. (1986) A link between macroscopic phenomena and molecular chirality: crystals as probes for the direct assignment of absolute configuration of chiral molecules. *Topics in Stereochem.,* **16**, 1–83

Allen, F. H., Kennard, O. and Taylor, R. (1983) Systematic analysis of structural data as a research technique in organic chemistry. *Acc. Chem. Res.,* **16**, 146–153

Bennema, P. and van der Eerden, J. P. (1987) Crystal graphs, connected nets, roughening transition and the morphology of crystals. In *Morphology of Crystals* (ed. Sunagawa, I.) Terra Scientific, Tokyo, Chap. 1

Bennema, P., Liu Xiang Yang, Lewtas, K., Tack, R. D., Rijpkema, J. J. M. and Roberts, K. J. (1992) Morphology of orthorhombic long chain normal alkanes. *J. Crystal Growth,* in press

Berkovitch-Yellin, Z. (1985) Toward an ab initio derivation of crystal morphology. *J. Am. Chem. Soc.,* **107**, 8239–8253

Bladon, P. and Breckinridge, R. (1988) *The Molecular Modelling Package INTERCHEM,* University of Strathclyde, Glasgow

Blake, C. C. F. and Small, R. W. H. (1972) The crystal structure of benzamide. *Acta Crystallogr., Sect. B,* **28**, 2201–2206

Boistelle, R. (1980) Defect structures and growth mechanisms of long-chain normal alkanes. *Curr. Topics Mater. Sci.,* **4**, 413–480

Born, M. (1923) *Atom Theorie des Festen Zustandes (Dynamik der Kristallgitter),* 2nd edn, Leipzig, Berlin

Bravais, A. (1866) *Etudes Crystallographiques,* Gauthiers Villars, Paris

Bruno, G. and Randaccio, L. (1980) A refinement of the benzoic acid structure at room temperature. *Acta Crystallogr., Sect. B,* **36**, 1711–1712

Buckingham, R. A. and Corner, J. (1947) Tables of second virial and low-pressure Joule–Thomson coefficients for intermolecular potentials with exponential repulsion. *Proc. R. Soc.,* **189**, 118–129

Burton, W. K., Cabrera, N. and Frank, F. C. (1951) The growth of crystals and the equilibrium structure of their surfaces. *Phil. Trans. R. Soc. London,* **243**, 299–358

Chang, S. and Hamilton, A. D. (1988) Molecular recognition of biologically interesting substrates: synthesis of an artificial receptor for barbiturates employing six hydrogen bonds. *J. Am. Chem. Soc.,* **110**, 1318–1323

Chemical Design Ltd (1988) *CHEM-X Molecular Modelling Software,* Chemical Design Ltd, Oxford

Clydesdale, G. (1991) The development of a predictive approach for modelling the polymorphic stability and crystal habit of normal alkanes and other molecular crystals. Ph.D. Thesis, University of Strathclyde, Glasgow

Clydesdale, G. and Roberts, K. J. (1991) The structural stability and morphology of even number *n*-alkanes crystallising in the homologous series $C_{18}H_{38}$–$C_{28}H_{58}$. American Institution of Chemical Engineers, *AIChE Symp. Ser. 284*, **87**, 130–137

Clydesdale, G., Docherty, R. and Roberts, K. J. (1991a) HABIT – a program for predicting the morphology of molecular crystals. *Comput. Phys. Commun.*, **64**, 311–328

Clydesdale, G., Roberts, K. J. and Docherty, R. (1991b) A predictive approach to modelling the morphology of organic crystals based on crystal structure using the atom–atom method. In *Proceedings 3rd European Conference on Crystal Growth* (ed Lorinczy, A.) Trans. Tech. Publishers, Budapest, 234–243

Clydesdale, G., Roberts, K. J. and Docherty, R. (1992a) Modelling the morphology of molecular crystals in the presence of disruptive tailor-made additives. *J. Crystal Growth*, in press

Clydesdale, G., Docherty, R. and Roberts, K. J. (1992b) The computer program CRYSTLINK, unpublished results

Cox, J. D. and Pilcher, G. (1970) *Thermochemistry of Organic and Organometallic Compounds*, Academic Press, New York

Davey, R. J. (1987) Looking into crystal chemistry. *Chem. Eng.*, **443**, 24–27

Derissen, J. L. and Smit, P. H. (1978) Intermolecular interactions in crystals of carboxylic acids. IV. Empirical interatomic potential functions. *Acta Crystallogr., Sect. A*, **34**, 842–853

Desiraju, G. R. (1989) *Crystal Engineering – The Design of Organic Solids*, Elsevier, Amsterdam

Desiraju, G. R. and Jagarlapudi, A. R. P. (1987) C–H···O interactions and the adoption of 4 Å short-axis crystal structures by oxygenated aromatic compounds. *Chem. Soc. Perkin Trans. 2*, 1195–1202

Dewar, M. J. S. and Thiel, W. (1979) Ground states of molecules. 38. The MNDO method. Approximations and parameters. *J. Am. Chem. Soc.*, **99**, 4899–4907

Docherty, R., Roberts, K. J. and Dowty E. (1988) MORANG – a computer program aid in the analysis of the morphology of crystals. *Comput. Phys. Commun.*, **51**, 423–430

Docherty, R. and Roberts, K. J. (1988) Modelling the morphology of molecular crystals; application of anthracene, biphenyl and β-succinic acid. *J. Crystal Growth*, **88**, 159–168

Docherty, R. (1989) Modelling the morphology of molecular crystals. Ph.D. Thesis, University of Strathclyde, Glasgow

Docherty, R., Clydesdale, G., Roberts, K. J. and Bennema, P. (1991) Application of Bravais–Friedel–Donnay–Harker, attachment energy and Ising models to predicting and understanding the morphology of molecular crystals. *J. Phys. D: Appl. Phys.*, **24**, 89–99

Donnay, J. D. and Harker, D. (1937) A new law of crystal morphology extending the law of Bravais. *Am. Mineral.*, **22**, 446–447

Dowty, E. (1980) Computing and drawing crystal shapes. *Am. Mineral.*, **65**, 465–471

Etter, M. C. (1991) Hydrogen bonds as design elements in organic chemistry. *J. Phys. Chem.*, **95**, 4601–4618

Friedel, M. G. (1907) Etudes sur la loi de Bravais. *Bull. Soc. Fr. Mineral.*, **30**, 326–455

Gavezzotti, A. (1991) Generation of possible crystal structures from the molecular structure for low-polarity organic compounds. *J. Am. Chem. Soc.*, **113**, 4622–4629

Gavezzotti, A. and Desiraju, G. R. (1988) A systematic analysis of packing energies and other packing parameters for fused-ring aromatic hydrocarbons. *Acta Crystallogr., Sect. B*, **44**, 427–434

Gibbs, J. W. (1875) *Trans. Acad. Connecticut*, **3**; see also (1928), *The Equilibrium of Heterogenous Substances* (Scientific Papers, Vol. 1), and (1906), *Collected Works*, Longmans Green, New York

Gilmer, G. H. and Jackson, K. A. (1976) In *Crystal Growth and Materials* (ed. Kaldis, E. and Scheel, H. J.) North Holland, Amsterdam, 80–114

Hartman, P. (1973) Structure and morphology. In *Crystal Growth: An Introduction* (ed. Hartman, P.) North Holland, Amsterdam, 367–402

Hartman, P. and Bennema, P. (1980) The attachment energy as habit controlling factor. I. theoretical considerations. *J. Crystal Growth*, **49**, 145–156

Hartman, P. and Perdok, W. G. (1955) On the relations between structure and morphology of crystals. I. *Acta Crystallogr.*, **8**, 49–52

International Tables For X-Ray Crystallography (1983) Vol. A, Riedel, Dordrecht

Jackson, K. A. (1958) Mechanism of growth. In *Liquid Metals and Solidification*, American Society for Metals, Cleveland, OH, 174–186

Jetten, L. A. M. J. (1983) Crystallization processes: experiment and interpretation. Ph.D. thesis, University of Nijmegen

Jones, J. E. (1924) On the determination of molecular fields. – 1. From the variation of the viscosity of a gas with temperature. *Proc. R. Soc., Sect. A*, **106**, 441–477

Kitaigorodskii, A. I. (1973) *Molecular Crystals and Molecules*, Academic Press, New York

Kitaigorodskii, A. I., Mirskaya, K. V. and Tovbis, A. B. (1968) Lattice energy of crystalline benzene in the atom–atom approximation. *Sov. Phys. Crystallogr.*, **13**, 176–180

Larson, M. A. (1991) Population balance: the contribution of Alan D. Randolph. *AIChE Symp. Ser. 284*, **87**, 1–4

Lehmann, M. S., Koetzle, T. F. and Hamilton, W. C. (1972) Precision neutron diffraction structure determination of protein and nucleic acid compounds. I. The crystal and molecular structure of the amino acid L-alanine. *J. Am. Chem. Soc.*, **94**, 2657–2660

Leviel, J. L., Auvert, G. and Savariault, J. M. (1981) Hydrogen bond studies. A neutron diffraction study of the structures of succinic acid at 300 and 77 K. *Acta Crystallogr.*, **37**, 2185–2189

Lifson, S., Hagler, A. T. and Dauber, P. (1979) Consistent force field studies of intermolecular forces in hydrogen-bonded crystals. 1. Carboxylic acids, amides, and the C=O···H hydrogen bonds. *J. Am. Chem. Soc.*, **101**, 5111–5121

Lippincott, E. R. and Schroeder, R. (1955) One-dimensional model of the hydrogen bond. *J. Chem. Phys.*, **23**, 1099–1106

Momany, F. A., Carruthers, L. M., McGuire, R. F. and Scheraga, H. A. (1974) Intermolecular potentials from crystal data. III. Determination of empirical potentials and application to the packing configurations and lattice energies in crystals of hydrocarbons, carboxylic acids, amines and amides. *J. Phys. Chem.*, **78**, 1595–1620

Némethy, G., Pottle, M. S. and Scheraga, H. A. (1983) Energy parameters in polypeptides. 9. Updating of geometrical parameters, nonbonded interactions, and hydrogen bond interactions for the naturally occurring amino acids. *J. Phys. Chem.*, **87**, 1883–1887

Nyburg, S. C. and Lüth, H. (1972) *n*-Octadecane: a correction and refinement of the structure given by Hayashida. *Acta Crystallogr., Sect. B*, **28**, 2992–2995

Nyburg, S. C. and Potworowski, J. A. (1973) Prediction of unit cells and atomic coordinates for the *n*-alkanes. *Acta Crystallogr., Sect. B*, **29**, 347–352

Pavloska, A. and Nenow, D. (1972) Les surfaces non-singulieres sur la forme d'equilibre du naphtalene. *J. Crystal Growth*, **12**, 9–12

Phillips, F. C. (1963) *An Introduction to Crystallography*, 3rd edn, Longmans Green, London

Ponomarev, V. I., Filipenko, O. S. and Atovmyan, L. O. (1976) Crystal and molecular structure of naphthalene at −150°C. *Sov. Phys. Crystallogr.*, **21**, 215–216

Randolph, A. D. and Larson, M. A. (1988) *Theory of Particulate Processes: Analysis and Techniques of Continuous Crystallisation*, 2nd edn, Academic Press, New York

Roscoe, G., Dickinson, R. G. and Raymond, A. L. (1923) The crystal struture of hexamethylene-tetramine. *J. Am. Chem. Soc.*, **45**, 22–29

Simard, M., Su, D. and Wuest, J. D. (1991) Use of hydrogen bonds to control molecular aggregation, self-assembly of three dimensional networks with large chambers. *J. Am. Chem. Soc.*, **113**, 4696–4698

Smit, P. H. and Derissen, J. L. (1978) Intermolecular Interactions in Crystals of Carboxylic Acids. IV Empirical Interatomic Potential Functions. *Acta Cryst,* **A34**, 842

Taylor, R. and Kennard, O. (1982) Crystallographic evidence for the existence of C–H···O, C–H···N, C–H···Cl hydrogen bonds. *J. Chem. Soc.,* **104**, 5063–5070

Trotter, J. (1961) The crystal and molecular structure of biphenyl. *Acta Crystallogr.,* **14**, 1135–1140

van der Eerden, J. P., Bennema, P. and Cherepanova, T. P. (1978) Survey of Monte Carlo simulations of crystal surfaces and crystal growth. *Progr. Crystal Growth Characterisation,* **1**, 219– 254

Weissbuch, I., Shimon, L. J. W., Addadi, L., Berkovitch-Yellin, Z., Weinstein, S., Lahav, M. and Leiserowitz, L. (1985) Stereochemical discrimination at organic crystal surfaces. 1: The systems serine/threonine and serine/allothreonine. *Israel J. Chem.,* **25**, 353–362

Williams, D. E. (1966) Nonbonded potential parameters derived from crystalline aromatic hydrocarbons. *J. Chem. Phys.,* **45**, 3770–3778

Williams, D. E. (1983) Computer Program PCK83, *Quantum Chemistry Program Exchange Program No. 481*

Wright, J. G. (1987) *Molecular Crystals,* Cambridge University Press, Cambridge

Wulff, G. (1901) XXV. Zur Frage der Geschwindigkeit des Wachsthums und der Aufloesung der Krystallflaechen. *Z. Krist. Min.,* **34**, 499–530

Appendix: List of symbols

(hkl) — Miller indices of crystal growth face

$\{hkl\}$ — Miller indices of symmetry-equivalent crystal growth faces

d_{hkl} — interplanar spacing for face (hkl), thus slice thickness of growth slice (hkl)

E_{cr} — lattice (or crystallization) energy of crystal

E_{sl} — slice energy of growth slice (hkl)

E_{att} — attachment energy of growth slice (hkl), sum of $E_{att(+)}$ and $E_{att(-)}$

$E_{att(+)}$ — attachment energy of growth slice (hkl) along the positive growth direction for pure system (equal to $E_{att(-)}$)

$E_{att(-)}$ — attachment energy of growth slice (hkl) along the negative growth direction for pure system (equal to $E_{att(+)}$)

U, V, W — multiples of the unit cell along the crystal directions a, b and c

Z — position of symmetry-related molecule in the unit cell

$E_{sl'}$ — slice energy with additive at centre of growth slice (hkl)

$E_{att'}$ — attachment energy of growth slice (hkl) containing additive onto pure surface (sum of $E_{att(+)'}$ and $E_{att(-)'}$)

$E_{att(+)'}$ — attachment energy of growth slice (hkl) containing additive onto pure surface in positive growth direction

$E_{att(-)'}$ — attachment energy of growth slice (hkl) containing additive onto pure surface in negative growth direction

$E_{att''}$ — attachment energy of pure growth slice (hkl) onto surface containing additive (sum of $E_{att(+)''}$ and $E_{att(-)''}$)

$E_{att(+)''}$ — attachment energy of pure growth slice (hkl) onto surface containing additive in positive growth direction

$E_{att(-)''}$ — attachment energy of pure growth slice (hkl) onto surface containing additive in negative growth direction

Δb — difference in incorporation energy (equals $E_{sl'} + E_{att'} - E_{cr}$)

Chapter 5

Aerosol formation

I. Colbeck

5.1 Introduction

The fundamental properties of aerosols have been studied for more than 100 years. In the nineteenth century, aerosol particles represented the smallest division of matter known. Many of the great scientists of the time, such as Tyndall, Lister, Kelvin, Maxwell, Aitken and Millikan, contributed to our understanding of aerosols. The term 'aerosol', meaning a suspension of solid or liquid particles in gas, originates from military research during World War I. The name is associated with Donnon (Whytlaw-Gray *et al.*, 1923), although first publication of the term was due to Schmauss (1920) where it was used as an analogue to 'hydrosol', a stable liquid suspension of solid particles. An aerosol consists of a relatively stable suspension of solid or liquid droplets in a gas or vapour. Aerosols are part of the colloid family.

Aerosol technology is of great interest in numerous scientific and engineering applications. The commercial production of carbon black and titanium oxide, controlling radioactivity released from nuclear reactor accidents, the spray drying of suspensions ranging from milk to detergents and combustion technology are only a few such applications. The safety aspects of particle behaviour are also important. There is a large body of information on dust explosions and damage to structures involving dust accumulation as an undesirable by-product (Hidy, 1984). In the microelectronic industries products are continually being developed with decreasing structural dimensions (Fissan, 1990; Liu, 1990). Hence great precautions are taken to ensure that particles are removed from ambient air so that these devices are not contaminated during manufacture.

The particle size of interest in aerosol behaviour ranges from molecular clusters of 0.001 μm to fog droplets and dust particles as large as 100 μm; a variation of 10^5 in size. There are various types of aerosol which are classified according to physical form and method of generation. The commonly used terms are:

dust – a solid particle formed by mechanical disintegration of a parent material, such as crushing, grinding and blasting
fume – solids produced by physicochemical reactions such as combustion, sublimation or distillation

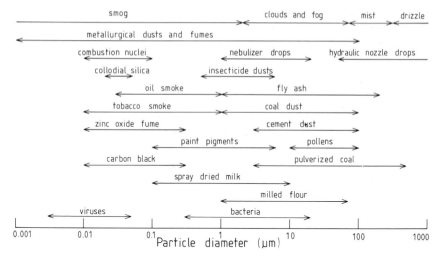

Figure 5.1   *Size range of aerosols in the atmosphere and industrial settings*

smoke – a visible aerosol produced by the disintegration of liquid or the condensation of vapour.

Figure 5.1 shows the typical size ranges for these and other types of aerosol.

The basic laws of aerosol behaviour apply to both undesirable aerosols (e.g. gas welding fumes) and intentionally generated aerosols. Aerosol systems evolve in time due to coagulation, condensation and evaporation, nucleation and chemical reactions.

5.2 Aerosol generation

A large amount of time and effort has gone into the investigation and development of aerosol-generation devices. Ideally one should have control of particle-size distributions; stability of operational performance for key periods of time; and control of volumetric output. The output from a generator is either monodisperse (all particles of one size) or polydisperse (many different sizes). Polydisperse aerosols have many industrial applications (e.g. agricultural spraying) and most environmental aerosols are polydisperse. It is much easier to produce polydisperse than monodisperse aerosols. The generation of a well-defined aerosol is an essential prerequisite for virtually any experimental work in the field of aerosol technology. Aerosols of very narrow size distributions have many applications in aerosol engineering, e.g. for size measurement, equipment calibration, for filtration efficiency testing, for lung inhalation studies and, lately, in the production of ceramic powders. For practical monodispersity to be achieved, the geometric standard deviation σ_g should be less than 1.2. For many applications monodisperse aerosols make it easier to design and perform experiments.

A wide variety of instruments for aerosol production have been described in the literature (Raabe, 1976; Willeke, 1980; Mitchell, 1986) and the techniques used are basically similar to the mechanisms of natural aerosol formation. However, no single technique can produce particles spanning the size range 0.01–100 μm, the range generally required for aerosol studies. The two main methods of aerosol formation are break-up mechanisms and condensation. Break-up mechanisms include the atomization of liquids and the dispersion of solids and powders. There is a lower limit for the size of particles produced by break-up processes since the energy necessary for atomization or dispersion increases with decreasing particle size. Polydisperse aerosols are normally produced by direct dispersion from a dust generator such as the fluidized bed (Marple *et al.*, 1978) and the Wright dust feed (Wright, 1950). Electrostatic charge is a problem common to all dry dispersion dust generators (Johnston *et al.*, 1987; Yeh *et al.*, 1988). The dispersion of powders is limited in terms of the smallest particle size that can be dispersed. For particles below 1 μm, dispersion by simple mechanical methods has often been regarded as impossible.

Atomization also generally produces a broad distribution of relatively coarse droplets, with the minimum particle size determined by a balance between the surface tension forces that resist droplet formation to pressure or other forces that attempt to disrupt the fluid surface. The common methods used to disperse liquids mechanically to form aerosols include air nebulizers (May, 1973; Kerker, 1975), spinning discs (Whitby *et al.*, 1965; May, 1966; Melton *et al.*, 1991), ultrasonic nebulizers (Lang, 1962; Denton and Swartz, 1974; Homma *et al.*, 1990), and vibrating orifice generators (Bergland and Liu, 1973). The last three techniques are not strictly based on atomization and are able to produce monodisperse aerosols. Full details of these methods may be found in Willeke (1980) or Hidy (1984).

Some methods use a combination of methods. The exploding wire technique is a hybrid of condensation and break up. A massive surge of thermal energy may be used to explode a metal wire resulting in ultrafine primary particles of 0.01–0.1 μm diameter. Aerosols produced by such methods consist of a long chain of smooth, spherical particles, whose size distribution is near log-normal. The high initial number concentrations results in rapid agglomeration and a highly polydisperse aerosol. Smaller primary particles are produced if the thermal energy is increased (Wegrzyn, 1976).

Laser ablation has been used controllably to produce nucleated silver particles (Shaw *et al.*, 1990). An ArF excimer laser is focused on a metal target. When the laser fluence is higher than a critical value, ablation occurs and atomic species are produced that expand out into a plume that interact with gaseous species present. On collision, the particles can thermalize and vapour phase condensation can occur. Particles smaller than 4 nm have been produced by this technique.

A number of gases and vapours react chemically with each other to form a solid or liquid. Particulate commodities such as carbon blacks and pigmentary titania are routinely made on a large scale by such reactions in turbulent flows.

Careful control of the mixing of the gases results in aerosols which have reproducible size and concentration characteristics. Particle growth is coagulation driven. The turbulent intensity, reactant concentration and initial pressure determine the optimal conditions for the manufacture of powders of narrow size distributions in turbulent flows (Pratsinis and Xiong, 1990).

Particles can be produced by the condensation of vapours on already existing nuclei (heterogeneous condensation) or at higher supersaturations, by instantaneous formation of vapour molecules; molecule clusters that grow to particles when their size exceeds a critical value (homogeneous condensation). Typically heterogeneous condensation is utilized. In such generators four, not necessarily separate, processes are common; nuclei production, vapour generation, vapour–nuclei mixing and condensation through controlled cooling of the carrier gas. The aerosol particles generated by this method are fairly monodisperse for particles in the submicrometre range. Examples of such generators are the Sinclair–LaMer generator (Sinclair and LaMer, 1949), the Rapaport and Weinstock generator (Rapaport and Weinstock, 1955), the Tu single stage generator (Tu, 1982) and that described by Liu and Lee (1975). Particle size is around $0.5–1.0$ μm and $\sigma_g = 1.2–1.4$. Recently, Japuntich *et al.* (1990) studied 17 different condenser systems. They found that monodispersity is very much dependent on the balance of forced and free convection in the condenser tube and the most successful condensates have low diffusion coefficients. They found a linear correlation between the Reynolds number (Re) or flow in the condenser and the Rayleigh number (Figure 5.2). Hence for a chosen Re the Rayleigh number may be found and for chosen temperature parameters the tube radius may be derived. Therefore, a condenser tube system which dimensionally and dynamically meets the criteria of mixed convection laminar flow may be optimized prior to aerosol generator construction. Liu and Lin (1990) have designed an ultrafine monodisperse aerosol generator which, in addition to stability of output, produces particles about 0.01 μm in size. The principle of the generator is the evaporation condensation process. Dioctyl sebacate is atomized and the aerosol is vapourized by a heater prior to cooling abruptly with a high flow rate of clean air (Figure 5.3). The impurities in the liquid serve as condensing nuclei.

Monodisperse, polystyrene latex spheres (PSL) in an aqueous solution are commercially available and the sizes range from 0.05 to 90 μm. Since the particles are already formed and need only be suspended in a support gas the latex aerosol generator is probably the simplest technique to use. The standard method utilizes a gas driven nebulizer to suspend the particles. Tajima *et al.* (1990) optimized the dimensions of such a nebulizer and concluded that the length of the nebulizing nozzle had an important effect on generated particle concentration. Yamamota *et al.* (1990) have gone one step further and have developed a generator which, in addition to a precisely machined nebulizer, maintains an optimum concentration of PSL for atomization. Aerosols in the range $0.08–3$ μm can be generated with constant output and reproducibility. Such a system requires ultraclean water since any small quantities of dissolved substances may appear in the aerosol, either coated on the latex sphere or as a small particle when the original droplet contained no latex.

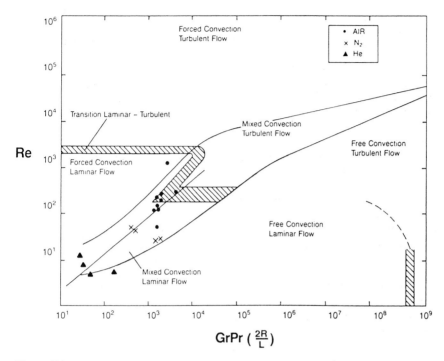

Figure 5.2   *Regimes of natural, forced and mixed convection for flow-through vertical tubes. Results of condenser tube analyses from 17 condensation aerosol generators. Re, Reynolds number; Gr, Grashoff number; Pr, Prandtl number; 2R/L, tube diameter/length ratio. (Courtesy of Academic Press, Japuntich et al., (1990))*

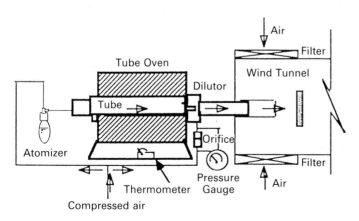

Figure 5.3   *Schematic diagram of ultrafine monodisperse aerosol generator (courtesy of Pergamon Press, Liu and Lin (1990))*

Aerosol technologies are well established in chemical process technology, accounting for the production of millions of tonnes of particles per year. Aerosol processes are applicable to a broad range of chemical compositions, yielding high product purity. Such processes can be expected to play a major role in aerosol generation if the physical processes that govern particle growth and structure are better understood. Materials with unique mechanical or electronic properties can be generated through synthesis of particles in the nanometre size range. Research is currently underway on the use of aerosol particles in the fabrication of electronic materials exhibiting quantum size effects (Flagan, 1990). On a related topic, silicon micromachining, the technology of creating microscopic physical structures in silicon or silicon dioxide using photolithographic and etching techniques employed in the fabrication of electronic integrated circuits, can produce particles of uniform size and shape (Hoover *et al.*, 1990; Kaye *et al.*, 1991). The controlled crystallization from colloidal sols or ionic solutions has enabled particles of various non-spherical shapes to be produced (Marshall *et al.*, 1990; Gowland *et al.*, 1991). This method is still a 'black art', and it is likely that specific particle shapes can only be reliably produced within a small size range using carefully defined formation chemistry.

5.3 Coagulation

The process whereby two aerosol particles collide, due to a relative velocity between them, and adhere or coalesce to form a single particle is called 'coagulation'. It is impossible to suppress coagulation completely and it acts to decrease particle number and increase particle size. However, it does not alter the mass of material in the coagulated particle. Due to the large specific surface area of aerosols, generally all contacts between particles result in a coagulation process and the particles cannot separate from each other. Hence, for aerosols coagulation is the most important interparticle phenomenon. At high number concentrations, as in smokes, the coagulation half-life can be in the order of seconds, while in low number concentrations coagulation can be negligible. Coagulation processes find application in a wide variety of fields including the pharmaceutical, food and ceramic industries. Particle coagulation may make powdered products more useful or it may make possible their further processing. It may be used to reduce dusting hazards and losses, to impart desirable flow properties or facilitate the recovery of fines.

For the comprehensive description of coagulation all processes have to be examined leading to non-zero relative velocities. This relative velocity can arise from a variety of physical causes:

Brownian motion – thermal motion of gas molecules
gravitational settling – different settling velocities of differently sized particle.
turbulence – acceleration of particles by turbulent convection
acoustic – motion induced by acoustic oscillations.

Coagulation may also be induced by electric forces between charged particles, or temperature and density gradients. Generally coagulation is a superposition of the various different processes.

5.3.1 Brownian coagulation

The theoretical description of aerosol coagulation is frequently restricted to spherical particles, which remain spherical after coagulation has occurred, i.e. liquid particles. The coagulation of solid particles is important and this leads to the formation of randomly shaped aggregates. Much work remains to be done on the theoretical treatment of this process. Analytical formulae are available for most coagulation processes assuming spherical particles, although only for limited regimes. Coagulation in monodisperse or nearly monodisperse aerosols has been studied experimentally. Again, for many applications, coagulation of particles with different size and chemical composition is of overriding importance.

The first theory of the physics of coagulation was published by Smoluchowski in 1916. He assumed that the particles were spherical, were all identical and that Brownian motion was responsible for their impacts.

The change with time, t, of the total number of particles, N, in a colloid undergoing coagulation is given by

$$\frac{dN}{dt} = -KN^2 \tag{5.1}$$

The coagulation coefficient K is given by

$$K = \frac{4kT}{3\eta} \tag{5.2}$$

where k is Boltzmann's constant, T is temperature and η viscosity. Hence K is dependent on the properties of the carrier medium, but independent of the size and density of the particles. Particle behaviour depends on the ratio of particle size (r) to the mean free path of the gas molecules (λ). This ratio is called the 'Knudsen number': $Kn = \lambda/r$. For aerosols with $Kn < 0.1$, i.e. the continuum regime, coagulation is given by eqn (5.1). As particle size decreases eqn (5.1) must be multiplied by the Cunningham slip correction C (Jennings, 1988). The slip correction represents the mechanism for transition from the continuum $r > \lambda$ to the molecular region $r < \lambda$. It increases as particle size decreases. Once particle size approaches gas-molecule dimensions (i.e. $Kn \rightarrow \infty$) coagulation is controlled by gas kinetic theory and the coagulation coefficient in this region, Kg, is:

$$Kg = \frac{4\pi r^2 v}{\sqrt{2}} \tag{5.3}$$

where v is the mean thermal velocity of the particle. The physical and thermodynamic properties of the particles and the carrier gas, respectively, jointly control coagulation under gas kinetic conditions.

It is evident that the coagulation coefficient increases for small Kn up to a maximum and decreases when changing to gas kinetic conditions. Fuchs (1964) attempted to resolve this anomaly between theories. He explained the correction with his concentration jump theory:

$$K = \frac{4kTC\beta_F}{3\,\eta} \tag{5.4}$$

where

$$\beta_F = \frac{1}{1 + G} \tag{5.5}$$

and

$$G = 4\,\frac{D}{r}\,\sqrt{\frac{\pi m}{8kT}}$$

D is the diffusion coefficient and m is the mass of the particle.

Davies (1979) concluded that the Fuchs factor was incorrect, because it was based on a false analogy between the coagulation of particles and the evaporation/condensation of gas molecules. He refers to an underestimation of K as Kn rises from 0.5 to 15. Few measurements have been made with Kn > 15, since at atmospheric pressure this requires particles to have radii less than 4.4 nm. Measurements which have been made give conflicting results (Fuchs and Sutugin, 1965; Rooker and Davies, 1979; Egilmez and Davies, 1982; Kim and Liu, 1984). Davies (1979) deduced a factor β_D which would interpolate smoothly between the gas kinetic and the continuum regime:

$$K = \frac{4\pi kTC\beta_D}{3\eta} \tag{5.6}$$

He argued from the partition of the thermal energy of a small particle between random and directed motion before colliding with another particle that β_D could be represented by

$$\beta_D = 1 + \left[\frac{8\,\exp\,(-9.03Pe)}{2Pe} \right]^{-1} \tag{5.7}$$

where Pe is the Peclet number given by

$$Pe = \frac{2vr}{D} \tag{5.8}$$

The corrections by both Fuchs and Davies are based on semiempirical theories. More experiments are required to clarify the situation, although the theory of Fuchs is often regarded as more reliable (Bunz, 1990). A modified version of Fuchs' theory has recently been derived (Meitzig, 1989):

$$K = \frac{4\pi kTC\beta_m}{3\eta} \tag{5.9}$$

$$\beta_m = \frac{1}{0.985 + M} \qquad (5.10)$$

where

$$M = \frac{4D}{r} \sqrt{\frac{\pi m}{8kT}} \left(\frac{2 + \dfrac{Kn}{25}}{1 + \dfrac{Kn}{100}} \right)$$

Figure 5.4 illustrates how the theories compare with experimental data. Here values of the coagulation coefficient are plotted against the Knudsen number.

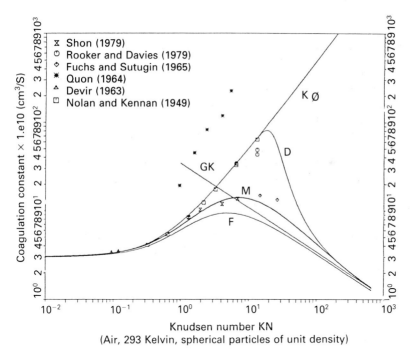

Figure 5.4 *Coagulation coefficient according to different authors: F, Fuchs (1964); D, Davies (1979); GK, gas kinetics; M, Meitzig (1989); K_o, Smoluchowski (1916) (courtesy of Pergamon Press, Bunz (1990))*

This can be misleading if measurements made at constant pressure (Shon *et al.*, 1980) are compared with those at constant particle size and varying gas pressure (Wagner and Kerker, 1977). Figure 5.5 shows why this is a problem – because identical values of Kn can be obtained by choosing various combinations of particle radius, *K* initially increases and after passing through a maximum

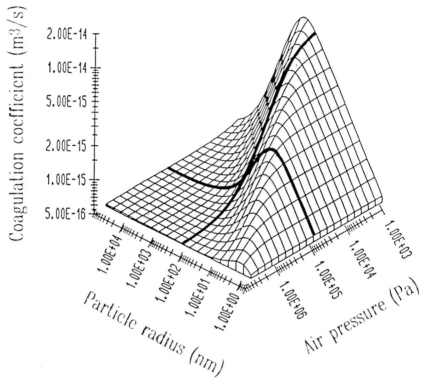

Figure 5.5 *Coagulation coefficient according to Fuchs (1964) for equal-sized particles shown versus particle radius and carrier gas pressure*

decreases again. At constant particle size, K increases with decreasing pressure but then approaches a constant limiting value.

There are many problems in performing reliable coagulation experiments. Since changes in particle-size distribution depend on the number concentration squared any spatial inhomogeneities of a coagulating aerosol will normally lead to an increase of the coagulation rate compared with a uniform aerosol. In addition, various simultaneously occurring effects such as diffusion, sedimentation, condensational growth, Ostwald ripening have to be taken into account. Wagner (1989) summarizes the requirements for accurate experimental data on aerosol data. Most possible errors in the measured data tend to cause an overestimation of the coagulation rate.

5.3.2 Coagulation of polydisperse particles

An important effect, ignored by Smoluchowski's coagulation theory, is the greater rate of coagulation of polydisperse aerosols. For the case of an aerosol with particles of two different sizes d_1 and d_2 ($d_1 < d_2$) the coagulation coefficient $K_{1,2}$ is given by

$$K_{1,2} = \pi(d_1 + d_2)(D_1 + D_2) \tag{5.11}$$

where D_1 and D_2 are the diffusion coefficients of the small and large particles. Since $d_1 < d_2$ and $D_2 < D_1$, we find

$$K_{1,2} = \pi d_2 D_1 \tag{5.12}$$

The two components of coagulation are the absorbing surface and the diffusing particle. Small particles have rapid diffusion but a small surface area, whereas big particles have a large surface area but slow diffusion. The combination of large absorbing surface of the big particles and the rapid diffusion of the small particle to that surface means that coagulation between dissimilar size particles is more rapid than that between same size particles.

Values of $K_{1,2}$ based on eqn (5.12) are given in Table 5.1, from which it can be seen that the greater the difference in particle size the greater the coagulation

Table 5.1 *Coagulation coefficients $K_{1,2} \times 10^{10}\,cm^3\,s^{-1}$ of unequal particles*

d_1 (μm)	d_2 (μm)						
	0.01	0.05	0.1	0.5	1.0	5.0	10.0
0.01	66						
0.05	100	15					
0.1	190	14	8.5				
0.5	850	42	14	4.0			
1.0	1700	79	24	4.3	3.5		
5.0	8300	380	110	18	6.1	3.1	
10.0	17000	750	220	22	10	3.4	3.0

coefficient. The size of the new particle differs little from the size of the larger particle. Hence fine particles in the polydisperse aerosol are quickly removed by the coarse ones. For a polydisperse aerosol the coagulation rate is time dependent since the size distribution and dispersity are constantly changing.

Several mathematical models have been developed to simulate aerosol dynamics, including coagulation. The methods differ mainly by their representation of the size distribution. The three major approaches are: (i) discrete representation, (ii) parameterized representation, and (iii) continuous representation (Seigneur *et al.*, 1986). Discrete methods are those in which the size domain is divided into discrete sections (Gelbard and Seinfeld, 1980; Warren and Seinfeld, 1985), parameterized models assume that the size distribution adheres to a known form depending on a small number of parameters (Whitby, 1981), and techniques that offer any degree of approximation are continuous (Tsang and

Brock, 1982 and 1983). Since the size distribution and dispersity are constantly changing, these methods are only valid at a particular time.

Allowing for this change, solutions indicate that a coagulating aerosol should, with time, reach the same steady-state size distribution, independent of the initial size distribution. Once this steady state is reached, gains by coagulation to the number of particles of a given size are equalled by losses by coagulation or sedimentation from that size. The self-preserving size distribution has been widely used in both theoretical and experimental aerosol studies (Friedlander and Wang, 1966; Clark and Whitby, 1967; Lai *et al.*, 1972). The time to reach the self-preserving form depends on the shape of the initial distribution. It has been shown that for initially monodisperse aerosols, the self-preserving form is reached after a time (Hidy and Lilly, 1965) approximately equal to $9\eta/kTN(t = 0)$. This theory has been used to explain the general shape of the observed size distributions of atmospheric aerosols.

Seigneur *et al.* (1986) have reviewed various mathematical models of aerosol dynamics. They reported that discrete models simulate coagulation well provided that a fine size resolution is used.

Parameterized approaches based on log normal size distributions tend to overestimate the rate of coagulation, while continuous representations provide an accurate solution but are computationally expensive. Each method had its merits and the specific technique to use will depend on the requirements of the application.

5.3.3 Molecular forces

It is well known that charged particles may experience either enhanced or retarded coagulation rates, depending on their charges. However, even electrically neutral particles are subjected to intermolecular forces which tend to enhance the apparent coagulation rate.

Van der Waals forces between electrically neutral and symmetrical molecules result from instantaneous dipoles produced by fluctuations in the electron clouds surrounding the nucleus. These instantaneous dipoles induce dipoles in neighbouring molecules. Fuchs (1964) suggested that the coagulation coefficient, given by eqn (5.4), should include a correction factor, to allow for molecular forces. For equal-sized water droplets at 293 K the correction factor is 1.27 and typically for most species the factor is in the range 1.25–1.50. This indicates that coagulation rates for equal-sized particles resulting from van der Waals forces cannot be expected to exceed 50%. In the continuum regime particle motion is mostly determined by the interaction of particles with the carrier gas and the influence of the interaction forces is small. The continuum regime enhancement factor for van der Waals forces has been presented by Friedlander (1977) and Schmidt-Ott and Burtscher (1982).

However, in the transition and free molecule regimes, the effect of van der Waals forces on coagulation rates is more pronounced. Fuchs and Sutugin (1965) found that the van der Waals forces enhanced coagulation by a factor of 2.19 in the free molecule regime. Graham and Homer (1973) measured the coagulation rate for liquid lead particles (diameter 10–20 nm) and showed that van der Waals forces resulted in a doubling of the rate, compared with non-interacting particles. A similar value of coagulation enhancement was obtained by experiments with sulphate aerosols with diameters in the range 5–40 nm (Brockmann *et al.*, 1982). Recently, Okuyama *et al.* (1984) reported the coagulation rates of ultrafine (diameters 5–40 nm) NaCl, $ZnCl_2$ and silver aerosols. They found coagulation enhancement factors, due to van der Waals force, in the range 1.2–3.7 for NaCl and $ZnCl_2$ particles and between 6 and 9 for silver particles. In addition, they concluded that the enhancement factor increases as particle size decreases.

The interaction potential may be calculated on the basis of the frequency-dependent electric polarizability of the particles, and hence it depends strongly on the material properties of the particles (Marlow, 1980a, 1981, 1982). A modified Fuchs formula has been suggested to take into account interaction potentials for aerosols in the transition regime. For water droplets coagulation may be enhanced by a factor of up to 2.5 (Marlow, 1980b). The enhancement is greatest where the radii of the coagulating particles is comparable and least where the ratio of the radii is greatest as well as where particle sizes approach the continuum regime. Viscous forces act to counter the effects of van der Waals forces. These arise because a particle in motion in a fluid induces velocity gradients in the fluid that influence the motion of other particles when they approach its vicinity. A fluid resists being squeezed out from between approaching particles, and hence viscous forces retard the coagulation rate from that in their absence. Alam (1987) has shown that van der Waals forces are largest for particles of equal size, whereas viscous forces are largest when the ratio of particle radii is large. For water, in the continuum regime, viscous forces dominate over van der Waals forces and coagulation is retarded (Alam, 1987). In the kinetic regime, coagulation is always enhanced due to the absence of viscous forces.

5.3.4 Coulomb forces

Most aerosol particles carry some electric charges with the effect on coagulation depending on the sign of their charges. Obviously, particles with similar charges will repel each other while those of opposite charges will attract each other. Fundamental studies of the coagulation rate of charged particles were made by Fuchs (1964). Assuming that the force between these particles is given by Coulomb's law, the correction term is given by

$$\beta = \frac{y}{e^y - 1} \tag{5.13}$$

where $y = q_1 q_2 / 4\pi\epsilon_0 (r_1 + r_2) kT$ and ϵ_0 is the dielectric constant of a vacuum.

For unipolar charged particles (i.e. like charges) $y > 0$, $\beta < 1$ and coagulation is retarded from that for pure Brownian motion. With bipolar aerosols (unlike charges) $y < 0$, $\beta > 1$ and coagulation is enhanced. Values of β for various values of y are shown in Table 5.2. Assuming equal numbers of positively and

Table 5.2 *Correction factor for Coulomb forces between particles of equal size*

y	β(*unipolar*)	β(*bipolar*)	β(*average*)
0.2	0.9033	1.1033	1.0033
0.4	0.8133	1.2133	1.0133
0.6	0.7298	1.3300	1.0299
0.8	0.6528	1.4528	1.0528
1.0	0.5820	1.5820	1.0820
2	0.3130	2.3130	1.3130
4	0.0746	4.0746	2.0746
6	0.0149	6.0149	3.0149
8	2.9×10^{-3}	8.0026	4.0028
10	4.5×10^{-4}	10.0005	5.0005

negatively charged particles, the average correction factor is just the arithmetic mean of the two values. It is evident from this table that for a weak bipolar aerosol, $y \ll 1$ there is little enhancement of coagulation since the increase brought about attraction is compensated by a similar decrease due to repulsion. In the case of a strongly bipolar aerosol $y \gg 1$, the increase of coagulation due to attraction greatly exceeds the decrease due to repulsion resulting in an increase in coagulation. Eliasson *et al.* (1987) have shown how, at least theoretically, a faster coagulation of smaller particles with larger ones can be achieved by using asymmetric bipolar coagulation. If the small particles are charged with one polarity and the large ones with the opposite polarity one can shift the particle distribution to larger radii. This process may be used to increase the efficiency of an electrostatic precipitator by removing fine particles from the off-gas in a stack.

Once two oppositely charged particles coagulate the absolute value of the aerosol charge decreases rapidly. In addition, although the coagulation rate for highly bipolar aerosols is much greater than equivalent neutral aerosols, this enhanced rate decreases very rapidly because of the annihilation of particle charge. Since the number concentration of a unipolar aerosol is always decreasing, due to electrostatic scattering, the rate of coagulation should be determined by the increase in the mean size of particles and is thus rather difficult to measure.

Electrostatically enhanced coagulation plays an important role in the coagulation of particles produced by flames (Whitby and Liu, 1967). Generally, fume and smoke particles produced by flames have a chain-like shape. Charged aerosols are more likely to form chain-like aggregates than are uncharged aerosols. Whenever chain-like aggregates are observed, it may be concluded that electrostatically enhanced coagulation has occurred during the history of the aerosol.

5.3.5 Gravitional coagulation

Collisions between particles can be induced by differences in motion associated with the action of external forces. In addition to Brownian coagulation, gravitational coagulation influences the behaviour of all particle-size distributions dispersed in any carrier gas. The theory developed is a continuum approximation which models the action of gravity. Approximate solutions are only available if the particles are similar in size or either very large or of intermediate size. For non-spherical particles, theoretical results can only be obtained for special geometries (e.g. ellipsoids). Gravitational coagulation is important for large particles ($d > 10$ μm) and for particles not too different in size (Bunz, 1990).

5.4 Industrial applications

Removal of particulates from industrial emissions is a major problem in pollution abatement. Particles in the size range 0.1–1 μm which make up a large fraction of the particulate emission are difficult to remove by conventional separation technology. Such particles pose a major health hazard since it is a contributor to respiratory disease. These particles are often toxic due to their ability to act as a vehicle for noxious agents (e.g. bacteria) or to their origin (e.g. chemically active elements produced during combustion processes). One method of removing particles in this size range is by techniques which enhance the coagulation rate. While numerous methods work on laboratory scale experiments, the vast majority would involve enormous energy requirements on the industrial scale.

5.4.1 Acoustic agglomeration

Sound vibrations can induce coagulation between particles. The first serious experiments on acoustic coagulation were performed 60 years ago. Since then, the agglomeration mechanism has been discussed extensively (Mednikov, 1965; Shaw, 1978). Acoustic coagulation has been used successfully for particle-size conditioning, e.g. as a preconditioner for devices like cyclones to remove particles from industrial off-gases (Chou *et al.*, 1981). However, industrial applications have been slow to develop due to lack of suitable, high intensity, high efficiency sound source. Regulations regarding particle emission and limiting the maximum allowable mass concentration are currently being introduced. Acoustic preconditioning offers a potential solution to this problem. It can also be used in any environment, for instance if the gas is chemically aggressive. Recent developments have led to a resurgence of interest in this topic for pollution control and in various types of scrubbers, filters and electrostatic precipitators and diffusion.

Acoustic coagulation is believed to be a combination of several processes occurring in the presence of a high intensity and acoustic field: orthokinetic, hydrodynamic, turbulent inertia, and turbulent diffusion. Orthokinetic interactions result from collisions between small particles highly influenced by vibrations and large particles which are virtually stationary. For a given sound wave, small particles have little inertia. Thus small particles tend to oscillate together with the gas medium, whereas large ones tend to remain stationary in the fluid. The coagulation coefficient for this effect is directly proportional to the ratio of the large particle concentration to the total concentration resulting in a quite strong influence of the form of the particle-size distribution. The hydrodynamic interaction results from mutual distortion of the flow field around interacting particles. As their name implies, both turbulent inertia and turbulent diffusion result from the fact that the acoustic field can induce turbulence if the sound pressure level exceeds a critical value, normally of the order of 160 dB. The agglomeration process occurs in small eddies and when the particles are completely entrained by the fluid the process may be treated in a similar manner to that of Brownian diffusion, i.e. turbulent diffusion. If the particles are not fully entrained then the relative velocities of the particles must be taken into account, i.e. turbulent inertia. The turbulent inertia coagulation rate is a function of particle radius to the fourth and, hence, is extremely sensitive to particle size. Figure 5.6 illustrates how each process is important in different regimes of particle size and to the sound level pressure.

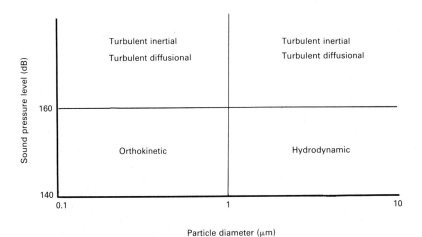

Figure 5.6 *Relative importance of acoustic agglomeration mechanisms (after Bunz (1990))*

There have been numerous reports of agglomeration and precipitation of aerosols in large-scale indoor experiments (Magill, 1989; Somers *et al.*, 1991; Gallego-Juarez *et al.*, 1991). Acoustic saturation at 20 kHz and 160 dB limits the useful field to a maximum distance of about 3 m from the emitters. Emitters at lower frequencies are now being developed. For larger scale application the energy requirements become prohibitive. The power dissipated at 20 kHz and 140 dB, for a mass loading of 1 g/m^{-3} and a particle diameter of 1 μm, is 9 W m^{-3}. The absorbed power can be considerably reduced by working at lower frequencies. Research is currently being undertaken in the use of acoustic agglomeration in large-scale open-air applications to combat such hazards as chemical explosions, fog clearance, pest control in agriculture, tunnel fires (Magill *et al.*, 1989; Gellego-Juarez *et al.*, 1991). To reach high intensity level over long distances, frequencies in the range 1–5 kHz or lower are required. However, fine particles will then be fully entrained in the sound field and will move backwards and forwards in phase; no agglomeration should take place. This problem may be overcome by the addition of seed aerosols with relatively large particle size. These seed particles act as scavenging centres and accelerate the agglomeration process. Hazardous gases may also be contained in a similar manner. Activated carbon or zeolite aerosol is sprayed into the gas to absorb it. These particles can then be subjected to sound waves, coagulation results and the 'heavy' particles containing the absorbed gas fall to the ground (Magill, 1989).

5.4.2 Other techniques

Wet scrubbing is a possible process for aerosol elimination. Such scrubbers are reliable, economical to operate and precipitate gases and aerosols simultaneously. However, particles between 0.01 and 1 μm in diameter are eliminated insufficiently because of the relative large wash droplets. Backhaus *et al.* (1988) have proposed a method to overcome this problem. Liquid droplets with diameters of 5–40 μm are injected into the aerosol. The aerosol is then taken up by these primary drops. The primary drops are transformed by a novel droplet agglomerator into secondary drops with diameters between 30 and 500 μm (Figure 5.7). The agglomerated lamellas collect the primary droplets and these form a liquid film. This film is dragged by shear forces of the gas flow to the backside of the lamellas and disintegrates into larger secondary droplets.

Recently, Colbeck and Hardman (1991) have reported enhanced coagulation rates of sodium chloride aerosols, primarily the result of thermophoretic forces. By subjecting aerosol to an intense light source, particles with diameters below 0.5 μm may be significantly influenced by photophoresis, while particles with diameters greater than 1 μm will be affected to a lesser extent. The photophoretic force may then be utilized to increase the relative velocity between small and large particles and hence alter the coagulation rate. Colbeck and Hardman (1991) found that the geometric mean diameter of a NaCl aerosol was doubled and the coagulation rate enhanced by a factor of approximately 100 when it was subjected to an intense light source. On further investigation, they concluded that this increase was due to thermophoresis (i.e. the light source was producing a temperature gradient in the apparatus) rather than photophoresis. While this method of increasing coagulation works in the laboratory, it would not be economically feasible on an industrial scale.

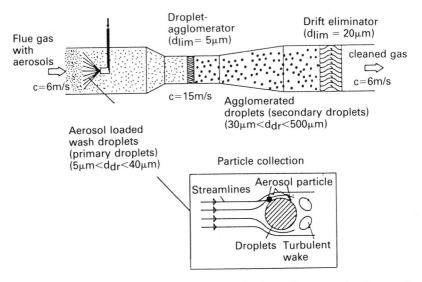

Figure 5.7 *Wet scrubbing with an integrated droplet agglomerator for the use of small droplets (courtesy of Pergamon Press, Backhaus et al. (1988))*

5.5 Conclusions

Aerosol technology is used in the manufacture of particulate commodities and materials for high technology application. Aerosol processes do not require the high liquid volumes and many tedious operations of wet chemistry. They may also be used in the production of highly pure materials. Depending on the actual process considered, a certain, and in many cases preferably monodisperse, size of particle is desired. Once formed the aerosol continually changes in size due to coagulation. Numerous models have been proposed to describe the development of the particle-size distribution. They perform reasonably well with respect to integral properties of the size distribution. Poor knowledge of specific physical phenomena (e.g. van der Waals forces and particle interception rates) does not allow precise model predictions from first principles. Advances in experimental techniques and theoretical techniques for studying coagulation of highly concentrated, non-spherical aerosols will greatly benefit aerosol formation technology. The basic concept of fractal geometry in the characterization of aerosol structure (Kaye, 1989; Colbeck and Harrison, 1991) should lead to great progress in studies of the dynamical behaviour of aerosol particles.

References

Alam, M. K. (1987) The effect of van der Waals and viscous forces on aerosol coagulation. *Aerosol Sci. Technol.*, **6**, 41–52

Backhaus, A., Conrads, M., Wurz, D. and Zimmerman, M. (1988) Development of a new agglomerator for the elimination of flue gas aerosol. *J. Aerosol Sci.*, **19**, 1373–1376

Bergland, R. N. and Liu, B. Y. H. (1973) Generation of monodisperse aerosol standards. *Environ. Sci. Technol.*, **7**, 147–153

Brockmann, J. E., McMurry, P. H. and Liu, B. Y. H. (1982) Experimental study of simultaneous coagulation and diffusional loss of free molecule aerosol in turbulent pipe flow. *J. Colloid Interface Sci.*, **88**, 522–529

Bunz, H. (1990) Coagulation workshop. *J. Aerosol Sci.*, **21**, 139–153

Chou, K. H., Lee, P. S. and Shaw, D. T. (1981) An investigation of high intensity acoustic agglomeration. *J. Colloid Interface Sci.*, **83**, 335–353

Clark, W. E. and Whitby, K. T. (1967) Concentration and size distribution, measurement of atmospheric aerosols and a test of the theory of self-preserving size distribution. *J. Atmos. Sci.*, **26**, 677–687

Colbeck, I. and Hardman, E. J. (1991) Thermophoretic and photophoretic motion of aerosols. *Powder Technol.*, **65**, 447–451

Colbeck, I. and Harrison, R. M. (1991) Aerosol science – from molecules to mists. *Chem. Ind.*, **7 Jan.**, 20–23

Davies, C. N. (1979) Coagulation of aerosols by Brownian motion. *J. Aerosol Sci.*, **10**, 151–161

Denton, M. B. and Swartz, D. B. (1974) An improved ultrasonic nebulizer system for the generation of high density aerosol dispersions. *Rev. Sci. Instrum.*, **45**, 81–83

Devir, S. E. (1963) Coagulation of aerosols. *J. Colloid Sci.*, **18**, 744–756

Egilmez, N. and Davies, C. N. (1982) Coagulation of fine aerosols. *Proc. R. Soc.*, **380**, 99–118

Eliasson, B., Egli, W., Ferguson, J. R. and Jodeit, H. (1987) Coagulation of bipolarly charged aerosols in a stack coagulator. *J. Aerosol Sci.*, **18**, 869–872

Fissan, H. (1990) Contamination of product surfaces by particles-reduction and control. In *Aerosols: Science, Industry, Health and Environment* (ed. Masuda, S. and Takahashi, K.) Pergamon Press, Oxford, 73–76

Flagan, R. C. (1990) Aerosol processes for material synthesis. In *Aerosols: Science, Industry, Health and Environment* (ed. Masuda, S. and Takahashi, K.) Pergamon Press, Oxford, 50–54

Friedlander, S. K. (1977) *Smoke, Dust and Haze*, Wiley, New York

Friedlander, S. K. and Wang, C. S. (1966) Self preserving particle size distribution for coagulation by Brownian motion. *J. Colloid Interface Sci.*, **22**, 126–132

Fuchs, N. A. (1964) *The Mechanics of Aerosols*, Pergamon Press, Oxford

Fuchs, N. A. and Sutugin, A. G. (1965) Coagulation rate of highly dispersed aerosols. *J. Colloid Sci.*, **20**, 492–500

Gallego-Juarez, J. A., Riera-Franco de Sarabia, E., Rodriguez-Corral, G., Magill, J., Richter, K., Fourcaudot, S. and Barraux, P. (1991) An acoustic system for particulate precipitation. *Environ. Pollut.*, **1**, 220–227

Gelbard, F. and Seinfeld, J. H. (1980) Simulation of multicomponent aerosol dynamics. *J. Colloid Interface Sci.*, **78**, 485–501

Gowland, R. J., Wilshire, B. and Clark, J. M. (1991) Production of novel particulate materials for aerosol characterization. In *Aerosols: Their Generation, Behaviour and Application*, The Aerosol Society, Loughborough, 217–222

Graham, S. C. and Homer, J. B. (1973) Coagulation of molten lead aerosols. *Discuss. Faraday Soc.*, **7**, 85–96

Hidy, G. M. (1984) *Aerosols: An Industrial and Environmental Science*, Academic Press, Orlando, FL

Hidy, G. M. and Lilly, D. K. (1965) Solutions to the equations for the kinetics of coagulation. *J. Colloid Sci.*, **20**, 867–874

Homma, K., Serita, F. and Takaya, M. (1990) Aerosol generation from particle suspensions by ultrasonic nebulizer. In *Aerosols: Science, Industry, Health and Environment* (ed. Masuda, S. and Takahashi, K.) Pergamon Press, Oxford, 231–234

Hoover, M. D., Casalnuovo, S. A., Lipowicz, P. J., Hsa, C. Y., Hanson, R. W. and Hurd, A. J. (1990) A method of producing non-spherical monodisperse particles using integrated circuit fabrication techniques. *J. Aerosol Sci.*, **21**, 569–575

Japuntich, D. A., Stenhouse, J. I. T. and Liu, B. Y. H. (1990) Conditions for monodispersity of heterogeneous condensation aerosols using dimensionless groups. *J. Colloid Interface Sci.*, **136**, 393–400

Jennings, S. G. (1988) The mean freepath in air. *J. Aerosol Sci.*, **19**, 159–166

Johnston, A. M., Vincent, J. H. and Jones, A. D. (1987) Electrical charge characteristics of dry aerosols produced by a number of laboratory dispensers. *Aerosol Sci. Technol.*, **6**, 115–127

Kaye, B. H. (1989) *A Random Walk through Fractal Dimensions*, V.C.H., Weinheim

Kaye, P. H., Hirst, E. and Clark, J. M. (1991) The manufacture of standard non-spherical particles by silicon micromachining. In *Aerosols: Their Generation, Behaviour and Application*, The Aerosol Society, Loughborough, 223–228

Kerker, M. (1975) Laboratory generation of aerosols. *Adv. Colloid Interface Sci.*, **5**, 105–172

Kim, C. S. and Liu, B. Y. H. (1984) Experimental studies of coagulation of free molecule aerosols. In *Aerosols* (ed. Liu, B. Y. H., Pui, D. Y. G. and Fissan, H.) Pergamon Press, Oxford, 923–925

Lai, F. S., Friedlander, S. K., Pich, J. and Hidy, G. M. (1972) Self-preserving particle size distribution for Brownian coagulation in the free molecule regime. *J. Colloid Interface Sci.*, **39**, 395–405

Lang, R. J. (1962) Ultrasonic atomization of liquids. *J. Acoustic Soc. Am.*, **34**, 6–8

Liu, B. Y. H. (1990) Aerosols and microelectronics. In *Aerosols: Science, Industry, Health and Environment* (ed. Masuda, S. and Takahashi, K.) Pergamon Press, Oxford, 5–9

Liu, B. Y. H. and Lee, K. W. (1975) An aerosol generator of high stability. *Am. Ind. Hyg. Assoc. J.*, **36**, 861–865

Liu, B. Y. H. and Lin, B. Y. (1990) An ultrafine DOS monodisperse aerosol particle generator for large scale filter testing. In *Aerosols: Science, Industry, Health and Environment* (ed. Masuda, S. and Takahashi, K.) Pergamon Press, Oxford, 263–266

Magill, J. (1989) Acoustic aerosol scavenging. In *Ultrasonics International Conference Proceedings*, Butterworth, Oxford, 313–320

Magill, J., Pickering, S., Fourcaudot, S., Gallego-Juarez, J. A., Riera-Franco de Sarabia, E. and Rodriguez-Corral, G. (1989) Acoustic aerosol scavenging. *J. Acoust. Soc. Am.*, **85**, 2678–2680

Marlow, W. H. (1980a) Lifshitz–van der Waals forces in aerosol particle collisions 1. Introduction: water droplets. *J. Chem. Phys.*, **73**, 6288–6295

Marlow, W. H. (1980b) Derivation of aerosol collision rates for singular attractive contact potentials. *J. Chem. Phys.*, **73**, 6284–6287

Marlow, W. H. (1981) Size effects in aerosol particle interactions: the van der Waals potential and collision rates. *Surface Sci.*, **106**, 529–537

Marlow, W. H. (1982) Lead aerosol Brownian collision rates at normal and elevated temperature: theory. *J. Colloid Interface Sci.*, **87**, 209–215

Marple, V. A., Liu, B. Y. H. and Rubow, K. L. (1978) A dust generator for laboratory use. *Am. Ind. Hyg. Assoc. J.*, **39**, 26–32

Marshall, I. A., Mitchell, J. P. and Griffiths, W. D. (1990) The formation of highly regular non-spherical particles by controlled crystal growth techniques. In *Aerosols: Their Generation, Behaviour and Application*, The Aerosol Society, Loughborough, 7–12

May, K. R. (1966) Spinning top homogeneous aerosol generator with shock proof mounting. *J. Sci. Instrum.*, **43**, 841–842

May, K. R. (1973) The collision atomizer: description, performance and application. *J. Aerosol. Sci.*, **4**, 235–243

Mednikov, E. P. (1965) *Acoustic Coagulation and Precipitation of Aerosols* (trans. from Russian by Larrick, C. V.) Consultants Bureau, New York

Meitzig, G. (1989) The use of different corrections for the Brownian coagulation function in aerosol behaviour modelling. *Report KFK 4606*, Kerforschungszentrum, Karlsruhe, 61–71

Melton, P. M., Harrison, R. M. and Burnell, P. K. P. (1991) The evaluation of an improved spinning top aerosol generator and comparison with its predecessor. *J. Aerosol Sci.*, **22**, 101–110

Mitchell, J. P. (1986) Aerosol generation for instrument calibration. *AEEW-R 2092*, HMSO, London

Nolan, P. J. and Kennan, E. L. (1949) Condensation nuclei from hot platinum: size, coagulation coefficient and charge distribution. *Proc. R. Ir. Acad.*, **A52**, 171–190

Okuyama, K., Kousaka, Y. and Hayashi, K. (1984) Change in size distribution of ultrafine aerosol particles undergoing Brownian coagulation. *J. Colloid Interface Sci.*, **101**, 98–109

Pratsinis, S. E. and Xiong, Y. (1990) Particle production in reactive turbulent flows. In *Aerosols: Science, Industry, Health and Environment* (ed. Masuda, S. and Takahashi, K.) Pergamon Press, Oxford, 267–270

Quon, J. E. (1964) Experimental determination of the coagulation rate constant for nuclei. *Int. J. Air Water Pollut.*, **8**, 335–368

Raabe, O. G. (1976) The generation of aerosols of fine particles. In *Fine Particles* (ed. Liu, B. Y. H.) Academic Press, New York, 55–110

Rapaport, E. and Weinstock, S. G. (1955) A generator for homogeneous aerosols. *Experientia*, **11**, 363–364

Rooker, S. J. and Davies, C. N. (1979) Measurement of the coagulation rate of a high Knudsen number aerosol with allowance for wall losses. *J. Aerosol Sci.*, **10**, 139–150

Schmauss, A. (1920) The chemistry of fog, clouds and rain. *Umschau, Munich*, **24**, 61–63

Schmidt-Ott, A. and Burtscher, H. (1982) The effect of van der Waals forces on aerosol coagulation *J. Colloid Interface Sci.*, **89**, 353–357

Seigneur, C., Hudischewskyj, A. B., Seinfeld, J. H., Whitby, K. T., Whitby, E. R., Brock, J. R. and Barnes, H. M. (1986) Simulation of aerosol dynamics: a comparative review of mathematical models. *Aerosol Sci. Technol.*, **5**, 205–222

Shaw, D. T. (1978) Acoustic agglomeration of aerosols. In *Recent Developments in Aerosol Science* (ed. Shaw, D. T.) Wiley Interscience, New York, 279–321

Shaw, D. T., Haugan, T., Xie, G. and Patel, S. (1990) Processing of ultrafine particles by laser ablation. In *Aerosols: Science, Industry, Health and Environment* (ed. Masuda, S. and Takahashi, K.) Pergamon Press, Oxford, 255–258

Shon, S. N., Kasper, G. and Shaw, D. T. (1980) An experimental study of Brownian coagulation in the transition regime. *J. Colloid Interface Sci.*, **73**, 233–243

Sinclair, D. and LaMer, V. (1949) Light scattering as a measure of particle size in aerosols. *Chem. Rev.*, **44**, 245–267

Smoluchowski, M. (1916) Drei vorträge über diffusion, Brownsche molekularbewegung und koagulation von kolloidteilchen. *Phys. Zeit.*, **XVII**, 557–599

Somers, J., Magill, J., Richter, K., Fourcaudot, S., Lagarge, P. and Barraux, P. (1991) Acoustic agglomeration of liquid and solid aerosols: a comparison of a glycol fog and titanium oxide. *J. Aerosol Sci.*, **22**, S109–S112

Tajima, N., Fukushima, N. and Sato, Y. (1990) Optimum design of a nebulizer for standard aerosol particle generation. In *Aerosols: Science, Industry, Health and Environment* (ed. Masuda, S. and Takahashi, K.) Pergamon Press, Oxford, 239–242

Tsang, T. H. and Brock, J. R. (1982) Aerosol coagulation in the plume from a crosswind time source. *Atmos. Environ.*, **16**, 2229–2235

Tsang, T. H. and Brock, J. R. (1983) Simulation of condensation aerosol growth by condensation and evaporation. *Aerosol Sci. Technol.*, **2**, 311–320

Tu, K. J. (1982) A condensation aerosol generator system for monodisperse aerosols of different physicochemical properties. *J. Aerosol Sci.*, **13**, 363–371

Wagner, P. E. (1989) Measurement of the coagulation function for fine particles – the influence of experimental conditions and observation techniques. *Report KFK 4606*, Kerforschungszentrum, Karlsruhe, 7–22

Wagner, P. E. and Kerker, M. (1977) Brownian coagulation of aerosols in rarefied gases. *J. Chem. Phys.*, **66**, 638–646

Warren, D. R. and Seinfeld, J. H. (1985) Simulation of aerosol size distribution evolution systems with simultaneous nucleation, condensation and coagulation. *Aerosol Sci. Technol.* **14**, 31–43

Wegrzyn, J. E. (1976) An investigation of an exploding wire aerosol. In *Fine Particles* (ed. Liu, B. Y. H.) Academic Press, New York, 253–274

Whitby, K. T. (1981) Determination of aerosol growth rates in the atmosphere using lumped mode aerosol dynamics. *J. Aerosol Sci.*, **12**, 172–178

Whitby, K. T. and Liu, B. Y. H. (1967) The electrical behaviour of aerosols. In *Aerosol Science* (ed. Davies, C. N.) Academic Press, London, 59–86

Whitby, K. T., Lundgren, D. A. and Peterson, C. M. (1965) Homogeneous aerosol generators. *Int. J. Water. Poll.*, **9**, 263–277

Whytlaw-Gray, R., Speakman, J. B. and Campbell, J. H. P. (1923) Smokes: Part I – a study of their behaviour and a method of determining the number of particles they contain. *Proc. R. Soc. A*, **102**, 600–615

Willeke, K. (1980) *Generation of Aerosols and Facilities for Exposure Experiments*, Ann Arbor Science, Ann Arbor, MI

Wright, B. M. (1950). A new dust-feed mechanism. *J. Sci. Instrum.*, **27**, 12–15

Yamamota, S., Aotoni, S., Fukushima, N. and Kousaka, H. (1990) Generation and application of uniform aerosol of polystyrene particles. In *Aerosols: Science, Industry, Health and Environment* (ed. Masuda, S. and Takahashi, K.) Pergamon Press, Oxford, 247–250

Yeh, H. C., Carpenter, R. L. and Cheng, Y. S. (1988) Electrostatic charge of aerosol particles from a fluidised bed aerosol generator. *J. Aerosol Sci.*, **19**, 147–151

Chapter 6

Bubble nucleation and growth

S. D. Lubetkin

6.1 Introduction

It may be an obvious point, but it is one worth making at the outset; the word 'bubble' is actually ambiguous. It is applied equally to gas surrounded by a bilayer of surfactant in a second gas phase, and to gas surrounded entirely by a liquid, with or without surfactant. Note that at very high bubble volume fractions (which are only achievable in practice in the presence of surfactant) this distinction might be lost. The present review deals only with bubbles consisting of gas in a continuous liquid phase. There are excellent monographs on soap bubbles (i.e. those of the first type) available, and those of Boys (1890) and Isenberg (1978) are recommended.

The size of bubbles is important in a number of ways. At the upper end of the size range, bubbles could in principle be of almost unlimited size, although there is a practical upper limit for aqueous systems of about 1 cm, beyond which hydrodynamic forces are likely to cause break up, as noted by Clift *et al.* (1978). Also, the larger the bubble, the less well it will approximate in shape to a sphere, with both gravitational and hydrodynamic distortions becoming increasingly important. A lower limit on the size of bubbles is set by the critical bubble size (as we will see below) which in turn is governed by a combination of (at least) three factors; these being, the solubility of the gas, and the mechanical equilibrium set by the Laplace pressure in combination with the local interfacial tension.

This review will chiefly be concerned with small bubbles. The term 'microbubble', which appears with some frequency in the literature, will be avoided here. Some authors (Hansen and Derderian, 1976; D'Arrigo, 1986) use this term to describe bubbles of about 30–100 μm (or even larger) sizes. Bubbles of gas up to about 1 mm diameter in water are still approximately spherical, but much above this size, the shape of a free bubble becomes much more complex, and may not even be axisymmetric. We will take this as a working upper size limit here.

Between about 0.1 μm and about 1 μm, there is a subrange of particular interest, where creaming is sufficiently slow for the bubbles to be 'kinetically stable' in the sense used in colloid science. Such suspensions seem to have been sadly neglected, despite their potential uses in manufacturing, and their intrinsically interesting properties. We will see that in many respects, suspensions (of colloidal size, or greater) of bubbles share common characteristics with suspensions of liquid droplets (emulsions), and that the chief differences arise from the relatively large compressibility and the reduced density and viscosity of the gaseous phase.

There is no definite volume fraction of bubbles at which it is obviously better to refer to the suspension as a foam. Those concerned with foams will probably have a mental picture of continuously interacting bubbles – for uniform, monosized bubbles, this state of affairs will begin at a volume fraction close to 70%, and will continue up to the $\geq 97\%$ commonly found with foams.

The literature is well served with reviews of foams; their stability and rheology (Akers, 1976; Dickinson, 1987), and it is not the purpose of the present chapter to add to that literature; rather the focus will be on the individual bubbles and their behaviour.

Where the formation of bubbles is accompanied by a decrease in free energy (e.g. in boiling, cavitation or the spontaneous formation of bubbles in supersaturated solutions) then the size of bubbles, as with the size of other suspensions, depends on both the nucleation and the growth steps. Where the bubbles are formed by attrition (i.e. the break up of larger bubbles) or by sparging or entrainment, the bubble size distribution will be governed by mechanical and hydrodynamic interactions, and nucleation (and possibly growth too) will be unimportant. To achieve control of the final particle size, each of these steps, and the many possible mechanisms leading to each will have to be controlled.

Controlling the size of small bubbles is not easy. Not only do bubbles suffer from all the same difficulties in this respect as do liquid drops or solid particles (such as Ostwald ripening), but also because of their compressibility, there is a fundamental problem in achieving monodispersity of the particle-size distribution. Regardless of how the bubbles are formed, under otherwise identical external conditions bubbles will be smaller the greater their depth in the liquid. The only exception to this general rule will be in cases where gravity may be ignored; for gas bubbles formed from liquids very close to the critical point, where the densities of the gas and liquid phases converge, or in microgravity environments (Karri, 1988), but these cases are highly specific, and probably not of great practical interest. Note, however, that for bubbles in aqueous media characteristic laboratory depths (say 10 cm) correspond to only an increase of about 1% in ambient pressure, and hence to a decrease of about 0.3% in the bubble radius.

A further difficulty specific to gas bubbles is that bubble nuclei (and indeed fully fledged bubbles) can be formed purely mechanically, for example by vigorous stirring and consequent entrainment or attrition of larger bubbles, in a system where such bubbles are not expected to arise thermodynamically. Such possibilities complicate the production of bubble suspensions, and the interpretation of measurements on such systems. There is an apparent similarity of attrition as a means of formation of bubbles, with the phenomenon of 'secondary nucleation' in crystal formation (Mason and Strickland-Constable, 1966).

If we attempt to treat a dilute bubble suspension as a classical colloidal system, we recall that to be considered in this way, the bubble diameter will have to be roughly between about 1 μm and about 1 nm. (Note that bubbles at the lower end of this size range are likely to be well below the critical bubble size, and hence unrealizable in practice.) In this size range, negative sedimentation or creaming under gravity will be relatively unimportant. The criterion for deciding when gravity may be ignored in this way is encapsulated in the Peclet number, and this will be more the fully discussed below. We note that even where gravity might be ignored from point of view of sedimentation, it will still have an impact on the hydrostatic head, and hence will have a small effect on particle size; these two aspects are therefore inextricably linked in a way that simply does not arise in the case of incompressible phases.

This chapter is divided into four sections, dealing with (i) the formation of bubbles, (ii) the growth of bubbles, (iii) the detachment of bubbles, and (iv) the role of surfactants in bubble formation and stability.

6.2 Formation of bubbles

We can broadly separate bubble formation into two categories: those cases where the free energy of the system is reduced by the appearance of bubbles, thus being spontaneous in the thermodynamic sense; and those where bubble formation is accompanied by a corresponding increase in the free energy.

6.2.1 Spontaneous bubble formation

Bubbles may be formed spontaneously in several ways, but note that bubbles can form spontaneously in a single-component system only by cavitation or boiling (ebullition), whereas in a two (or greater) component system supersaturation of a second component may arise, and hence under appropriate conditions bubbles may be formed as a result of the presence of a greater than equilibrium concentration of a dissolved gas, as well as by the two previously mentioned mechanisms. Note, however, that a dissolved gas may have an effect upon the boiling of the major component, even if it is itself insufficiently supersaturated to produce bubbles, as noted by Mori *et al.* (1976).

Conceptually, boiling in a single-component system, and the formation of gas bubbles of a second component, are similar. Where the second component does have an important influence (see Section 6.4 for the effect on surface tension and contact angle) this will be discussed in the context of specific alterations to an otherwise unified approach.

Bubbles may arise spontaneously from six quite separate and essentially independent sources, and in each case after their formation, growth to a macroscopic size may take place:

(a) homogeneous nucleation
(b) heterogeneous nucleation
(c) cavitation
(d) electrolysis
(e) Harvey nuclei
(f) pre-existing and colloidally stable free bubbles.

Cases (e) and (f) are somewhat different in kind from (a)–(d), inasmuch as in each of the latter eventualities, the new (bubble) phase already exists, and true nucleation is thus avoided. They are none the less important for that – in fact in practical situations such pre-existing gas phases may be impossible to avoid, and for this reason nucleation will often be of no great significance.

Homogeneous nucleation, single-component system

When bubbles are formed in the bulk of the liquid, well away from any surface (for example, the walls of the container, dust particles or other foreign bodies, or the free liquid surface), then the nucleation is said to be 'homogeneous'. The thermodynamic drive for the phase change is the excess of chemical potential of the liquid phase as compared with that of the vapour and, for the purposes of the present chapter, this drive may arise in two chief ways.

Alteration of temperature

In a single-component system, raising the temperature of the system at a constant pressure (usually close to but not exactly equal to atmospheric) sufficiently above the appropriately pressure-corrected boiling point will result in the formation of initially critical-sized bubbles, which will then rapidly grow. The pressure correction and the inequality to atmospheric pressure arise from the fact that there is a hydrostatic head to be considered. Usually, this correction is small, but in dense liquids, or at great depths, these could be significant. In this context, it is worth noting that, given a spatially uniform temperature for the liquid (in practice, convection will render the liquid thermally inhomogeneous), boiling should start near the top of the containing vessel where the hydrostatic pressure is least. The liquid surface itself may be the preferred site for nucleation, where the role of inhomogeneities will be important. Chief amongst these is likely to be the presence at the interface of surface active species, which might either promote or antagonize nucleation (see Section 6.5). A purist might further argue that such nucleation at the free liquid surface is not actually homogeneous in any case, and should for this reason be considered in the section on heterogeneous nucleation (see below).

Alteration of pressure

The vapour may be stabilized with respect to the liquid by a reduction of pressure. It is of course a commonplace that reducing the pressure above a liquid will cause a reduction in the boiling point. Similarly, reduction of the hydrostatic pressure in a liquid can cause cavitation, and this problem has a mature and extensive engineering literature (see e.g. Knapp *et al.*, 1970). For example, the power that can be transmitted by a screw (propeller) to the surrounding water before the reduction in hydrostatic pressure over the trailing edges causes cavitation (and collateral damage to the material of the screw) limits the efficiency of screw-driven vessels. Studies of the (negative) pressure required to cause breaking of a thread of liquid contained in a capillary tube have a venerable history, and have been used extensively to test theories of cavity nucleation. Most such experiments have actually revealed that the nucleation of cavities is not homogeneous, but takes place preferentially at surface imperfections, probably related to Harvey nuclei (see Section 6.2.3). Note that evacuating the head space above a liquid can only impose a (negative) pressure or tension of (at most) 1 atmos, whereas mechanically imposed tensions larger than this can be achieved by other procedures, as discussed in Section 6.2.3.

The thermodynamic/kinetic treatment of homogeneous, single-component system nucleation applies equally to the two above chief ways of supersaturating the system, and also, with minor modification, to the formation of bubbles in two (or more) component systems.

The theory of both homogeneous and heterogeneous nucleation has been covered extensively in the literature. Particularly comprehensive reviews have been given by Hirth and Pound (1963) and by Zettlemoyer (1969), while an earlier work by Dunning (1955) is still unsurpassed for clarity. More recently, Kashtiev has published several excellent articles covering both atomistic and so-called 'classical theories' (Kashtiev and Exerova, 1980; Kashtiev, 1982, 1984, 1985; Exerova *et al.*, 1983; Gutsov *et al.*, 1985; Trayanov and Kashtiev, 1986). It is noteworthy that bubble nucleation does not have an extensive review literature, and only rates a mention in the review of Hirth and Pound. In view of the large number of general nucleation reviews, only a superficial treatment is given here, sufficient for understanding the later complications.

In general, homogeneous nucleation occurs at higher supersaturations than for the heterogeneous case, and the theoretical prediction is that under conditions of low supersaturations, homogeneous nucleation is unlikely. The other way of expressing this is that the rate of heteronucleation for a given supersaturation is likely to be much greater than the corresponding homonucleation rate. Such a statement carries a risk of misinterpretation, however, since the heteronucleation rate will always be limited at sufficiently high rates, by the number of hetero-sites. In the case of homogeneous nucleation, the ultimate rate would not be limited in this way, though it would of course be limited by the diffusion of monomer units to the growing nucleus.

To emphasize this, consideration of the CO_2/water system shows that the conditions for homogeneous nucleation of CO_2 bubbles requires a super-saturation in excess of about 1000: this should be compared with a typical supersaturation of about 4 for commercially carbonated drinks, for example. The experimental finding is, of course, that many bubbles are indeed formed in such systems. Conditions for homogeneous nucleation are fairly infrequently met, and thus despite its theoretical importance, this type of nucleation is often not as practically significant as the heterogeneous case.

Before discussing the kinetic expressions for J, the rate of nucleation under various conditions, we introduce the concept of a 'critical' size. It is helpful first to examine the concept of critical size in the case of a liquid drop formed from a supersaturated vapour. A supersaturated vapour is thermodynamically less stable (and therefore has a higher chemical potential) than the liquid. The free energy of formation of a drop of the (new) liquid phase of radius r is composed of two terms – the 'surface' term and the 'bulk' term. The surface term expresses the fact that energy must be expended to form the drop/vapour interface of interfacial tension γ

$$\Delta G_{surface} = 4\pi r^2\gamma \tag{6.1}$$

The bulk term is due to the gain in free energy as a result of the greater stability of the liquid phase ΔG_v per unit volume of new phase

$$\Delta G_{bulk} = 4\pi r^3 \Delta G_v/3 \tag{6.2}$$

Overall, the free energy change is given by

$$\Delta G_{total} = 4\pi r^2 \gamma - 4\pi r^3 \Delta G_v/3 \tag{6.3}$$

Now, the Kelvin equation (eqn (6.4)) relates the curvature of the interface $(1/r)$ to the ratio of the vapour pressure P of a drop of radius r to that over a flat surface P_0, where $\alpha = P/P_0$:

$$2\gamma/r = kT \ln \alpha \tag{6.4}$$

and ΔG_v can be expressed in terms of α, since for a molecular volume of O

$$\Delta G_v = kT \ln \alpha/O \tag{6.5}$$

As can be seen from Figure (6.1), ΔG_{total} passes through a maximum at a size r^*, given by differentiating ΔG_{total} with respect to the radius, and setting the resulting expression equal to zero. This gives the maximum free energy barrier, ΔG^*, which has to be surmounted in the nucleation process. The size of the droplet r^* corresponding to this maximum in the free energy is called the critical drop size. If such a drop loses one molecule (monomer unit), it will have an increasing propensity to evaporate, since its curvature is now greater than $1/r^*$, and from eqn (6.4) this results in an increased vapour pressure; finally, the droplet will thus return to individual molecules. If, on the other hand, it gains a molecule, it becomes essentially free growing, since its curvature is now decreased, and its vapour pressure correspondingly reduced below that of the ambient supersaturated vapour. The equilibrium at the size r^* is thus unstable, and the formation of critical-sized droplets is the kinetic bottleneck to the appearance of the new liquid phase from the supersaturated vapour.

A very similar argument leads to the definition of the critical bubble size, and a full analysis of the problem is given by Blander and Katz (1975). In terms of P and p^*, the pressures in the liquid and the vapour pressure of the liquid (at the ambient pressure P), respectively, and the surface tension γ:

$$r^* = 2\gamma/(p^* - P) \tag{6.6}$$

In terms of measurable quantities, since $\alpha \approx p^*/P$, and $\sigma = \alpha - 1$, this gives approximately:

$$r^* = 2\gamma/\sigma P \tag{6.7}$$

At this size, the bubble is in mechanical equilibrium with the liquid, and the chemical potential of the vapour in the bubble is equal to that of its surrounding liquid.

In order to predict the rate of nucleation, we need to calculate (i) the number of critical-sized bubbles formed per unit time, and (ii) the rate at which they acquire one extra molecule, and thus become free growing. We start by considering the case of boiling (ebullition) in a single-component system. The most important parameter is (as before) the saturation ratio α, which for boiling

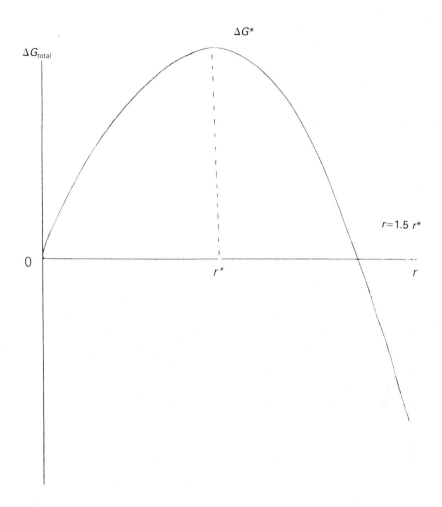

Figure 6.1   *The overall free energy (ΔG_{total}) for the formation of a cluster passes through a maximum with increasing size of the cluster. The size corresponding to this maximum free energy requirement occurs at the radius r*, and since the free energy decreases in both directions away from this maximum, the drop will either grow or evaporate; the equilibrium is unstable. The quantity ΔG^* (corresponding to r*) is thus the maximum free energy barrier in the nucleation process, and constitutes the kinetic 'bottleneck' to the formation of the new phase. Once such a critical nucleus has been formed by fluctuations (whose probability decreases in proportion to exp ($-\Delta G/kT$)), the stability of the mother phase (supersaturated vapour in this case) is threatened. The point at which the curve crosses the x axis, which occurs at $r = 1.5r^*$, corresponds to the cluster size at which the surface and bulk terms exactly balance – it has no great significance insofar as nucleation theory is concerned*

is a measure of the degree to which the liquid is superheated above its normal boiling point under the prevailing conditions of pressure, both applied (often atmospheric) and hydrostatic.

The rate of nucleation J (i.e. the number of bubbles formed per second per cubic centimetre of solution) is given by

$$J = C \exp [-16\pi\gamma^3/3kT(p^* - P)^2] \tag{6.8}$$

In this equation, p^* is the partial pressure of the vapour in the critical bubble, and is equal to $P + 2\gamma/r$, where P is the applied hydrostatic head. C is a quantity which varies relatively slowly with σ, the supersaturation, defined as above, $\sigma = \alpha - 1$, and T. The value of C can usually be considered to be approximately constant for small ranges of supersaturation. In those instances where such approximations are not valid, a fuller account must be taken of the parameters going to make up C. According to Döring (1937, 1938), C is given by

$$C = [6\gamma/\pi m(3 - b)]^{1/2} \cdot \exp (-\Delta H_{vap}/kT)n_0 \tag{6.9}$$

where γ is the interfacial tension between the liquid and its vapour at the appropriate temperature, m is the molecular mass, ΔH_{vap} is the enthalpy of vaporization per molecule, and n_0 is the number of molecules per unit volume (1 cm^3 in the present context). The quantity $b = (p^* - P)/p^*$ introduces a problem, since for $b > 3$ this leads to a negative square-root term. Döring thus restricted his equation to cases where $b < 3$. This problem was surmounted by Hirth and Pound (1963) who give as an alternative to eqn (6.9) for C:

$$C = \Delta G^*/(3\pi kTi^{*2})^{1/2} \cdot p^*/(2\pi mkT)^{1/2} \cdot 4\pi r^{*2} \cdot n_0 \tag{6.10}$$

This can also be expressed in other terms which emphasize the similarities with eqn (6.9)

$$C = \Delta G^*/(3\pi kTi^{*2})^{1/2} \cdot n_s v \exp (-\Delta H_{vap}/kT) \cdot 4\pi r^{*2} \cdot n_0 \tag{6.11}$$

where n_s is the number of molecules per square centimetre in the liquid surface, and v is the vibrational frequency of the liquid molecules.

In order to compare experimental results with eqn (6.8), together with expressions (6.9), (6.10) or (6.11) for the pre-exponential, a means of evaluating p^* is needed. Hirth and Pound (1963) showed that the transcendental eqn (6.12) provides the required expression

$$kT \ln (p/p^*) = O(p^* - P) \tag{6.12}$$

Homogeneous nucleation, two-component system

Let us now briefly consider the case of a two-component system, where the second component is a dissolved gas. With small modification, the equations developed above are used again. There are two main modifications required to the theory, these being that the effective pressure in the critical bubble nucleus is increased by the presence of the dissolved gas, and the modification of the surface tension by the dissolved gas.

Firstly, the pressure inside the bubble p^* is now composed of the sum of the (partial) vapour pressure of the liquid p_s together with the partial pressure of the dissolved gas p_v

$$p^* = p_s + p_v \qquad (6.13)$$

For weak gas solutions, it can be shown (Ward *et al.*, 1970) that if c_2 is the concentration of dissolved gas, and c_{2e} is the corresponding concentration over a flat surface at the applied pressure P, then

$$p_s/P = c_2/c_{2e} \qquad (6.14)$$

The same authors showed that $p_v = \varphi p_{vap}$, where φ is given by

$$\varphi = \exp [O(P - p_{vap})/kT - c_2/c_1] \qquad (6.15)$$

here O is the molecular volume of the solvent and c_1 is the solvent concentration in moles per unit volume (*ca.* 55.5 M for water). With these modifications, the nucleation equation reads

$$J = C' \exp [-16\pi\gamma^3/3kT(Pc_2/c_{2e} + \varphi p_{vap} - P)^2] \qquad (6.16)$$

where use has been made of eqns (6.14) and (6.15). Now, C' is not strictly a constant (as before), but includes N, the number of molecules per unit volume of solvent, a factor B (=1 in the present discussion), the surface tension γ, and the molecular mass of the gas species m. The interested reader is referred to Ward *et al.* (1970, 1986) for details. Lubetkin and Blackwell (1988) showed that at low supersaturations, where $\varphi p_{vap} \ll Pc_2/c_{2e}$, then the bracketed denominator in eqn (6.16) simplifies to give

$$J = C' \exp [-16\pi\gamma^3/3kT(\sigma P)^2] \qquad (6.17)$$

where σ is the saturation ratio, $\alpha = c_2/c_{2e}$ (see eqns (6.7) and (6.16)), minus one; thus $\sigma = \alpha - 1$. Equation (6.17) is now in a form to be directly compared with experimental data.

Secondly, the adsorption of the dissolved second component will in general alter the interfacial tension. Since this quantity, γ, appears to the third power in the exponential, it will exercise a considerable influence on the rate of nucleation. This question is addressed in detail in Section 6.4.

Heterogeneous nucleation, single-component system

Bubbles formed at a surface (e.g. the walls of the container, specks of dust or other inhomogeneities), rather than in the bulk of the solution, are said to be nucleated 'heterogeneously'. Usually, heterogeneous nucleation of bubbles is easier (i.e. occurs at lower values of σ) than for the homogeneous case. The degree to which it is easier varies according to two main parameters: (i) the contact angle θ of the gas/solution/solid interface; and (ii) the geometry of the nucleation site.

The contact angle θ

As the contact angle increases away from zero (where nucleation is exactly as difficult as for the homogeneous case), it becomes easier for bubbles to be formed. Mathematically, this increased ease of formation is expressed by the appearance of a factor $f(\theta)$ in the exponential term in eqn (6.1), and the equation now reads

$$J = C'' \exp[-16p\gamma^3 f(\theta)/3kT(p^* - P)^2] \tag{6.18}$$

C'' now includes a further two terms, which depend upon the contact angle. The first is simply $\{f(\theta)\}^{-1/2}$, and the second, $1/2\{1 + \cos(\theta)\}$. As before, we note that changes in the pre-exponential are usually negligible in their effect on J when compared with changes in the exponential term, and this is the justification for treating the pre-exponential as a constant.

Thus $f(\theta)$ is given by

$$f(\theta) = [2 + 3\cos(\theta) - \cos^3(\theta)]/4 \tag{6.19}$$

This function changes from 1 when $\theta = 0°$ (i.e. 'perfect wetting') to 0 when $\theta = 180°$ and, correspondingly, the free energy barrier to the formation of bubbles changes from a size as large as that for the homogeneous case when $\theta = 0°$ to zero when $\theta = 180°$.

Conceptually, the easiest way to understand this reduction in free energy needed for nucleation is to note that, as the contact angle increases away from zero, so the critical bubble will adopt the shape of a spherical cap, whose volume is governed by the contact angle, as shown in Figure (6.2).

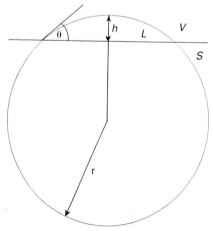

Figure 6.2   *The spherical cap model of a nucleus of a partially wetting liquid forming on a solid surface. The maximum height of the cap h is a function of the sphere radius r and the contact angle θ. Eliminating h from the expression for the volume of the cap gives the function f'(θ)·V = [2 − 3 cos (θ) + cos ³(θ)]V/4, where V is the volume of the whole sphere, of radius r. The function f'(θ) goes from 1 for θ = 180° (i.e. non-wetting), to zero for θ = 0° (i.e. perfect wetting). Note that this is the opposite of the case for bubbles, where the function f(θ) goes from 0 to θ = 180° (i.e. non-wetting), to 1 for θ = 0° (i.e. perfect wetting), as expressed by eqn (6.19)*

The geometry of the nucleation site

This problem has not yet been fully investigated, but an analysis is available for certain well-defined geometries (e.g. spherical and conical depressions and spherical and conical protrusions), for nucleation in two-component systems (see below). For details of the effects in the case of nucleate boiling, see the review by Cole (1974).

Heterogeneous nucleation, two-component system

Equation (6.17) can be modified to take account of variations in the contact angle θ and in the geometry of the nucleation site for the case of heterogeneous nucleation in two-component systems. Essentially, it is found that nucleation becomes considerably easier in certain cases, and in particular for conical pits. A conical pit of half-angle β, introduces a factor in the exponential term (and also into the pre-exponential term, which we are ignoring in the present analysis) which reduces the barrier to nucleation. If we take the effect of the contact angle and the conical pit together, the factor can be called $f(\theta,\beta)$, and is defined as

$$f(\theta,\beta) = [2 - 2 \sin (\theta - \beta) + \cos (\theta) \cos^2 (\theta - \beta)/\sin (\beta)]/4 \qquad (6.20)$$

and the modified nucleation equation becomes

$$J = C''' \exp [-16\pi\gamma^3 f(\theta,\beta)/3kT(\sigma P)^2] \qquad (6.21)$$

With these modifications, and for some values of γ, θ, β and σ, eqn (6.21) predicts reasonable rates of formation of bubbles under conditions of super-saturation easily attainable in practice. As a concrete example, for the CO_2/water system, supersaturations of around 4 are sufficient to cause nucleation in cavities of half-angle of about 5°, with contact angles of about 90° (Wilt, 1986). The factor $f(\theta,\beta)$ becomes negative when $\theta - \beta > 90°$, and the treatment breaks down (Carr, 1993). The factor $f(\theta,\beta)$ used by Wilt is chosen to give a very small numerical value (*ca.* 10^{-5}), and it is chiefly this which results in predictions of reasonable rates of nucleation at low supersaturations. Small changes in either θ or β result in such large (negative) values of the exponential term in eqn (6.21) that it becomes impossible to evaluate the full expression for J. The whole basis of the analysis above has been criticized (Lubetkin, 1989a, b), and this is referred to below (Section 6.4).

Single-component nucleation experiments – boiling

This field is really too large for a satisfactory treatment here, and the interested reader is referred to Cole's (1974) review. Briefly, early experiments (see e.g. Wismer, 1922) attempting to measure superheat limits on bulk samples of liquids gave sometimes confusing and irreproducible results. The most likely explana-tion for such apparent experimental failures is that Harvey nuclei (and more generally heterogeneous nucleations sites of unknown detailed geometry and contact angle) were promoting nucleation or bubble formation. These early experiments all share the feature that the superheat limits are not as large as theoretically expected. Deliberately roughened surfaces were employed by Corty and Foust (1955), while Clark *et al.* (1959) used macroscopically polished surfaces for their high speed photographic investigations, and identified individual 'nucleation' sites.

More promising from the point of view of interpretation in terms of homogeneous nucleation are experiments on the boiling of individual droplets. Such a method had long been used in studies of crystallization (see e.g. Sander and Damkohler, 1943; Turnbull, 1952). Amongst those using this approach were Trefethen (1957), Wakeshima and Takata (1958), Skripov and Sinitsyn (1964, 1968) (see also Skripov and Ermakov, 1964; Sinitsyn and Skripov, 1968; Blander *et al.*, 1971). The results of these experiments appear to agree much more closely with expectations.

Two-component nucleation experiments

Reports of experiments on bubble nucleation in supersaturated gas solutions (other than electrochemically evolved gases) are rare, although there is a substantial qualitative and semiquantitative literature on the gas content of carbonated drinks (see e.g. Findlay and King, 1913). More recently, Lubetkin and Blackwell (1988), and Lubetkin (1989a, b) have reported measurements on the rate of nucleation of bubbles of CO_2 from supersaturated water solutions, and attempted to compare the results with the classical heterogeneous nucleation theory of Wilt (1986) with rather variable results. The interpretation of the data shows that the nature of the hetero surface, and in particular the degree to which it is hydrophilic or hydrophobic, has important consequences. This is discussed more fully in Section 6.4.

Cavitation

Cavitation is the process of bubble formation by reduction in pressure. The reduction may be effected by mechanical means, placing the liquid under tension, or by acoustic means, where rarefaction waves similarly introduce negative pressures. It is of interest to note that the elementary step in the growth to supercritical size of the cavity nucleus is not (as might be supposed) the transfer of liquid molecules to the vapour in the cavity, but the transport of vacancies from liquid to growing cavity (Hirth and Pound, 1963).

The question of the further growth of the bubbles once formed appears not to have been studied, and will not be further addressed here, except to note that if the bubble has been produced by a transitory change in pressure (as might be expected in the case of cavitation produced by acoustic means) there is no obvious drive for further growth – indeed, it is often the extremely rapid collapse of the cavities which liberates enough mechanical energy to cause the damage found on ships' screws. Clearly, in a solution that is supersaturated, but insufficiently so to give spontaneous nucleation, cavitation could potentially provide supercritical nuclei, and the further growth of these nuclei would then be assured, assuming that transport of the growth units was adequately fast to outpace incipient bubble collapse.

Cavitation experiments have been conducted using a variety of techniques. The earliest attempts were made by completely filling glass tubes with liquids, sealing and then cooling them; from the known thermal expansivities of the glass and the liquid, the observed rupture could be related to a calculated tension

(Meyer, 1911; see also Kenrick *et al.*, 1924). The direct application of negative pressures by bellows (Vincent, 1941, 1943) and centrifugal methods have been used more recently (Briggs, 1950, 1951a, b, 1953, 1955).

As noted earlier, experiments generally give fracture tensions much lower than those calculated from theoretical equations, and this is illustrated by Table 6.1, which shows the calculated and experimental tensile strengths of various liquids. This large discrepancy between expectation and experiment leads to the conclusion that cavitation (in common with many other nucleation phenomena) is normally heterogeneous, unless special precautions are taken.

Table 6.1   *Calculated and experimental strengths of various liquids[a]*

Liquid	Theoretical fracture pressure (atmos)	Observed fracture pressure (atmos)
Water	−1380	−270
Chloroform	−318	−290
Benzene	−352	−150
Acetic acid	−325	−288
Aniline	−625	−280
Carbon tetrachloride	−315	−275
Mercury	−23100	−425

[a] Courtesy of Hirth and Pound (1963).

Electrolytic and chemical bubble generation

Electrolytic bubble generation

This is a wide topic and really deserves a review in itself, and many have been published; for a recent example, see Sides (1986). However, in general terms, electrolytic bubble generation is not markedly different from heteronucleation and growth on surfaces other than those of electrodes. It should be remarked, however, that bubbles will in general be charged, and the electrode surface itself will carry a charge, the sign and magnitude of which will vary as the electrochemical conditions are altered. This introduces a variable interaction between the hetero surface and the growing nucleus/bubble not met in other cases of heterogeneous nucleation.

This specific interaction will alter both the thermodynamics (through changes in the electrochemical potentials of the various species and phases) and the kinetics (by altering diffusion rates, for example). Furthermore, it is to be expected that the size of the bubbles upon detachment from the electrode will be strongly influenced by subtle changes in composition of the electrode surface, and by the electrostatic interaction. Some of these issues have been addressed in the literature (Brandon and Kelsall, 1985; Brandon *et. al.*, 1985).

The chief advantage of using an electrolytic method for nucleation and growth of bubbles is that one has effective control over both the kinetics (by manipulating the current) and the thermodynamics (by similarly maintaining desired voltage differences). From the point of view of both understanding and process control, these are powerful advantages, and they have been fully exploited in industrial practice, notably in the chloralkali process. Given the economic importance of the chloralkali process, it is unsurprising that chlorine should be one of the three most extensively studied species, the others being hydrogen and oxygen.

While studies of the growth kinetics of electrogenerated bubbles are relatively common (see e.g. Glas and Westwater, 1964; Verhaart *et al.*, 1980), few appear to have been undertaken on bubble nucleation kinetics. Indirect evidence obtained from the transient current behaviour, particularly on microelectrodes, has been used to differentiate between the importance of mass transport and diffusion, but these experiments lack in either temporal or spatial resolution. Recently, a new technique has become available which shows considerable promise in this regard. The electrochemical quartz crystal microbalance (EQCM) has great sensitivity to changes in attached mass (of the order of nanograms per square centimetre) and has been shown to be capable of detecting mono-molecular layers of adsorbed species. A bubble arising at such a surface has an instantaneous effect on the apparent attached mass. The first report of this method shows that information may be obtained both on the nucleation and the subsequent growth of bubbles (Carr *et al.*, 1989), and the subsequent report of Gabrielli *et al.* (1991) confirmed the promise of this novel method.

Chemical bubble generation

The supersaturated gas/liquid solution can of course be produced by a chemical reaction which does not involve the passage of an electric current. Thermal decomposition reactions producing gases are the most obvious type to exploit, and current practice uses carbonates, nitrates, hydrogen peroxide, and diazo derivatives. Thermal decomposition can be replaced by acid–base reaction, or other chemically induced gas releasing processes. In urethane foam production, the carbon dioxide gas released in the reaction of the isocyanates may be used to produce the foam. Careful control of the initial particle size of the chosen reactant(s) provides the possibility of complete control of the final bubble size, but the thermal instability of many of the obvious candidate reactants may inhibit milling to the appropriate size.

Harvey nuclei

The formation of macroscopic bubbles may take place without the need for a nucleation step, if there are pre-existing sources of bubbles. Here and in the subsequent sections, we consider three such sources: Harvey nuclei and pre-existing bubbles whose size is sufficiently small to prevent creaming under gravity, and entrained or sparged bubbles. The distinction between these is not simply a matter of where the pre-existing bubble is located (although for Harvey nuclei this must be at a solid–liquid interface, by definition (Harvey *et al.*, 1944a, b, 1945, 1947; see also Bankoff, 1958), but includes possible differences in size of the bubble source. In the case of Harvey nuclei, the source is any gas-filled (usually) re-entrant cavity of such geometry and size that the entrapped gas is stable to displacement by the surrounding liquid phase. The size requirement will be discussed in more detail below, but in all events has to be larger than the critical bubble nucleus size under the prevailing conditions. It is not a requirement that the trapped gas be chemically the same as the dissolved gas or the vapour of the liquid, and this permissible difference in chemical composition can be of significance, since any container or indeed adventitious solid particle will usually have previously been in contact with air, and air-filled cavities will then potentially serve as Harvey nuclei for the formation of bubbles in the system. Perhaps the most familiar everyday example is the use of antibumping granules during distillation, where the deliberate (if unstated) aim is to introduce as many Harvey nuclei as possible.

The thermodynamic and to a lesser degree the hydrodynamic requirements for the stability of Harvey nuclei have been studied by Cole (1974), while the behaviour of sites of known geometry during boiling of water were tested experimentally by Griffith and Wallis (1960). In the case of bubble formation from a supersaturated gas solution, similarly well-defined sites were examined by Lubetkin (1989b).

Pre-existing bubbles

Most liquids which have been subject to shaking or stirring in a container will contain bubbles. Naturally, because of the generally large density difference between liquids and vapours or gases, creaming will take place on a time-scale determined by bubble diameter. For a given liquid/vapour combination, the importance of creaming will be regulated by the Peclet number, which is defined as the ratio of the gravitational force to the Brownian force on the bubble:

$$P_e = \Delta\rho vr/kT \tag{6.22}$$

where $\Delta\rho$ is the density difference between liquid and bubble, v is the bubble volume, r its radius, k is the Boltzmann constant, and T the absolute temperature. Creaming will only be important where $P_e \gg 1$, which for water at room temperature will be satisfied for bubbles of rather more than 1 μm in diameter. Obviously, bubbles much smaller than this size will not cream on reasonably short time-scales, and as they become smaller still, they will effectively not cream at all. Such behaviour is well known for colloidal dispersions of solids and liquids, and the potential long-term (kinetic) stability of such suspensions hardly needs re-emphasis here. The issue is not settled by such an analysis, however. Stability against creaming or sedimentation is no guarantee of stability in other respects, and so it is here; bubbles are as subject to dissolution and coalescence as are liquid droplets or solid particles. These issues are addressed in Section 6.4. For the present, suffice it to say that bubble suspensions are frequently less concentrated than other conventional colloidal systems, so that Brownian collision, and hence coalescence, is less important. For the case of supersaturated gas solutions at least, there is no problem of dissolution for supercritical-sized bubbles. Furthermore, surfactants may play an even more important role in stabilising bubbles than in other colloidal systems, and this aspect is dealt with more fully in Section 6.5.

In the case of aqueous media, it is known that bubbles may be very persistent. The original work of Harvey *et al.* (1944a, b, 1945, 1947) and Bankoff (1958) demonstrated this well, although in these cases, the bubbles were not free, but trapped at solid–liquid interfaces. Recently, D'Arrigo (1986) has reviewed the evidence for such stable bubbles, particularly in natural waters, and has at least partially identified the natural surfactants which are responsible for their stability. It is highly likely that unless very careful precautions are taken, that most water will contain a large number of bubbles, and that, statistically, a proportion of these will satisfy the Peclet criterion, and be kinetically stable over long periods. The bubbles considered by D'Arrigo are generally much larger than this non-creaming size (say between about 30–100 μm in diameter), and his chief interest is in the stability of such relatively large bubbles against dissolution.

A sudden change in conditions in such a stable bubble-containing system (reduction of pressure, increase in temperature etc.) would then give rise to a surge of bubbles, as the pre-existing, submicroscopic bubbles rapidly grew to a visible (or detectable) size. The appearance would be of a rapid burst of nucleation, but such an impression might be false: no nucleation would be needed, and none might have taken place; this could only be decided on the basis of whether the imposed change was sufficient to cause nucleation on its own account, and by whether the rate of bubble appearance dropped with time more rapidly than the rate of relief of supersaturation apparently warranted.

From a practical standpoint, the interesting question is how long might such pre-existing bubbles last? Unfortunately, there seems to be no clear answer to this question at present. By analogy with colloidal suspensions of solids or liquids, however, we may speculate that the life-times may be very long indeed. Faraday's original gold sols are still stable at the Royal Institution in London, nearly 150 years after their preparation. It is relatively easy to give an answer to the related question of how to get rid of the bubbles; the application of a sufficiently great pressure will collapse all bubbles in the system, the magnitude of the pressure required will of course depend on the size of the bubble to be removed. Rather more discouragingly, these pressures are often very large, and in excess of 1000 atmos. Worse still, there is always the prospect that upon reduction of the applied pressure, that the released gas will again nucleate to form a fresh crop of microbubbles.

6.2.2 Non-spontaneous bubble formation

Sparging
The term 'sparging' is intended to cover the insertion of gas bubbles directly into a liquid by pumping gas (usually through a frit, filter or bubble column) into the bulk liquid. In principle, this should be the most controllable and reliable method of producing monosized bubbles, and furthermore the theory is reasonably well understood. The thermodynamic factors ruling bubble production at an orifice are the subject of an extensive literature. We note that the mathematics of the size and shape of a bubble forming at an orifice are identical (with an appropriate sign

reversal) to the treatment of a liquid drop similarly formed. Reviews and books in this area abound (a useful book in this context is that by Hartland and Hartley, (1976)), and no more than a cursory summary will be given here. The hydrodynamic factors are as usual less readily accessible, although the excellent book by Clift *et al.* (1978) will be found useful.

When inertial and viscous effects are unimportant, and when the process is conducted close to equilibrium, the size of the bubble formed at an orifice can readily be calculated from the force balance about any horizontal plane through the bubble:

$$2\pi x \gamma \sin (\phi) = V\Delta\rho g + \pi x^2 \Delta p \tag{6.23}$$

where x is the horizontal distance to the bubble axis at an angle ϕ to that axis, V is the bubble volume, $\Delta\rho$ is the density difference between gas and liquid, and g is the acceleration due to gravity. The pressure drop Δp across the interface at any point is given by

$$\Delta p = 2\gamma/b + g\Delta\rho z \tag{6.24}$$

Making use of the Laplace equation relating the principal radii of curvature R_1 and R_2 to the difference in pressure Δp across the curved interface and to the interfacial tension γ,

$$\Delta p = \gamma(1/R_1 + 1/R_2) \tag{6.25}$$

gives

$$\gamma(1/R_1 + 1/R_2) = 2\gamma/b + g\Delta\rho z \tag{6.26}$$

This equation can be written in dimensionless form using a parameter β, related to the bubble shape, and given by

$$\beta = \Delta\rho g b^2/\gamma \tag{6.27}$$

b is the radius of curvature at the apex of the bubble. Combining these parameters into a single equation gives

$$[1/(R_1)/b] + [\sin (\phi)/(x/b)] = \beta(z/b) + 2 \tag{6.28}$$

This is the form of equation solved numerically (by hand) by Bashforth and Adams (1883). More conveniently, the required quantities can be expressed in terms of two characteristic diameters of the bubble d_e and d_s, where d_e is the diameter at the equator of the bubble and d_s is the diameter at a distance d_e from the apex. The ratio d_s/d_e is often called S, which is experimentally determined, and defining a new quantity H given by

$$H = \beta(d_e/b)^2 \tag{6.29}$$

it follows that

$$\gamma = g\Delta\rho d_e^2/H \tag{6.30}$$

Equation (6.30) is exact, and tabulated values of S as a function of $1/H$ are readily available. Knowing (experimentally) the value of S, the tables give the appropriate value of H, and thus from eqn (6.30) γ can be determined. This is perhaps the most convenient method for relating bubble size and shape to the interfacial tension.

A much simplified treatment of the detachment volume (often referred to as 'Tate's law') is available, but unfortunately the proportion of gas remaining attached to the surface after detachment is a less tractable quantity. Tables of empirical correction factors allowing for this amount are available (see e.g. Boucher and Evans, 1975; Boucher *et al.*, 1976; Boucher and Kent, 1978; Boucher, 1980).

In most cases of sparging, equilibrium calculations are inadequate to predict the detachment volume. Usually, the dynamics of the fluid–fluid interface have to be accounted for, since rapid bubble formation is usual in sparging. This is a more difficult problem, and while empirical correlations are available (Perry *et al.*, 1984), little is known theoretically. Three regimes can be distinguished: the single bubble regime, the intermediate regime, and the jet regime. The single bubble regime can be reasonably satisfactorily described by the thermodynamic analyses presented above. It is simple to show that the frequency, $\bar{\omega}$, of bubble detachment for a volumetric flow rate of gas of Q, and an orifice diameter of D in a liquid/gas combination characterized by densities ρ_l and ρ_g, respectively, is given by

$$\bar{\omega} = Qg(\rho_l - \rho_g)/\pi D\gamma \tag{6.31}$$

In the intermediate regime, represented by Reynolds numbers of about 200–2100, the frequency of bubble detachment increases more slowly than predicted by eqn (6.31), and the bubbles themselves tend to increase in size. The jet regime is characterized by turbulence, and the appearance of 'jets' from the orifice. Despite their name, these 'jets' actually consist of many small bubbles (typically between about 200 μm and 4 mm in diameter). There is no adequate theoretical description of this regime, and even empirical correlations are not good.

Entrainment

Entrainment covers those situations where gas is enveloped by the liquid at the interface between the two phases, and most usually occurs where the liquid is in violent motion. Common examples are the entrainment of air into water in fountains or by wave action or other turbulent motion in oceans, rivers and lakes. Clearly, in such circumstances, the violence of the motion involved makes control of the bubble size difficult, and for this reason this method of bubble formation will not be considered in detail here. Cascade systems, where the jet of liquid is directed at an appropriate velocity into a pool of the liquid, can successfully entrain large quantities of air. An approximate expression for the volumetric ratio of gas to liquid Q_g/Q_l thus entrained is given by

$$Q_g/Q_l = 0.0316(V^2\rho_l L/\gamma g)^{1/2} \tag{6.32}$$

where V is the jet velocity, ρ_l is the liquid density, γ is the surface tension, and L is the length of the jet.

Agitational methods are common in industrial practice, and are one of the chief means of introducing bubbles into fluid systems. Mechanical agitators may work by a combined mechanism involving both entrainment and attrition (see below). The final bubble size produced by a mechanical agitator is strongly correlated with the average tip speed of the impeller, but the time taken to achieve this final size is dependent upon the throughput of the agitator.

Attrition

Mechanical disruption of the gas–liquid interface will usually result in the entrapment of gas. Of course, the lifetime of bubbles so formed will depend on the factors discussed below but, assuming a sufficiently long survival, these bubbles will be subject to the same mechanical/hydrodynamic processes responsible for the formation of the original, larger bubbles, and break up is likely to follow. These processes, which we refer to as 'attrition', are complex and there is a substantial, if mainly empirical, chemical engineering literature available (Perry *et al.*, 1984).

An elegant and revealing experiment was performed by Prins (1976), who used a rotating wire cage to measure foam volumes as a function of the tip speed of the cage, at various surfactant concentrations. In all cases, the foam volume plots showed very sharp maxima at a velocity which increased with the surfactant concentration. This velocity was identified with the critical value, beyond which the agitation caused foam breakage. It is likely that, in a general way, this sort of behaviour would also apply to more dilute bubble suspensions, although there are good reasons for believing that dilute suspensions would not behave identically. Clearly, mechanical break up of bubbles will share common features with the similar phenomenon in emulsions, a subject which has been extensively reviewed by Tadros and Vincent (1983) and by Walstra (1983).

6.3 Growth of bubbles

Whatever the means by which a supercritical bubble is formed, it will tend to grow; growth will be fostered by the same conditions of supersaturation that gave rise to the nucleation event. Only in the case of a nucleation pulse, rapidly followed by a return to equilibrium, would further growth not take place (as noted above, such conditions may be found in acoustic cavitation). Even then if the cavity formed was sufficiently large that its Peclet number was substantially greater than unity, bouyancy would result in subsequent growth.

It is convenient to consider bubble growth at a surface and free bubble growth separately, even though they clearly have many attributes in common.

6.3.1 Bubble growth at a surface

The primary sources for information on the growth of bubbles at surfaces are from the electrochemical literature (see e.g. Westerheide and Westwater, 1961) and from the heating engineering literature. Broadly speaking, two regimes can be distinguished during the growth. Firstly, there is a short period, typically about 10 ms, where the growth is controlled by surface tension forces, inertia, viscosity and pressure, followed by growth essentially dominated by diffusion, either of matter or energy. In the case of electrogenerated bubbles, a further regime may occur, in which gas is directly 'injected' into the bubble, and the growth rate is governed by purely Faradaic considerations.

There is a substantial literature on the diffusion-controlled growth (Frank, 1950; Tao, 1978) and it appears to be common ground that the growth rate is given by variants of Scriven's (1959) equation, which relates the radius r as a function of time t to a growth coefficient β (which is supersaturation dependent) and the diffusion coefficient D:

$$r(t) = 2\phi\beta(Dt)^{1/2} \qquad (6.33)$$

This equation assumes spherical symmetry, and ignores the (partial) obscuration of the electrode (in electrogenesis), and the interference in the diffusion field in diffusion-controlled growth in a two-component system. In the case of boiling, a further possible complication arises in that thermal gradients at the surface will differ from those found at some distance from it.

The same equation, modified to allow for hydrostatics, will apply to the growth of free bubbles (see below). It is worth noting that despite a considerable literature, the effects of surfactants, including the dissolved gas itself, have not been properly studied. In the case of electrolytic bubble growth, the work of Brandon and Kelsall (1985) and of Brandon *et al.* (1985) is significant. They used both anionic and cationic surfactants during studies of hydrogen, oxygen and chlorine bubble growth, and measured the electrophoretic mobility of the electrogenerated bubbles of hydrogen, oxygen and chlorine. The studies were carried out on microelectrodes, and this has an important bearing on the rate of growth, but not on the detachment.

In surfactant-free solutions, Brandon *et al.* (1985) showed that the bubbles had a point of zero charge (PZC) at pH 2–3. Below pH 2, they were positively charged, and above pH 3 they were negatively charged. Classical equations developed for solid particles were inadequate for the calculation of the electrophoretic mobility of the bubbles. Addition of ionic surfactants tended to rigidify the interface (see Section 6.3.2), and at monolayer coverage the agreement with theory for solid particles was restored. This is in broad agreement with the Frumkin–Levitch theory (Frumkin and Levitch, 1947).

Studies by Brandon and Kelsall (1985) showed how important was the electrostatic interaction between the charged bubble and the electrode. In the case of hydrogen evolved at a cathode, at pH < 2, this was a strongly attractive interaction, and bubble departure diameters were correspondingly increased. For pH > 3, the interaction was repulsive, and departure diameters were greatly reduced. For oxygen and chlorine on anodes, the change was in the opposite sense. Surfactants screened these interactions to a greater or lesser extent, and reduced the dependence of departure diameter on pH.

6.3.2 Free bubble growth

Basically, the same factors control the growth of the free bubble as in the similar case of a bubble attached to a surface, but with the additional influence that as the bubble rises through the liquid, the hydrostatic pressure is gradually reduced, and this aids the growth of the bubble. This problem does not seem to have been studied, and it may be important. Since the rate of growth of the bubble by diffusion is a function of the interfacial area, significantly increased rates of growth might be expected as a result of this positive feedback mechanism. Few reports have been found in the literature (but see Shafer and Zare, 1991) where measurements have been made of free bubble growth.

It is noteworthy that even the apparently simple question of how fast a *constant size* bubble will rise under various conditions is not fully answered. Theoretical analysis for non-rigid (fluid) spheres (Hadamard, 1911; Rybczynski, 1911) of the low Reynolds number case predicts that rise will be 50% faster than the Stokes' law rate.

From Stokes' law, the terminal velocity V_T of a rigid sphere is given by

$$V_T = 2r^2 \Delta \rho g / 9 \mu \tag{6.34}$$

where r is the radius, and μ is the viscosity.

From the Hadamard–Rybczynski treatment, V_T is given by

$$V_T = 2r^2 \Delta \rho g [(2 + 3k)/(1 + k)]/3 \tag{6.35}$$

Here, the new variable k is the viscosity ratio, and is zero for gas bubbles.

Experimental tests of this theory are hard to come by, and the reason is that small amounts of contaminant at the bubble surface radically affect its rise, and the Stokes' law prediction for rigid spheres is recovered, although the mechanism involved in this restoration of apparent rigidity is unclear; Frumkin and Levich (1947) and Levich (1970) ascribe it to Gibbs–Marangoni effects at the bubble interface.

It should be apparent that even if a definitive solution to the terminal velocity problem was at hand, there are still complex non-linear problems to solve. Firstly, the terminal velocity is itself a function of bubble size, which must by definition increase as the bubble rises. Secondly, the rate of growth by diffusion is a function of bubble size, increasing with the square of the radius, but the increase in radius resulting from unit increase in the number of moles of gas in the bubble only results in a change in the radius to the one-third power.

Considerable simplification is achieved by ignoring the increase in size attributable to the reduction in hydrostatic head as the bubble rises. Note that for bubbles in water, a rise of the order of 10 cm results in a reduction of the hydrostatic head of only about 1%, and hence an increase in the radius of about 0.3%. With this assumption, and the assumption that Stokes' law holds, the equation of motion (and hence growth) can be solved easily: two recent treatments have appeared in the literature (Hansen and Derderian, 1976; Shafer and Zare, 1991). Hansen deduces that

$$r^2 = BP^{-1}(P_i - P)Dt \tag{6.36}$$

Where B is the Bunsen coefficient (the volume of gas dissolved by unit volume of the liquid), P_i is the initial pressure of the dissolved gas, P is the final pressure (usually atmospheric), D is the diffusion coefficient, and t is the time. The prediction is that the radius will increase in proportion to $t^{1/2}$. Note that this is in accord with both the Frank (1950) and Scriven (1959) theories.

Shafer and Zare (1991), by a different argument, reach the equation

$$r = r_0 + v_r t \tag{6.37}$$

where r_0 is the initial radius of the bubble, and v_r is the rate of increase of the bubble radius. Note that this (apparently) linear equation is actually only linear if v_r is a constant, and this is not expected to be generally true. When v_r is indeed constant, then this equation predicts that r will increase in direct proportion to t.

6.4 The detachment of bubbles

6.4.1 Quasi-equilibrium treatment

We have seen (Section 6.2.2) that nucleation at a surface is generally much more facile than that in the homogeneous bulk of the liquid. For this reason, unless the supersaturation is very large, bubbles will normally be formed at surfaces. The question then arises, what effect does the surface have upon subsequent nucleation and growth, and upon the rate of bubble release? If there is a mechanism for growth, then bubbles formed on surfaces will eventually become detached, rise through the solution, and finally burst at the interface between the solution and air. We have seen (above) that the growth rate during the rise through the solution is not well understood, and although attempts have been made to allow for the partial interruption of the diffusion field by the surface (see e.g. Buehl and Westwater, 1966), the theory is incomplete. The same is true but with still greater force for the kinetics of the detachment process. The thermodynamics and hydrostatics present essentially no problem, and despite the general lack of analytical solutions, numerical computations of arbitrary precision and wide applicability have been available since 1883 (Bashforth and Adams, 1883) together with more recent refinements (Boucher and Evans, 1975; Hartland and Hartley, 1976). It is important to recognize that these solutions have but little relevance to the *kinetics* of detachment, since they are all based upon the assumption of (quasi) equilibrium.

In the simplest model, bubbles become detached from a surface when the buoyancy force is greater than the surface tension acting around the periphery of the bubble foot (Lubetkin, 1989b). Actually, this ignores the fact that in general, necking of the bubble takes place, and detachment leaves a proportion of the gas behind, and perhaps more seriously ignores inertial effects. The controlling factors are likely to be the rate of growth of the bubble β and the contact angle θ. The larger θ becomes, the more difficult the detachment process becomes, and the greater the size of the bubble produced in the bulk liquid after detachment finally occurs. Similarly, the greater β becomes, the larger the bubble will grow as it rises. The literature seems not to discuss the interplay of these factors which, together with the nucleation rate, are likely to be determining parameters in the size and number of bubbles formed by nucleation/growth processes.

Consider a generally flat solid surface which has a certain density per unit area of nucleation sites n_s, each equally potent. If nucleation produces a bubble on such a site, and the bubble has a radial growth rate ρ, then the bubble on this site will grow over an adjacent site with a characteristic time τ_g of approximately $1/\rho(n_s)^{1/2}$. Now, one can similarly calculate a characteristic time for nucleation τ_n, since the nucleation frequency is given by J^{-1}. If τ_g is short compared to τ_n, and hence the ratio $\tau_g/\tau_n = M$ is less than unity, then most potential nucleation sites will be obscured before they can give rise to a bubble, and the observed rate of bubble production will be correspondingly reduced. However, if $M = \tau_g/\tau_n$ is greater than unity, then essentially the full rate predicted by J will be observed.

In the preceding paragraph, the number of bubbles produced *at the surface* was considered. Usually, observation relies on the bubbles released from the surface, which then rise through the solution. Release depends on the surface characteristics and, in particular, on the contact angle θ at the three phase line between solid/liquid/gas. As before, note that there will be some characteristic time τ_d required for the detachment process. We have already noted the characteristic nucleation time τ_n. The dimensionless ratio τ_d/τ_n has been designated L (Lubetkin, 1989a, b). If $L > 1$, then detachment dominates the overall kinetics of bubble release into the solution, while if $L < 1$ nucleation is dominant.

A simplistic treatment of the relative importance of nucleation and detachment gives rise to a theoretical equation for L:

$$L = J \sin(\theta)\{6\gamma[2 - \cos(\theta)]/g\Delta\rho\}^{1/2}/2\beta \qquad (6.38)$$

Now, the nucleation rate and the growth rate are likely to be coupled, in the sense that large values of one will generally imply similarly large values of the other. For this reason, the first of the dimensionless ratios (τ_g/τ_n) will probably not be a sensitive function of the chief experimental variable σ. For $L = \tau_d/\tau_n$, however, there is good reason to expect almost complete decoupling of the two times going to make the ratio L. For example, for a given set of conditions of supersaturation, a low contact angle will reduce the rate of nucleation, but will speed the detachment, and conversely for a high contact angle. For this reason, it is expected that very wide variations in the relative importance of nucleation and detachment will be found experimentally, and this has (to some extent) been confirmed (Lubetkin, 1989a, b). The significance of these findings is that predictions of bubble nucleation rate using equations like eqn (6.21) above will generally be in error, and this error may be extremely large, particularly when $L \gg 1$.

If $L \ll 1$, we expect broad agreement with classical nucleation theory. Given that complete wetting ($\theta = 0$) is quite a common state of affairs, we might expect that very often $L \ll 1$, and nucleation would dominate the kinetics of bubble formation. This is indeed so, but it is important to recognize that when $\theta = 0$, then heterogeneous nucleation ceases to have any energy advantage over homogeneous nucleation – in fact the two mechanisms merge, as may be seen by comparing eqns (6.8) and (6.18). The conditions which favour easy release of bubbles therefore correspond with those of greatest difficulty in the nucleation step, and vice versa.

Experimentally, the problems relate to the possible effects of pressure on θ and γ, and the determination of β for the systems of interest. The latter quantity has received attention, particularly from the electrochemists, while the former seem to have been almost ignored. Jho *et al.* (1978) have measured the change in surface tension for solutions of various gases, including CO_2, in water. A significant dependence of the surface tension on pressure was found. It should be made clear that this dependence is not actually as a direct result of the applied pressure – because of the very low compressibility of the liquid phase, such effects are very small. It is the effect of the enhanced solubility of the gas at elevated pressure which causes measurable changes in γ. The magnitude of the change in γ is a matter of considerable importance for the nucleation of bubbles. Note that γ appears to the third power in the exponential term in eqn (6.21). Very roughly speaking, changes of 1% or so in the exponential give rise to changes of the order of a power of 10 in the rate of nucleation *J*.

Experiments by Lubetkin and Akhtar (1994) to measure accurately the effect of pressure on the surface tension of the water/CO_2 interface gave values of $d\gamma/dp$ substantially greater than those of Jho *et al.* In the experiments of Jho *et al.*, a very short time was allowed for equilibration (about a few minutes) after an alteration in the applied pressure of gas. It has been demonstrated that the equilibration process for the CO_2/water system is rather slow (Lubetkin and Blackwell, 1988) and times of hours (if not days) rather than minutes are required. The sensitivity of the rate of nucleation to the surface tension is of great significance, and Table 6.2 illustrates just how large is the change wrought by the experimentally determined values of γ when compared with the calculated values used by Wilt (1986). The table shows the ratio of the J/J_0, J_0 being the rate

Table 6.2   *The calculated values of γ derived from $d\gamma/dp = -0.7\,mNm^{-1}/atm$ used by Wilt (1986) and the measured values of γ (Lubetkin and Akhtar, 1994) show the dramatic effect on the nucleation rate of changes in γ. Note that the calculations here ignore changes in σ.*

p/p_0	γ (Wilt)	γ (L&A)	J/J_0 (L&A)
1	73.4	70.1	5.1×10^{178}
2	72.7	67.3	4.0×10^{69}
3	72.0	64.3	6.8×10^{41}
4	71.3	62.3	2.7×10^{26}
5	70.6	61.1	2.3×10^{17}
6	69.9	60.0	1.7×10^{12}
7	69.2	59.2	7.3×10^{8}
8	68.5	58.2	6.4×10^{6}
9	67.8	57.8	1.3×10^{5}
10	67.1	56.7	1.6×10^{4}

L&A, Lubetkin and Akhtar (1994); Wilt (1986).

calculated from the Wilt equation (eqn (6.21)) using $\theta = 94°$, $\beta = 5°$, and Wilt's estimates for the surface tension, while ignoring the pre-exponential constant C'''. The dramatic increase in the nucleation rate, especially at low supersaturations, is obvious. Note that in the region of supersaturation characteristic of carbonated drinks (say about $p/p_0 = 4$) this factor alone would introduce an increase in rate of about 10^{26} as a result of the lower surface tension measured for this system under these conditions. However, this calculation does not account for alterations in the contact angle as the partial pressure of CO_2 is increased.

Furthermore, in the Lubetkin and Akhtar experiments, the contact angle was also measured, and found to vary considerably as the pressure was increased. As noted above, altering the contact angle term in eqn (6.21) produces very large changes in the rate of nucleation. It was impractical to calculate the exponentials, since the magnitude of the exponential terms was so large that overflow conditions occurred during computation. This implies of course that the ratios J/J_0 (L&A) would be altered by very large factors; since the measured contact angles were all well below the values assumed by Wilt, the values of the factor $f(\theta,\beta)$ are all substantially larger, typically by about a factor of 10^4, thus resulting in a dramatic decrease in the predicted rate of nucleation for these lower contact angle surfaces.

6.4.2 Kinetic treatment

The analysis presented above is based essentially on an equilibrium detachment model. Very often, the detachment process is not well represented by such a quasi-equilibrium treatment, and this will be particularly true when bubbles are formed rapidly. This is likely in most industrial practice (in sparging, for example), or in growth of bubbles at large supersaturations. Under these conditions, it is not permissible to ignore hydrodynamic effects, and account must be taken of inertia and viscous drag on the detachment.

Analyses of this problem have appeared (Prisnyakov, 1970; Karri, 1988). The description here follows that of Karri, and establishes a force balance equation between the drag force F_d and the surface tension force F_s on the one hand, and the inertial force F_i, the 'pressure' force F_p and the bouyancy force F_b on the other:

$$F_d + F_s = F_i + F_p + F_b \tag{6.39}$$

F_d is given by

$$F_d = C_d \rho_l [dR/d\tau] \pi R_s^2 \tag{6.40}$$

where C_d is the drag coefficient, ρ_l is the liquid density, R_s is the radius of the bubble foot, and τ is the time.

F_s is given by the product of $2\pi R_s$ (the periphery of the bubble foot) with the surface tension γ and the sine of the contact angle θ:

$$F_s = 2\pi R_s \gamma \sin(\theta) \tag{6.41}$$

F_i is given by the product of the bubble volume (here corrected to allow for the incomplete spherical cap, by the inclusion of the factor $f(\theta)$ from eqn (6.19)), the density of the liquid ρ_l, the departure radius R_d, and with the second derivative of the radius with time:

$$F_i = -4/3[f(\theta)\pi R_d^3 \rho_l (d^2R/d\tau^2)] \tag{6.42}$$

Similarly, F_p can be expressed in terms of a dynamic excess internal pressure Δp_v:

$$F_p = [2g/R_d + \Delta p_v]\pi R_d^2 \tag{6.43}$$

Δp_v is given by

$$\Delta p_v 4\pi R_d^2 = -4/3(\pi R_d^3 \rho_l)[d^2R_d/d\tau^2] + 1/2(C_d \rho_l \pi R_d^2)[dR_d/d\tau]^2 \tag{6.44}$$

F_b is simply the product of the volume and density difference terms:

$$F_b = 4/3[f(\theta)\pi R_d^3 (\rho_l - \rho_v)g] \tag{6.45}$$

These equations can be solved either analytically (in part) or numerically (in part) to predict bubble departure diameters where inertia is important.

Finally, some mention should be made of sharp corners and edges, and the effect they may have upon detachment. The three-phase contact line between gas, liquid and solid rarely moves smoothly when a force is applied. The stick–slip phenomenon is the rule rather than the exception (Huh and Scriven, 1971; Winterton and Blake, 1983), and this phenomenon is particularly noticeable when rough or inhomogeneous surfaces are involved. The three-phase contact line can become pinned by such microtopographical features, and if this contact line is the foot of a growing bubble, then the diameter of the foot cannot increase while it is thus pinned. Noting that very often, the microcavities which are expected to be the preferred sites for nucleation (including Harvey 'nuclei') will have just such surface discontinuities at their mouths, it might be concluded that such behaviour will be common in bubble nucleation. Given that in simplistic terms, the diameter of the bubble foot controls the detachment size of the bubble, then it is clear that this could result in early detachment, and hence reduced size of the bubbles rising through the liquid. There appears to be no literature on the possible effects of pinning on bubble size.

6.5 The role of surfactants in bubble formation and stability

6.5.1 Formation of bubbles

At the most elementary level, the creation of a bubble of radius r, in a liquid where the interfacial tension is γ, requires (at least) the expenditure of $4\pi r^2 \gamma$ joules of energy. Clearly, the smaller γ is, the smaller this energy requirement and, since surfactants reduce the interfacial energy, they are expected to increase the ease of bubble formation. While quite correct, this simple approach ignores many important questions. The reversible work of formation of the new interface is very much less than the irreversible work in almost all practical situations. For instance, work has to be expended in moving the liquid, and for the rapid formation of relatively large bubbles (by sparging, for example), this may involve a substantial head of liquid. The stretching of the interface during its formation may involve viscoelastic effects, and thus thermal losses, and in the presence of surfactants (and particularly where such surfactants are polymeric, or liquid crystalline), these losses may be large.

Thus, in a general way, it is possible to assert that the efficiency of bubble formation will be very low indeed, and will be worst in the cases of mechanical means of bubble generation such as sparging or entrainment. However, even though surfactants may reduce the efficiency of bubble formation through mechanisms such as surface elasticity, they are nonetheless a practical necessity. The reason for this is that surfactants are important for bubble stability. A recent review by D'Arrigo (1986) has highlighted the importance of surfactants in stabilizing bubbles in natural waters.

6.5.2 Stability of bubbles in open systems

When bubbles reach a free liquid surface, they may burst rapidly, or may persist for very long times. The distinction drawn in the opening remarks of this chapter, between bubbles surrounded by gas and those surrounded by liquid, are important here. The conditions for stability of a bubble, the whole or part of which is surrounded by a bilayer in a gaseous phase are entirely different from those of a bubble totally immersed in a liquid. The former problem has been extensively studied, and has a rich literature, while the latter seems to have been relatively neglected. The reason for this state of affairs is not hard to determine; while the same factors are important for the persistence of emulsions (Ostwald ripening, coalescence), it is more difficult to study such effects in bubbles as a direct result of their strong tendency to cream. At the outset, it was decided not to attempt to treat the subject of foams, and the persistence of bubbles which have reached a free liquid surface properly belongs in that domain.

The lifetimes of bubbles still surrounded by liquid are a legitimate matter for the present chapter. Apart from reaching the free surface and bursting (which has been excluded) what mechanisms are available for bubble disappearance? The answer is that this can only occur by dissolution or by coalescence. In a bubble population, dissolution will inevitably result in the growth of the larger bubbles in the distribution and this is of course Ostwald ripening. Note that in mixed gas bubbles, where the solubilities of the two gases are very different, that such uniform trends in the median bubble size may perhaps unexpectedly not be found, and bimodal distributions may develop, as noted by Kabal'nov *et al.* (1987) for two-component emulsions. The generally appreciable solubilities of gases in liquids (and particularly where there is a chemical interaction, e.g. CO_2/water, SO_2/water and NH_3/water), and the relative ease with which gases penetrate the surface layers of the bubble means that Ostwald ripening is usually faster in bubble suspensions than in emulsions or solid suspensions.

Coalescence, on the other hand, is likely to be unimportant, at least in dilute bubble suspensions. There are two main reasons for this. Firstly, for bubbles at the low number concentrations considered in this chapter, collisions are intrinsically unlikely. Secondly, where surfactants are present, which is the normal state of affairs, then coalescence is greatly inhibited; more so than for a solid–solid collision or even a liquid–liquid collision. The reason for this is that the ease of deformation of the interface for fluid–fluid collisions is greatly enhanced when compared with a collision of identical energy in a solid–solid system, and this allows the system to dissipate the energy of collision efficiently. When the fluids are gases, then added to the ease of deformation is the compressibility of the bubble. Clearly, when bubbles are forced together (as, for example, in a foam) then neither of these mechanisms will be so efficient, and coalescence becomes an important means of foam collapse.

Opposing the general resistance of bubbles to coalescence is the fact that the van der Waals attractions between bubbles will usually be as great as those between liquid drops in the appropriate gas; this is a direct result of the symmetry of the expressions for the Hamaker constants in the two cases (Everett, 1988). In the absence of surfactants, bubbles upon sufficiently close approach will strongly attract one another, and coalescence will inevitably follow, since there are no repulsive forces available to stabilize the interface.

It is well known that surfactants stabilize soap bubbles (i.e. those bubbles surrounded by a gas), and the criteria are discussed in the reviews on foams alluded to earlier. The significance of surfactants in altering the kinetics of coalescence, Ostwald ripening and diffusive loss in bubbles in a liquid phase are neither so well understood, nor widely reported, although there is an extensive literature on the diffusion of gases through interfaces, including the effect of adsorbed surfactant layers, as a means of reducing water losses from reservoirs. A useful review is to be found in LaMer (1962), and in particular on evaporation resistance in Barnes and LaMer (1962) and Rosano and LaMer (1956).

6.5.3 Stability of bubbles in closed systems

Ward *et al.* (1982) showed that in a closed system, a single bubble may have no equilibrium state, or one or two equilibrium sizes, depending on the amount of gas present. In the case where there are two possible equilibrium sizes, the larger is the stable size. This prediction (which was experimentally verified) is in stark contrast to the situation in an open system, where only one equilibrium size is possible, and that is an unstable equilibrium. When many bubbles are present, a similar analysis shows that the equilibrium size is dependent on the number of bubbles per unit volume: as this number density increases, so the stable bubble size decreases. For the water/nitrogen system, a single bubble must exceed 207 μm, while at a number density of 10^7 bubbles/cm^3 this critical size has dropped to 3.47 μm. Numerical calculations by Wilt (1986) give similar results for the water/carbon dioxide system, and show that if such pre-existing and stable bubbles were present, they could act as 'nuclei' thus producing a crop of bubbles, possibly at supersaturations even below that theoretically needed for heterogeneous nucleation.

References

Akers, R. J. (ed.) (1976) *Foams*, Academic Press, London

Bankoff, S. G. (1958) Entrapment of gas in the spreading of a liquid over a rough surface. *AIChE J.*, **4**, 24–26

Barnes, G. T. and LaMer, V. K. (1962) The evaporation resistances of monolayers of the long-chain fatty acids and alcohols and their mixtures. In *Retardation of Evaporation by Monolayers: Transport Processes* (ed. LaMer, V. K.) Academic Press, New York, 9–33

Bashforth, F. and Adams, J. C. (1883) *An Attempt to Test the Theories of Capillary Action*, Cambridge University Press, Cambridge

Blander, M. and Katz, J. L. (1975) Bubble nucleation in liquids. *AIChE J.*, **21**, 833–848

Blander, M, Hengstenberg, D. and Katz, J. L. (1971) Bubble nucleation in n-hexane, n-pentane + hexadecane mixtures, and water. *J. Phys. Chem.*, **75**, 3613–3619

Boucher, E. A. (1980) Capillary phenomena: properties of systems with fluid/fluid interfaces. *Rep. Prog. Phys.*, **43**, 497–546

Boucher, E. A. and Evans, M. J. B. (1975) Pendent drop profiles and related capillary phenomena. *Proc. R. Soc. London, Ser. A*, **346**, 349–374

Boucher, E. A. and Kent, H. J. (1978) Capillary phenomena VII. Equilibrium and stability of pendent drops. *J. Colloid Interface Sci.*, **67**, 10–15

Boucher, E. A., Evans, M. J. B. and Kent, H. J. (1976) Capillary phenomena II. Equilibrium and stability of rotationally symmetric fluid bodies. *Proc. R. Soc. London, Ser. A*, **349**, 81–100

Boys, C. V. (1890) *Soap Bubbles and the Forces which Mould them* (reprinted (1960) in the Science Study Series) Heinemann, London

Brandon, N. P. and Kelsall, G. H. (1985) Growth kinetics of bubbles electrogenerated at microelectrodes. *J. Appl. Electrochem.*, **15**, 475–484

Brandon, N. P. and Kelsall, G. H., Levine, S. and Smith, A. L. (1985) Interfacial electrical properties of electrogenerated bubbles. *J. Appl. Electrochem.*, **15**, 485–493

Briggs, L. J. (1950) The limiting negative pressure of acetic acid and benzene in relation to temperature. *Science*, **112**, 427–430

Briggs, L. J. (1951a) The limiting negative pressure of acetic acid, benzene, aniline, carbon tetrachloride and chloroform. *J. Chem. Phys.*, **19**, 970–972

Briggs, L. J. (1951b) The limiting negative pressure of five organic liquids, and the two phase system, water–ice. *Science*, **113**, 483–490

Briggs, L. J. (1953) The limiting negative pressure of mercury in Pyrex glass. *J. Appl. Phys.*, **24**, 488–490

Briggs, L. J. (1955) Maximum superheating of water as a measure of negative pressure. *J. Appl. Phys.*, **26**, 1001–1003

Buehl, W. M. and Westwater, J. W. (1966) Bubble growth by dissolution: influence of contact angle. *AIChE J.*, **12**, 571–576

Carr, M. W. (1993) Ph.D. thesis, Bristol University, Bristol

Carr, M. W., Hillman, A. R., Lubetkin, S. D. and Swann, M. J. (1989) Detection of electrolytically generated bubbles using a quartz crystal microbalance. *J. Electroanal. Chem.*, **267**, 313–320

Clark, H. B., Strenge, P. S. and Westwater, J. W. (1959) Active sites for nucleate boiling. *Chem. Eng. Prog. Symp. Ser. 29*, **55**, 103–110

Clift, R., Grace, J. R. and Weber, M. E. (1978) *Bubbles, Drops and Particles*, Academic Press, New York

Cole, R. (1974) Boiling nucleation. *Adv. Heat Transfer*, **10**, 85–166

Corty, C. and Foust, A. S. (1955) Surface variables in nucleate boiling. *Chem. Eng. Prog. Symp. Ser.*, **17**, 1–12

D'Arrigo, J. S. (1986) *Stable Gas-in-liquid Emulsions*, Elsevier, Amsterdam

Dickinson, E. (ed.) (1987) Food emulsions and foams. *Royal Society of Chemistry Special Publication No. 58*, Royal Society of Chemistry, London

Doring, W. (1937) Die uberhitzunggrenze und zerreissfestigkeit von flussigkeiten. *Z. Phys. Chem.*, **36**, 371–386

Doring, W. (1938) Berichtigung zu der arbeit: Die uberhitzunggrenze und zerreissfestigkeit von flussigkeiten. **38**, 292–294

Dunning, W. J. (1955) Theory of crystal nucleation from vapour, liquid and solid systems. In *Chemistry of the Solid State* (ed. Garner, W. E.) Butterworths, London, 159–183

Everett, D. H. (1988) *Basic Principles of Colloid Science*, Royal Society of Chemistry, London, 170

Exerova, K., Balinov, B. and Kashtiev, D. (1983) Nucleation mechanism of rupture of black films. II. Experiment. *J. Colloid Int. Sci.*, **94**, 45–53

Findlay, A. and King, G. (1913) Rate of evolution of gases from supersaturated solutions. Part 1. Influence of colloids and of suspensions of charcoal on the evolution of carbon dioxide. *J. Chem. Soc. Trans.*, **103**, 1170–1193

Frank, F. C. (1950) Radially symmetric phase growth controlled by diffusion. *Proc. R. Soc. London Ser. A*, **201**, 586–599

Frumkin, A. and Levich, V. G. (1947) *Zh. Fiz. Khim.* **21**, 1183–1189

Gabrielli, C., Huet, F., Keddam, M. and Torresi, R. (1991) Investigation of bubble evolution with a quartz crystal microbalance. *J. Electroanal. Chem.*, **297**, 515–522

Glas, J. P. and Westwater, J. W. (1964) Measurements of the growth of electrolytic bubbles. *Int. J. Heat Mass Transfer*, **7**, 1427–1443

Griffith, P. and Wallis, J. D. (1960) The role of surface conditions in nucleate boiling. *Chem. Eng. Prog., Symp. Ser. No. 30*, **56**, 49

Gutsov, I., Kashtiev, D. and Avramov, I. (1985) Nucleation and crystallization in glass-forming melts: old problems and new questions. *J. Non-cryst. Solids*, **73**, 477–499

Hadamard, J. S. (1911) Mouvement permanent tent d'une sphere liquide et visquese dans un liquide visqueux. *Compte Rend Acad. Sci.*, **52**, 1735–1738

Hansen, R. S. and Derderian, E. J. (1976) Problems in foam origin, drainage and rupture. In *Foams* (ed. Akers, R. J.) Academic Press, London, 1–15

Hartland, S. and Hartley, R. W. (1976) *Axisymmetric Fluid Liquid Interfaces*, Elsevier, Amsterdam

Harvey, E. N., Barnes, D. K., McElroy, W. D., Whiteley, A. H., Pease, D. C. and Cooper, K. W. J. (1944a) Bubble formation in animals I. Physical factors. *Cell Comp. Physiol.*, **24**, 1–22

Harvey, E. N., Whiteley, A. H., McElroy, W. D., Pease, D. C. and Barnes, D. K. (1944b) Bubble formation in animals II. Gas nuclei and their distribution in blood and tissues. *J. Cell. Comp. Physiol.*, **24**, 23–34

Harvey, E. N., Barnes, D. K., McElroy, W. D., Whiteley, A. H. and Pease, D. C. (1945) Removal of gas nuclei from liquids and surfaces. *J. Am. Chem. Soc.*, **67**, 156–157

Harvey, E. N., McElroy, W. D. and Whiteley, A. H. (1947) On cavity formation in water. *J. Appl. Phys.*, **18**, 162–172

Hirth, J. P. and Pound, G. M. (1963) Condensation and evaporation – nucleation and growth kinetics. *Prog. Mater. Sci.*, **11**.

Huh, C. and Scriven, L. E. (1971) Hydrodynamic model of steady movement of a solid/liquid/fluid contact line. *J. Colloid Interface Sci.*, **62**, 85–101

Isenberg, C. (1978) *Science of Soap Films and Soap Bubbles*, Tieto Clevedon

Jho, C., Nealon, D., Shogbola, S. and King, Jr, A. D. (1978) Effect of pressure on the surface tension of water: Adsorption of hydrocarbon gases and carbon dioxide on water at temperatures between 0 and 50°C. *J. Colloid Interface Sci.*, **65**, 141–154

Kabal'nov, A. S., Pertzov, A. V. and Shchukin, E. D. (1987) Ostwald ripening in two-component disperse phase systems: application to emulsion stability. *Colloid Surfaces*, **24**, 19–32

Karri, S. B. R. (1988) Dynamics of bubble departure in microgravity. *Chem. Eng. Commun.*, **70**, 127–135

Kashtiev, D. (1982) On the relationship between nucleation work, nucleus size and the nucleation rate. *J. Chem. Phys.*, **76**, 5098–5102

Kashtiev, D. (1984) The kinetic approach to nucleation. *Crystal. Res. Technol.*, **19**, 1413–1423

Kashtiev, D. (1985) Nucleation at changing density of monomers. *Crystal. Res. Technol.*, **20**, 723–731

Kashtiev, D. and Exerova, K. (1980) Nucleation mechanism of rupture of black films. I Theory. *J. Colloid Interface Sci.*, **77**, 501–511

Kenrick, F. C., Gilbert, C. S. and Wismer, K. L. (1924) The superheating of liquids. *J. Phys. Chem.*, **28**, 1297–1307

Knapp, R. T., Daily, J. W. and Hammitt, F. G. (1970) *Cavitation*, McGraw-Hill, New York

LaMer, V. K. (ed.) (1962) *Retardation of Evaporation by Monolayers: Transport Processes*, Academic Press, New York

Levich, V. G. (1970) *Physicochemical Hydrodynamics*, Prentice-Hall, New York

Lubetkin, S. D. (1989a) Measurement of bubble nucleation rates by an acoustic method. *J. Appl. Electrochem.*, **19**, 668–676

Lubetkin, S. D. (1989b) The nucleation and detachment of bubbles. *J. Chem. Soc. Faraday Trans. 1*, **85**, 1753–1764

Lubetkin, S. D. and Akhtar, M. (1994) *J. Colloid Interface Sci.*, in press

Lubetkin, S. D. and Blackwell, M. R. (1988) The nucleation of bubbles in supersaturated solutions. *J. Colloid Interface Sci.*, **126**, 610–615

Mason, R. E. A. and Strickland-Constable, R. F. (1966) Breeding of crystal nuclei. *Trans. Faraday Soc.*, **62**, 455–461

Meyer, J. (1911) *Zur Kenntnis des negativen Druckes in Flussikeiten*, W. Knapp, Halle

Mori, Y., Hijikata, K. and Nagatani, T. (1976) Effect of dissolved gas on bubble nucleation. *Int. J. Heat Mass Transfer.*, **19**, 1153–1159

Perry, R. H., Green, D. W. and Maloney J. O. (ed.) (1984) Liquid–gas systems. In *Chemical Engineers' Handbook*, 6th edn, McGraw-Hill, New York, Section 18

Prins, A. (1976) Dynamic surface properties and foaming behaviour of aqueous surfactant solutions. In *Foams* (ed. Akers, R. J.) Academic Press, London, 51–71

Prisnyakov, V. F. (1970) Breakoff of vapour bubbles from a heating surface. *Inzhen.-Fiz. Zh.*, **19**, 1435–1441 (trans. Consultants Bureau, New York)

Rosano, H. L. and LaMer, V. K. (1956) The rate of evaporation of water through monolayers of esters, acids and alcohols. *J. Phys. Chem.*, **60**, 348–353

Rybczynski, W. (1911) A novel method for measuring the interfacial tension under applied pressure: CO_2/water system. *Bull. Int. Acad. Pol. Sci. Lett., Cl. Sci. Math. Nat., Ser. A*, 40–46

Sander, V. A. and Damkohler, G. (1943) Ubersattigung bei der spontanen keimbildung in wasserdampf. *Naturwissenschaften*, **39/40**, 460–465

Scriven, L. E. (1959) On the dynamics of phase growth. *Chem. Eng. Sci.*, **10**, 1–13

Shafer, N. E. and Zare, R. N. (1991) Through a beer glass darkly. *Phys. Today*, **Oct.**, 48–52

Sides, P. J. (1986) Phenomena and effects of electrolytic gas evolution. In *Modern Aspects of Electrochemistry*, Vol. 18 (ed. White, R. E., Bockris, J. O.'M. and Conway, B. E.) Plenum Press, New York, 303–354

Sinitsyn, E. N. and Skripov, V. P. (1968) Kinetics of nucleation in superheated liquids. *Russ. J. Phys. Chem.*, **42**, 440–443 (trans. Consultants Bureau, New York)

Skripov, V. P. and Ermakov, G. V. (1964) Pressure dependence of the limiting superheating of a liquid. *Russ. J. Phys. Chem.*, **38**, 208–213 (trans. Consultants Bureau, New York)

Skripov, V. P. and Sinitsyn, E. N. (1964) Experiments with a superheated liquid. *Sov. Phys. Usp.*, **7**, 887–889 (trans. Consultants Bureau, New York)

Skripov, V. P. and Sinitsyn, E. N. (1968) Nucleation in superheated liquids and surface tension. *Russ. J. Phys. Chem.*, **42**, 167–169 (trans. Consultants Bureau, New York)

Tadros, Th. F. and Vincent, B. (1983) Liquid/liquid interfaces. In *Encyclopedia of Emulsion Technology*, Vol. 1, *Basic Theory* (ed. Becher, P.) Marcel Dekker, New York, 1–56

Tao, L. N. (1978) Dynamics of growth or dissolution of a gas bubble. *J. Chem. Phys.*, **69**, 4189–4194

Trayanov, A. and Kashtiev, D. (1986) Growth shape of crystallites on a substrate. A Monte Carlo simulation. *J. Cryst. Growth*, **78**, 399–407

Trefethen, L. (1957) Nucleation at a liquid–liquid interface. *J. Appl. Phys.*, **28**, 923–924

Turnbull, D. (1952) Kinetics of solidification of supercooled mercury droplets. *J. Chem. Phys.*, **20**, 411–424

Verhaart, H. F. A., De Jonge, R. M. and van Stralen, S. J. D. (1980) Growth rate of a gas bubble during electrolysis in supersaturated liquid. *Int. J. Heat Mass Transfer*, **23**, 293–302

Vincent, R. S. (1941) Measurement of tension in liquids by means of a metal bellows. *Proc. Phys. Soc. London*, **53**, 126–140

Vincent, R. S. (1943) The viscosity tonometer – a new method of measuring tension in liquids. *Proc. Phys. Soc. London*, **55**, 41–48

Wakeshima, H. and Takata, K. (1958) On the limit of superheat. *J. Phys. Soc. Jpn.*, **13**, 1398–1403

Walstra, P. (1983) Formation of emulsions. In *Encyclopedia of Emulsion Technology*, Vol. 1, *Basic Theory* (ed. Becher, P.) Marcel Dekker, New York, 57–127

Ward, C. A., Balakrishnan, A. and Hooper, F. C. (1970) On the thermodynamics of nucleation in weak gas–liquid solutions. *J. Basic Eng.*, **85**, 695–704

Ward, C. A., Tikuisis, P. and Venter, R. D. (1982) Stability of bubbles in a closed volume of liquid–gas solution. *J. Appl. Phys.*, **53**, 6076–6084

Ward, C. A., Tikuisis, P. and Tucker, A. S. (1986) Bubble evolution in solutions with gas concentrations near the saturation value. *J. Colloid Interface Sci.*, **113**, 388–398

Westerheide, D. E. and Westwater, J. W. (1961) Isothermal growth of hydrogen bubbles during electrolysis. *AIChE J.*, **7**, 357–362

Wilt, P. M. (1986) Nucleation rates and bubble stability in water–carbon dioxide solutions. *J. Colloid Interface Sci.*, **112**, 530–538

Winterton, R. H. S. and Blake, T. D. (1983) Dynamic effects in contact angle hysteresis, applied to boiling incipience. *Int. Comm. Heat Mass Transfer*, **10**, 525–531

Wismer, K. L. (1922) The pressure–volume relation of superheated liquids. *J. Phys. Chem.*, **26**, 301–315

Zettlemoyer, A. C. (ed.) (1969) *Nucleation*, Marcel Dekker, New York

Chapter 7

Emulsions and droplet size control

E. Dickinson

7.1 Introduction

An emulsion is traditionally defined as an opaque, heterogeneous system of two immiscible liquid phases ('oil' and 'water') with one of the phases dispersed in the other as droplets of microscopic or colloidal size. According to which phase makes up the droplets, there are two kinds of simple emulsions: oil-in-water and water-in-oil. While in theory the essence of an emulsion is that it should contain *liquid* particles, the boundary between emulsions and solid particle suspensions is not entirely distinct, since an emulsion may still be called an emulsion even if the dispersed phase becomes partly or completely crystallized (e.g. on cooling). Some emulsions contain additional solid or liquid dispersed phases. In a multiple emulsion, one of the constituent components is itself an emulsion. So, an emulsion made by dispersing a W/O emulsion in water is called a water-in-oil-in-water emulsion.

The transient emulsion made by the vigorous agitation of two immiscible pure liquids is very unstable. It reverts rapidly to bulk oil and water. In order to make a stable emulsion, there must be surface-active material present to protect the newly formed droplets against immediate recoalescence. An *emulsifier* is a single chemical component, or mixture of components, having the capacity for facilitating emulsion formation and preventing recoalescence through a combination of surface activity and structure formation at the newly formed oil–water interface. Wherever possible, it is useful to distinguish the terms 'emulsifier' and 'stabilizer'. A *stabilizer* is a component, or mixture of components, added to an emulsion to confer long-term stability, possibly by a mechanism involving adsorption at the oil–water interface, but not necessarily so. Small-molecule surfactants and high-molecular-weight polymers are commonly used in both roles, though the former tend to make better emulsifiers and the latter better stabilizers.

To make an emulsion containing droplets in the colloidal size range requires the application of a large amount of mechanical energy. The two bulk phases may be brought together in several different ways: water injected into oil, oil injected into water, alternate additions of oil and water, or the simultaneous addition of oil and water. Common sense dictates that an O/W emulsion is more readily prepared by adding oil to water, and vice versa for a W/O emulsion. The major factor affecting the droplet-size distribution is the intensity of mechanical agitation. Whether one gets a W/O or an O/W emulsion depends on the relative amounts of oil and water and on the nature of the emulsifier. Depending on its chemical structure, an emulsifier will tend preferentially to stabilize either a water droplet in oil or an oil droplet in water. The continuous phase is usually the one in which the emulsifier is most soluble (Bancroft's rule).

This chapter is concerned with the principles and practice of droplet formation in simple *macro*emulsions stabilized by surfactants, polymers or finely divided particles (pickering stabilization). Aspects of emulsion stability will be discussed only insofar as they pertain to emulsification processes and the properties of freshly made emulsions. Information relating to the long-term stability of emulsions may be found elsewhere (Becher, 1965; Kitchener and Mussellwhite, 1968; Dickinson and Stainsby, 1982; Tadros and Vincent, 1983). The formation of *micro*emulsions, or any other types of thermodynamically stable surfactant-containing mesophase structures, will not be considered here. Also excluded from consideration here are so-called 'emulsifiable concentrates' which lead to emulsions on dilution in water with little or no agitation. Such systems are said to exhibit spontaneous emulsification as a result of the very low interfacial tension. Further information on these matters may be found in the articles by Moilliet (1976), Becher (1985) and Evans *et al.* (1986).

7.2 Techniques of emulsification

7.2.1 Methods of making particles

Colloids are traditionally made in one of two ways: by dispersion (i.e. breaking down particles of macroscopic dimensions), or by condensation (i.e. building up particles from the atomic to the colloidal scale). Prior to highlighting the particular features of emulsion droplet formation, it is useful first to review briefly the technical aspects of general solid particle formation by these two techniques.

The *dispersion method* usually involves the comminution of coarse material by brute force – commonly mechanical grinding, though ultrasonic and electrical dispersion methods are also employed. A ball mill, with metal balls rolling over and dropping down on one another, yields particles down to about 50 μm, but much finer dispersions require the use of a *colloid mill*. The principle of the colloid mill is the production of intense shearing forces between a stator surface and a rotor running at high speed (about 10^2 s^{-1}). Larger particles are ruptured on violently striking the stator surface under the influence of turbulent flow. The intense heat generated by friction is removed by a cooling jacket. The particle-size distribution reaches a steady state after prolonged milling, owing to the tendency of small particles to agglomerate under the combined influences of the applied mechanical forces and the induced interparticle attractive colloidal interactions. In dry milling, the presence of inert diluent (e.g. glucose in sulphur) aids dispersion by reducing the probability of the fine particles (e.g. sulphur) from sticking to one another. In wet milling, small particles are prevented from aggregating by small-molecule surfactants or polymer stabilizers. Surfactants also directly facilitate dispersion by penetrating microfissures in deformed crystalline materials and, thereby, weakening the macroscopic structure (the Rebinder effect). On the commercial scale, the dispersion method has wide application in the production of such varied colloidal materials as drilling muds, food emulsions, mineral slurries, pigment dispersions and pharmaceutical pastes.

In the *condensation method*, particles are nucleated from a supersaturated solution, grown to colloidal size, and then stabilized in some way. Particles may appear as the result of a precipitation reaction; or, alternatively, the sudden cooling of a solution, if it leads to the normal lowering of solubility, may throw out the solute in colloidal form. The initial rate of nucleation depends mainly on the degree of supersaturation. Highly dispersed precipitates are produced when the nucleation rate is high and the particle growth rate is low. Particle growth is determined by the mass transport rate to the particle surface and the kinetics of particle aggregation. Colloids are usually made in dilute solution because a high particle concentration favours coagulation and a high reactant concentration leads to particle growth beyond the colloidal size range. Condensation methods are used industrially in emulsion polymerization and in the production of photographic emulsions.* In emulsion polymerization, latex particles are formed by a condensation process involving polymerization of monomer (e.g. styrene) dispersed as individual molecules, swollen surfactant micelles and emulsion droplets in an aqueous medium. The short-chain growing oligomers produced in the early stages have a strong tendency towards micellization, but as oligomers become polymers the growing micelles are transformed into small colloidal particles composed of phase-separated polymer. A monodisperse latex can be made if all the nuclei are formed at the start, or if, instead of the system being allowed to nucleate spontaneously, it is seeded with a large number of small latex particles (see Chapter 8).

It is by dispersion rather than by condensation that emulsion droplets are normally formed. Compared with the milling of solid particles, it is much easier to disrupt large deformable liquid drops into small droplets by the vigorous application of mechanical energy. A colloid mill can produce droplets down to about 2 μm, and various types of high-pressure homogenizers (see below) readily yield droplets up to 10 times smaller than this in the presence of an efficient emulsifying agent. By contrast, the condensation method is not suitable for emulsion formation as it is a rare occurrence in nature for insoluble liquid droplets to form spontaneously in another liquid medium. When liquid–liquid phase separation does occur (e.g. on cooling a solution below its critical consolute temperature), the transition is a second-order cooperative process with no supersaturation, in sharp contrast to the precipitation of solid material which is usually first order. For these reasons, we concentrate exclusively on droplet formation by dispersion techniques in the rest of this chapter.

---

* The use of the word 'emulsion' here is somewhat unfortunate. A photographic emulsion is, in fact, a light-sensitive *dispersion* of solid silver halide particles in a gelatin gel dispersion medium.

7.2.2 The high-pressure homogenizer

Of the many devices which have been designed and built for the manufacture of emulsion droplets, by far the most efficient and widely used technique is that of high-pressure homogenization. The prototype was constructed by Auguste Gaulin at the turn of the century for the homogenization of milk (*'fixer la composition des liquides'*). A pressure of more than 200 atmos was used by Gaulin to force milk through a bundle of capillary tubes against one end of which pressed a concave valve (Phipps, 1985). Over the past 100 years, valve design has been improved in order to maximize efficiency and minimize wear, but the overall result and methodology is the same. That is, the fat globules of fresh milk (average diameter 3–4 μm) are broken down into much smaller droplets (average diameter <1 μm) by driving them at high pressure through a narrow slit or hole. The primary purpose of homogenization is to make milk less susceptible to creaming (Dickinson and Stainsby, 1982).

A modern Gaulin-type homogenizer consists of a high-pressure displacement pump and a homogenizing valve of the 'drop and lift' type (Phipps, 1975). Two alternative valve designs are illustrated schematically in Figure 7.1. The spring-

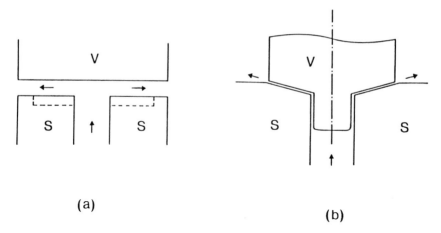

(a)

(b)

Figure 7.1 *Cross-sections of two designs of high-pressure homogenizing valve:* (a) *plane-faced valve;* (b) *conical valve. Liquid flows between valve (V) and seating (S) in the directions indicated by the arrows*

loaded valve may rest on a flat seating at the end of a tube (Figure 7.1a), while some valves have mating surfaces which are conical about a central axis with fluted locating guides entering the seating orifice (Figure 7.1b). The pump usually has three pistons (but occasionally five or seven) in order to minimize pressure fluctuations. During operation, a coarse pre-mix emulsion is pumped at a constant rate to the centre of the seating; this lifts the valve, and the mixture of oil and water passes radially through the slit aperture (15–300 μm width) to emerge as finely dispersed droplets. Pressure loadings on the valve typically lie in the range 4–40 MPa (i.e. 40–400 bar). As the homogenization pressure is increased, the valve lift is reduced, the slit width becomes narrower, and the resulting droplets become smaller. Reducing the mean droplet diameter d_{32} from 3.5 to 0.7 μm during milk homogenization increases the total oil–water interface by a factor of 5 and the total number of droplets by a factor of 125 (i.e. 5^3). Figure 7.2 shows a set of typical droplet-size distributions for cow's milk homogenized at various pressures. Similar droplet-size distributions are obtained for mineral oil-in-water emulsions stabilized by a wide range of water-soluble surfactants.

Emulsification by high-pressure homogenization takes place very rapidly. The average passage time t_p of a small fluid element through the valve is equal to the slit volume divided by the flow-rate Q, i.e.

$$t_p = h(r_2^2 - r_1^2)\pi/Q \tag{7.1}$$

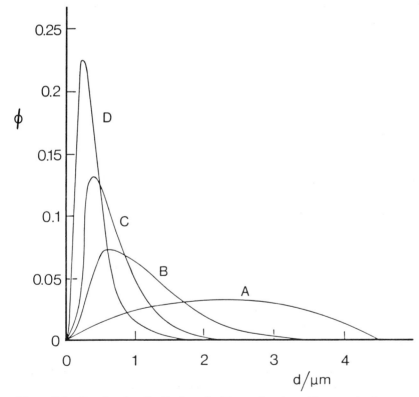

Figure 7.2 *Droplet-size distribution of milk as a function of homogenization pressure. The dispersed phase volume fraction* ϕ *per 0.1 μm size increment is plotted against the droplet diameter* d *for various pressures: (A) 25 bar; (B) 80 bar; (C) 180 bar; (D) 350 bar (after Mulder and Walstra (1974))*

where h is the gap width (see Figure 7.1a), and r_1 and r_2 are the inner and outer slit radii, respectively. Typical working parameter values for a Manton–Gaulin commercial homogenizer are $Q = 1$ m^3 h^{-1}, $r_1 = 5$ mm, $r_2 = 10$ mm and $h = 100$ μm; from eqn (7.1) these values give a passage time of $t_p \approx 0.1$ ms. This corresponds to a mean flow velocity across the slit of about 50 m s^{-1} and a maximum local flow velocity which is probably considerably higher. The average rate of energy dissipation per unit volume is

$$\bar{\epsilon} = p_h/t_p \tag{7.2}$$

where p_h is the homogenization pressure. Taking $p_h = 20$ MPa and $t_p = 0.1$ ms, eqn (7.2) gives an average power density of $\bar{\epsilon} = 2 \times 10^{11}$ W m^{-3}. This large power input leads to a heating of the emulsion by several degrees after passage through the homogenizer.

Droplet fragmentation into the submicrometre size range is achieved by a combination of intense laminar and turbulent flow (Mulder and Walstra, 1974; Walstra, 1983). At the inlet to the homogenizing valve, the Reynolds number Re is given by

$$Re = Q/2\pi r_1 \eta \tag{7.3}$$

where η is the viscosity. The change from laminar flow down the inlet pipe to turbulent flow at the slit entrance occurs at a value of Re which depends on the width h and on whether the inlet boundary is sharp edged or rounded (Phipps, 1985). Very roughly, for the general case, one can say that a value of Re $\leqslant 100$ is indicative of laminar flow and that a value of Re $\geqslant 1000$ implies turbulent flow. For a large Manton–Gaulin commercial homogenizer, the estimated Reynolds number is well in excess of 10^4 and so we can be sure that the flow in the inlet region is intensely turbulent. Reynolds numbers are lower in laboratory-scale homogenizers, but are still generally in the turbulent regime in the inlet region. In the slit itself, however, the turbulence is probably largely 'squeezed out' owing to the strong convergence of the flow. As it passes through the slit, therefore, the liquid is subjected to very high rates of strain, both in shear and in elongation. This means that droplet disruption in the homogenizer is due to the combined effects of severe laminar flow in the slit and local pressure fluctuations from turbulent eddies in the inlet region (see below). In addition, where the flow velocity in the slit is near its maximum, negative pressures may occur leading to limited droplet disruption by cavitation.

The most important control variable affecting droplet size is the homogenization pressure p_h. During power dissipation as heat in the valve, potential energy from the valve loading pressure is first converted into mechanical kinetic energy of fluid elements and then into thermal kinetic energy of the molecules. Assuming that these two processes are separable in time, we can estimate the maximum local fluid velocity u by roughly equating the mechanical kinetic energy to the potential energy, i.e.

$$p_h \approx \frac{1}{2}(\rho u^2) \tag{7.4}$$

where ρ is the fluid density. As u is inversely proportional to t_p, we see from eqns

(7.2) and (7.4) that the average power density is given by

$$\bar{\epsilon} \approx p_h^{1.5} \tag{7.5}$$

Instantaneous values of ϵ will vary throughout the homogenizing valve. For a particular value of ϵ, there will be a critical droplet diameter d_c such that droplets larger than d_c are disrupted whereas smaller ones are not. According to the theory of isotropic turbulence (see below), d_c is given by

$$d_c \approx \gamma^{0.6} \epsilon^{-0.4} \tag{7.6}$$

where γ is the interfacial tension. From eqns (7.5) and (7.6), the predicted dependence of the droplet size on the homogenization pressure is

$$d_c \approx p_h^{-0.6} \tag{7.7}$$

A distribution of droplet sizes (see Figure 7.2) is produced because of variations in ϵ throughout the homogenizer valve and also because of droplet recoalescence and other mechanisms of droplet disruption. Nevertheless, generally good agreement is found between eqn (7.7) and the dependence of the average droplet diameter (d_{32} or d_{43}) on the homogenization pressure. Figure 7.3 shows plots of

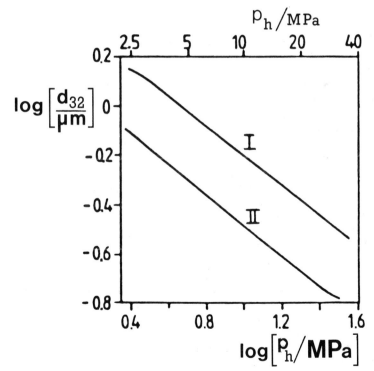

Figure 7.3 *Influence of homogenization pressure on droplet size for different valve homogenizers (I and II). The logarithm of the average volume/surface diameter d_{32} is plotted against the logarithm of the homogenization pressure p_h (after Walstra (1975))*

$\log_{10}d_{32}$ against $\log_{10}p_h$ for two different valve homogenizers as determined experimentally by Walstra (1975). The data in Figure 7.3 illustrate two important points relating to high-pressure homogenization. Firstly, the slopes of the lines are very close to –0.6, which is consistent with a droplet disruption mechanism primarily involving turbulent flow (see eqn (7.7)). Secondly, the intercepts are quite different, illustrating the point that substantially different droplet sizes may be obtained from two different homogenizing valves, or even from the same valve after extended usage and wear.

In addition to homogenization pressure, there are several other factors having an influence on the efficiency of emulsification: the dispersed phase volume fraction, the ratio of surfactant to dispersed phase, the viscosities of the oil and aqueous phases, and the temperature of homogenization. Increasing the oil volume fraction or the oil/emulsifier ratio leads to larger droplets because of the associated increase in the rate of coalescence of newly formed droplets. The encounter time of droplets during emulsification is inversely proportional to the volume fraction, and the probability that two colliding droplets will coalesce decreases with the surface coverage of emulsifier which in turn is determined by the total amount of emulsifier available per unit area. Some experimental data illustrating this effect are given in Figure 7.4 for oil-in-water emulsions of

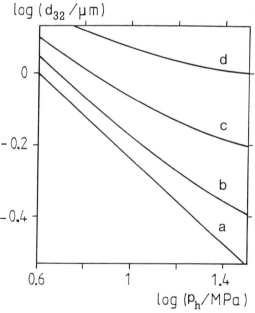

Figure 7.4 *Influence of oil volume fraction and homogenization pressure on droplet size of emulsions containing a constant protein concentration (0.4 wt%) in the aqueous phase. The logarithm of the average volume/surface diameter d_{32} is plotted against the logarithm of the homogenization pressure p_h for various dispersed phase volume fractions: (a) ≤ 0.05; (b) 0.2; (c) 0.3; (d) 0.4 (after Walstra (1993))*

increasing volume fraction made with a constant proteinaceous emulsifier content in the aqueous phase. We note the large deviation from eqn (7.7) at the high oil/emulsifier ratios. Increasing the oil volume fraction at constant oil/emulsifier ratio also leads to larger droplets due to the lowering of the encounter time and some suppression of turbulent eddies, but the effect is usually less than that of increasing the oil/emulsifier ratio at constant volume fraction, unless the viscosity or the emulsifier content is high. For a given set of homogenization conditions, increasing the amount of emulsifier usually leads to smaller droplets up to a certain limiting emulsifier concentration beyond which there is no significant improvement.

The effect of the continuous phase viscosity on the droplet-size distribution is generally slight in accordance with predictions of the theory of isotropic turbulence (see below). On the other hand, the effect of increasing the (internal) dispersed phase viscosity η_i is normally to increase the resulting emulsion droplet size. The experimental value of the exponent m in

$$d_{43} \approx \eta_i^m \tag{7.8}$$

is given by Walstra (1974) as $m = 0.37 \pm 0.02$, whereas the isotropic turbulence theory gives $m = 0$, though this theory can be refined somewhat (Davies, 1985) to give a theoretical dependence on η_i. In qualitative terms the influence of η_i on droplet size can be explained by a consideration of the relative time-scales for droplet deformation and small eddy dissipation in turbulent flow. The smallest eddies produce the smallest droplets but they have the shortest lifetimes. Increasing the dispersed phase viscosity increases the deformation time, and so this increases the size of eddies which last long enough to cause the required deformation. Hence, as η_i increases, the droplets are broken by the larger eddies, and so the resulting droplets are themselves larger. In addition to this, any disruption by laminar shear flow is strongly affected by the dispersed phase viscosity if it is significantly larger than the continuous phase viscosity (see below). Raising the temperature usually increases the efficiency of homogenization mainly through its influence on η_i. Milk taken straight from the refrigerator will not homogenize at all because the fat globules are solid (milk fat is about 60% crystallized at 5°C).

A distinct advantage of high-pressure homogenization for industrial use is that fine emulsions can be made rapidly and reproducibly in a continuous one-pass process. Recycling the primary emulsion repeatedly through the apparatus does lead to a reduced average droplet size and a narrower distribution, but such a procedure is generally not convenient or economic on the industrial scale, though it is sometimes used in laboratory or pilot-plant studies. The formation of a finer emulsion by repeated homogenization is consistent with the stochastic nature of droplet disruption in turbulent flow – there are significant fluctuations in local flow conditions as different regions of the sample pass through the valve. Commercial equipment commonly incorporates a two-stage homogenizing head, i.e. with two valves mounted in series. The purpose of the second stage is to disrupt aggregates of droplets (so-called 'homogenization clusters') formed during the first stage. The second valve operates at low pressure (say 20% of the total); there is no significant homogenization during the second stage.

High-pressure homogenization can be carried out on a small scale by reducing the liquid volume passing through the valve. In the mini-homogenizer of Dickinson *et al.* (1987) a single pass through the valve takes a few seconds. This small-scale one-stage homogenizer can make a 10–15 ml sample of oil-in-water emulsion with a droplet-size distribution similar to that for a two-stage Manton–Gaulin homogenizer which produces a minimum of several litres of emulsion. The main disadvantage of the valve homogenizer for small-scale use is the difficulty of making a pre-mix of the oil/water/emulsifier mixture without incorporation of air, since the presence of dispersed air leads to unreliable operation of the valve. This problem can be overcome by careful use of a specially designed pre-mixing cell, or by making the emulsion by high-pressure jet homogenization.

7.2.3 The jet homogenizer

The principle of jet homogenization (Burgaud *et al.*, 1990) is the bringing together of previously unmixed oil and aqueous phases into an homogenization chamber by passage at high velocity through a fixed small hole. Intermixing and emulsification of the phases occurs as the emerging liquid 'jet' impinges violently at the surface of a flat plate within the chamber. Though it requires a high pressure to drive the liquids through the hole, the jet homogenizer differs from the high-pressure valve homogenizer in having no moving parts in the emulsification region. Its advantage over the Gaulin-type homogenizer for small-scale use is that it requires no pre-mixing of the oil and aqueous phases.

A schematic diagram of a laboratory jet homogenizer system is given in Figure 7.5. Two parallel cylindrical chambers contain the oil and aqueous phases to be emulsified, and the emulsion volume fraction is determined by their relative cross-sections. The liquid phases are expelled from their cylinders by synchronously moving pistons driven by a pneumatic ram. The liquid flows rapidly along capillary lines before passing into the homogenization chamber (see Figure 7.6). The homogenization pressure (i.e. pressure in the chambers containing the oil and aqueous phases) is controlled by the thrust from an air cylinder. The droplet-size distribution of the emulsion emerging from the homogenizing head depends on the liquid jet velocity and the dimensions of the inlet and outlet holes of the homogenization chamber. Emulsion droplet sizes similar to those produced with a Gaulin-type valve homogenizer require typical homogenization pressures of 200–300 bar producing jet speeds of the order of, say, $200\,\mathrm{m\,s^{-1}}$. A feature of the design illustrated in Figures 7.5 and 7.6 is the accessibility and ease of detachment and assembly of the main cylinder block and homogenizing head. This facilitates filling, cleaning and thermostatting of the homogenizer. The emulsion volume fraction is fixed by the ratio of piston cross-sectional areas, and so it is necessary to have several homogenization units available with a range of piston radius ratios in order to make emulsions with a range of phase volume fractions.

As well as avoiding problems of air incorporation, the jet homogenizer has one other advantage over the high-pressure valve homogenizer. The narrow gap

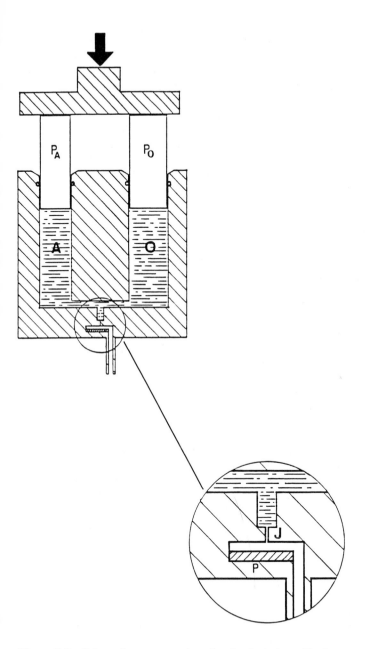

Figure 7.5 *Schematic representation of a simple design of jet homogenizer. Pistons P$_A$ and P$_O$ move synchronously in the direction indicated by the arrow. Aqueous and oil phases are driven from compartments A and O, respectively, through hole J to impinge on fixed plate P (from Castle* et al. *(1988))*

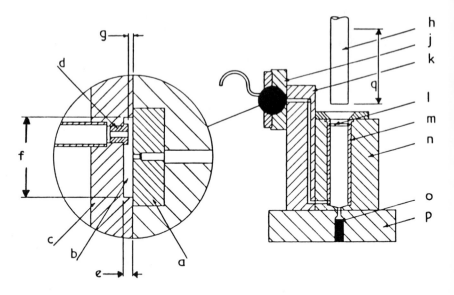

Figure 7.6   *Sectional scale drawing of homogenizing unit of laboratory scale jet homogenizer.* (a) *Jet disc incorporating jet hole;* (b) *homogenization chamber;* (c) *impact plate;* (d) *outlet insert;* (e) *chamber depth (ca. 0.6 mm);* (f) *chamber diameter (ca. 9 mm);* (g) *outlet gap (ca. 0.4 mm);* (h) *piston;* (j) *detachable head;* (k) *capillary tube block;* (l) *O-ring;* (m) *cylinder liner;* (n) *cylinder block;* (o) *pressure transfer piston;* (p) *base;* (q) *piston length (ca. 5 cm)*

through which the liquids are forced is fixed and permanent, and so it does not suffer from wear-related problems to the same extent as a valve homogenizer. Because of this, the long-term reproducibility of droplet-size distributions generated by jet homogenization is excellent. A similar principle is adopted in the so-called 'microfluidizer' of Korstvedt *et al.* (1984), which also has no moving parts. A particular feature of the laboratory-scale microfluidizer is its ease of temperature control.

7.2.4 Alternative methods of emulsification

There are many other ways of making emulsions apart from high-pressure homogenization (Walstra, 1983). Various types of mixer are commonly employed in the kitchen, laboratory or pilot plant. Turbulent flow is achieved with a high-speed stirrer or wisk in a tank with baffles near the walls to avoid stratification. Droplet diameters down to about 5 μm can be produced. Simple mixers are often used to pre-mix oil and aqueous phases prior to feeding into a high-pressure homogenizer or colloid mill. In making emulsions by simple mixing, addition of the dispersed phase slowly to the continuous phase during emulsification often gives the best results. This is because small droplets are most easily produced after small additions of dispersed phase when the viscosity is low and the amount of available emulsifier (dissolved in the continuous phase) is high. Once a dilute emulsion of quite fine droplets is established it becomes easier to break down further added dispersed phase because of the hydrodynamic action of the original droplets.

In a colloid mill, the liquids are dispersed in a narrow gap ($\leqslant 0.1$ mm) between a stator surface and a rotor running at high speed (up to 20 000 r.p.m.). Very strong shear flows (up to 10^7 s^{-1}) are generated, and turbulence is enhanced by roughening the stator and rotor surfaces with serrations. Droplet sizes produced by colloid mills are smaller than those produced by mixers, but not as small as those produced by high-pressure homogenization.

High-intensity ultrasound is a convenient way of inducing emulsification in the laboratory, although the spread of droplet sizes produced by ultrasonic devices is rather wide and sometimes bimodal. One such device is the so-called 'liquid whistle' in which a jet of liquid mixture impinges on the edge of a thin metal blade whose resonant vibrations (10–40 kHz) set up powerful disruptive oscillations in the liquid jet. Droplet disruption proceeds by a cavitation mechanism (see below). Very small droplets can be produced by ultrasonic emulsification, but the reproducibility is rather poor. Presumably some large droplets are fragmented by imploding cavities while others are disturbed hardly at all.

A new technique of emulsion droplet formation which shows considerable commercial potential is called 'membrane emulsification' (Nakashima *et al.*, 1991). The method involves using a low stable pressure (about 1 or 2 bar) to force a dispersed phase (oil or water) to permeate through a strong membrane having a uniform pore-size distribution. Average droplet size and droplet-size distribution of the resulting emulsion is controlled primarily by the size and uniformity of the micropores. It seems that the most reliable performance is obtained with glass membranes having cylindrical interconnected uniform micropores which are fabricated by microphase separation of a CaO/Al$_2$O$_3$/ B$_2$O$_3$/SiO$_2$ mixture. A particular feature of membrane emulsification is that it can be made to produce emulsions with a much greater degree of monodispersity than is normally possible with high-pressure valve or jet homogenization. In addition, it can be made to produce fine emulsions (<1 µm diameter) using less surfactant and much less energy than is required with these other methods. The main disadvantage of membrane emulsification is that it is a rather slow process, especially for making concentrated emulsions. Problems can also arise with contamination of the membrane and blocking of the micropores, and careful cleaning is necessary if small particles, proteins or strongly adsorbing polymers are present. The membrane must be wetted with the continuous phase prior to contact with the dispersed phase, which means that the membrane must be predominantly hydrophilic (hydrophobic) in order to make an oil-in-water (water-in-oil) emulsion. Techniques for regenerating hydrophilic or hydrophobic glass membranes which have become contaminated have also been developed (Nakashima *et al.*, 1991).

7.3 Theory of emulsification

7.3.1 Energy requirements

In thermodynamic terms, the energy required to convert a system of two immiscible liquids into a dispersion of emulsion-sized droplets is rather modest. To increase the surface area by an amount ΔA requires a free energy change:

$$\Delta G = \gamma \Delta A \tag{7.9}$$

where γ is the interfacial tension. Let us consider the formation of oil droplets of radius 1 μm in water at a volume fraction of 0.1. For pure oil + water with $\gamma = 50$ mN m^{-1}, the energy change is $\Delta G \approx 10^4$ J m^{-3}. For oil + water + surfactant with $\gamma = 5$ mN m^{-1}, the energy change is $\Delta G \approx 10^3$ J m^{-3}. In practice, however, the actual amount of mechanical energy necessary to make such an emulsion in the presence of a good emulsifier would be of the order of 10^6 J m^{-3}. This means that, thermodynamically speaking, emulsification is an extremely inefficient process. The vast additional excess of energy is required for inducing the break up of large droplets into small ones.

In order to disrupt a droplet of radius a, it is necessary to apply an external pressure gradient of the order of

$$dp/dr \approx \Delta p/a = 2\gamma/a^2 \tag{7.10}$$

where Δp is the Laplace pressure. The pressure gradient predicted from eqn (7.10) is $dp/dr \approx 10^{10}$ Pa m^{-1}. The local stress differences required to produce such a gradient may arise from intense laminar flow (shear or extensional) or from inertial effects (turbulence or cavitation). From eqn (7.10) we see that the lower the interfacial tension, the lower are the stress differences required to produce droplet disruption. This is one of the roles of the emulsifier during emulsification. The other, probably more important, role is to adsorb rapidly at the oil–water interface in order to protect the newly formed droplets against immediate recoalescence.

7.3.2 Droplet disruption in laminar flow

Complex laminar flow patterns may be considered locally to be composed of three kinds of flow: rotational flow, shear flow and elongational flow. Pure rotational flow causes no droplet deformation; it simply rotates in the flow field. Pure elongational flow causes deformation but no rotation. Pure shear flow causes both deformation and rotation. The relative amounts of shear and elongational flow during emulsification depend on the type of equipment being used. Droplet disruption by shear flow is important in the colloid mill, and droplet disruption by elongational flow is important in the high-pressure valve or jet homogenizers.

In simple shear flow, a spherical liquid droplet is distorted into a prolate ellipsoid as illustrated in Figure 7.7. Under steady-state conditions, the change in droplet shape depends on a balance between interfacial tension forces and forces due to normal surface stresses from the viscous shear flow (Taylor, 1934). Weak flows extend and compress the droplet in the principal directions of positive and negative straining, and in Couette flow this means that the prolate ellipsoid is orientated at 45° to the flow direction. The extent of deformation is expressed by the parameter

$$R = (l_{maj} - l_{min})/(l_{maj} + l_{min}) = Da\eta(19\eta_i + 16\eta)/16\gamma(\eta_i + \eta) \qquad (7.11)$$

where l_{maj} and l_{min} are the lengths of the major and minor axes, D is the shear rate, η is the shear viscosity of the continuous phase, and η_i is the shear viscosity of the internal droplet phase. Above a certain degree of deformation ($R \gtrsim 0.5$), the droplet bursts into smaller droplets. The detailed behaviour at large deformations depends on the viscosity ratio η_i/η. When the oil and water viscosities are of similar magnitude, a neck appears at the middle of the droplet, and this breaks into two large fragments and several smaller ones. For $\eta_i/\eta \ll$ 1 the droplet assumes a sigmoidal shape and small satellite droplets are gradually

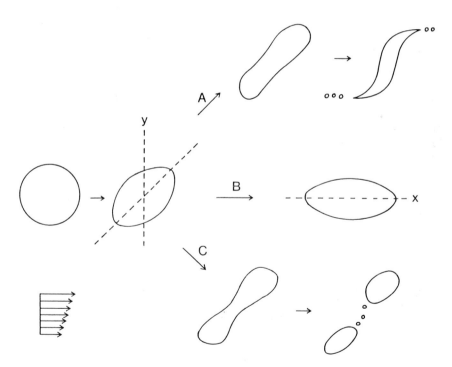

Figure 7.7   *Deformation and disruption of a liquid drop in shear flow in the x–y plane. The ellipsoidal deformed drop is orientated at 45° to the flow direction at low rates of shear. The type of deformation/disruption at high rates of shear depends on the ratio of the dispersed phase viscosity η_i to the continuous phase viscosity η. (A) $\eta_i \ll \eta$; (B) $\eta_i \gg \eta$; (C) $\eta_i \approx \eta$ (after Dickinson and Stainsby (1982))*

eroded from the pointed ends (Gopal, 1968). In contrast, for $\eta_i/\eta \gg 1$, the droplet reaches its maximum deformation in the direction of the flow ($\phi = 90°$) due to the tensile nature of the first normal stress difference.

Droplet disruption occurs when the external viscous stresses acting on the droplet exceed the Laplace pressure. That is, the droplet bursts when the dimensionless Weber number

$$\text{We} = Da\eta/\gamma \tag{7.12}$$

exceeds a critical value of the order of unity. Equation (7.12) indicates that shear rates of the order of $D \approx 10^7 \text{ s}^{-1}$ are necessary in order to produce emulsion droplets in water in the submicrometre size range. Such shear rates are difficult to produce in practice, and so it is generally not possible to make very fine emulsions by simple shear flow alone. In the absence of surfactant, the critical Weber number diverges to infinity as $\eta_i/\eta \rightarrow 4$ (see Figure 7.8). and so theory predicts that there is no droplet break up at all for $\eta_i/\eta > 4$ (Walstra, 1993). The explanation is that for such a highly viscous dispersed phase the droplet cannot deform as quickly as the velocity gradient can cause deformation; the droplet does deform slightly, of course, but mainly it just rotates at a rate of $D/2$. Figure 7.8 shows a comparison of the theoretical dependence of the critical Weber number on η_i/η with some experimental data obtained for emulsions made in a colloid mill (Ambruster, 1990). Agreement is mostly rather good, although break up was found to occur experimentally at higher values of the viscosity ratio (up to $\eta_i/\eta = 10$) than predicted theoretically. Possible reasons for this are the presence

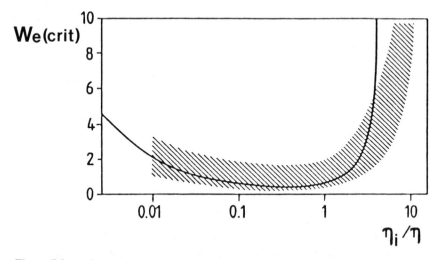

Figure 7.8   *Influence of viscosity ratio on the condition for the onset of droplet disruption in simple shear flow. The critical Weber number We(crit) is plotted against the ratio of the dispersed phase viscosity η_i to the continuous phase viscosity η. The hatched area refers to experimental data obtained with a colloid mill (after Walstra (1993))*

of other types of flow in the equipment (elongational or even turbulent flow) or the presence of an interfacial tension gradient at the droplet surface which hinders the development of flow within the droplet (Walstra, 1993). The theory assumes that the tangential stress is continuous across the oil–water interface, but this will generally not be the case in real systems due to the adsorbed emulsifier, and so in practice droplets of high viscosity can be disrupted in shear flow, although less effectively than low viscosity droplets.

In elongational flow, viscous droplets are pulled out into long threads, and these disintegrate by various hydrodynamic instability mechanisms (see below). The main limiting factor is that, whereas shear flow is readily produced on a continuous basis, elongational flow (e.g. during high-pressure homogenization) is transient. This means that, although in theory the flow rate is sufficient to disrupt the droplets, it may not occur in practice because the velocity gradient does not last long enough. For the values of η_i/η covered in Figure 7.8, the critical Weber number is constant at about 0.3, and so we see that elongational flow is much more effective at droplet disruption for $\eta_i/\eta \gg 1$.

7.3.3 Droplet disruption in turbulent flow

In turbulent flow, such as occurs in the high-pressure homogenizer, the droplets experience inertial* as well as viscous forces. Overall liquid motion is in eddies much larger than the droplets which transmit kinetic energy to small eddies in which most of the viscous dissipation takes place. As described in standard textbooks (Levich, 1962; Davies, 1972), turbulent flow is characterized by a local velocity v which fluctuates chaotically about a time-averaged value $\bar{v}$. In the direction of flow, $\bar{v}$ is equal to the macroscopic flow velocity u; at other angles θ to the flow direction we have

$$\bar{v} \approx u \cos \theta \tag{7.13}$$

Deviations from $\bar{v}$ are characterized by the root-mean-square average:

$$v_{rms} = \langle (v - \bar{v})^2 \rangle^{1/2} \tag{7.14}$$

The flow is said to be isotropic if v_{rms} is the same in every direction.

The disruption of droplets in turbulent flow has been described theoretically assuming that the flow is locally isotropic (Kolmogorov, 1941; Hinze, 1955; Shinnar, 1961). There is a spectrum of eddy sizes. It is the small energy-bearing eddies which lead to the formation of the smallest droplets. For the case where the droplets are larger than the smallest eddies of size δ, Kolmogorov (1941) has shown that the diameter of the largest droplet remaining unbroken, d_c, is proportional to $\epsilon^{-2/5}$ (see eqn (7.6)), where ϵ is the rate of energy dissipation. This means that, in order reduce the average droplet size from 5 to 0.5 μm, it is necessary to increase the power input to the homogenizer by a factor of approximately 300. If the system is very viscous or the droplets very small ($d_c \ll \delta$), viscous deformation may still be the overriding factor in strongly turbulent

---

* By 'inertial' force is meant the force associated with a fluid element obtained by multiplying mass by instantaneous acceleration (Newton's law of motion).

flow. Under these conditions, d_c is proportional to $\epsilon^{-1/2}$, although the difference from eqn (7.6) may be difficult to confirm experimentally.

According to Walstra (1983), a plot of $\log_{10}d_{43}$ against $\log_{10}\bar{\epsilon}$ gives a slope of –0.4 for a high-speed mixer (the Ultra Turrax) and for an ultrasonic device, in good agreement with the theory of isotropic turbulence. For the same total expenditure of energy, the stirrer is much less effective than the ultrasonic device, which in turn is substantially less effective than the high-pressure homogenizer. The reason for this is that the homogenizer dissipates the input energy in a shorter time than the ultrasonic device and in a much shorter time than the stirrer. Though the stirrer produces turbulent flow, most of the energy is dissipated in large eddies which cannot break up the small droplets. Another reason why high-pressure homogenizers are especially effective at emulsification is that the power density is proportional $p_h^{3/2}$, where p_h is the homogenization pressure. This leads to a larger slope (of –0.6) in the plot of $\log_{10}d_{43}$ against $\log_{10}\bar{\epsilon}$ (see Figure 7.3) than for the other devices.

7.3.4 Droplet disruption by cavitation

Cavitation is the phenomenon of sudden formation and subsequent collapse of small vapour bubbles in a liquid. A high-velocity liquid may produce a local negative pressure which leads to the formation of a cavity at that point by a process of heterogeneous nucleation. Controlled cavitation may be generated by sound waves of ultrasonic frequency (say 20 kHz). Cavity collapse proceeds on a time-scale shorter than 1/frequency. The imploding cavity generates a high-intensity microscopic shock wave and, if a large droplet is in the close vicinity, part of the dispersed phase is sucked towards the shrinking void, as shown in Figure 7.9. Due to hydrodynamic instability (see below), the jet of dispersed phase liquid breaks up into small droplets as it moves to fill the void.

Cavitation is probably the main mechanism of droplet disruption in ultrasonic emulsification. It also possibly occurs to a limited extent in high-pressure or jet homogenization, especially where there is some back pressure at the exit to the homogenization chamber.

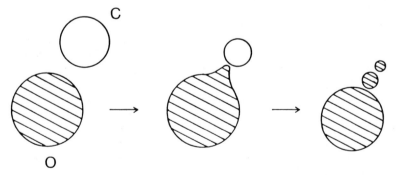

Figure 7.9   *Illustration of the mechanism of emulsification by cavitation as a vapour cavity C collapses in the vicinity of an oil droplet O*

7.3.5 Hydrodynamic instability of films and jets

An essential feature of many types of emulsification process is the break up of liquid films and jets into small droplets under the influence of viscous and inertial forces. Large droplets may be drawn into threads (e.g. during elongational flow), and the threads may be subject to varicose deformation (i.e. alternate contractions and swellings along the length). When the wavelength of the sinusoidal deformation exceeds the circumference of the unperturbed cylinder, the jet is unstable. Other ways in which jets are formed are by direct injection of one liquid into another (jet homogenization) or by disruption of plane interfaces in high-speed mixers.

Any disturbance at the interface of a planar film or a cylindrical jet will cause ripples to develop. These capillary waves are damped by high viscosity liquids and by adsorbed emulsifier, but if an external source acts continuously at a suitable frequency (as in an ultrasonic device) standing waves will be produced. If the amplitude is high enough, the interface may break down into small droplets. Under certain conditions in the absence of a constant frequency source, a short local perturbation may produce a disturbance that spontaneously grows in amplitude. At some point in the growth, deep 'fingers' of one liquid penetrate into the other, and this hydrodynamically unstable situation leads to a shredding of small droplets.

The Rayleigh–Taylor instability occurs when the interface is accelerated from the (less dense) oil phase into the (more dense) aqueous phase (Gopal, 1968). For the case in which the tension is constant in time and space (i.e. in the absence of surface-active material), the interface is unstable if the wavelength λ of the disturbance satisfies the inequality

$$(2\pi/\lambda)^2\gamma < a\Delta\rho \tag{7.15}$$

where a is the acceleration perpendicular to the interface, and $\Delta\rho$ is the density difference between the phases. This type of instability is important during the formation of a coarse emulsion by shaking or stirring, but its relevance to the formation of fine emulsions remains unproven.

When the two phases at a planar oil–water interface move with different tangential velocities, a Kelvin–Helmholtz instability may occur. The relative translational kinetic energy of the two liquids induces any wavy disturbances to grow in amplitude, and this causes the liquids to mix together. Droplet shredding takes place at relatively low shear velocities in laminar flow, especially if the tension is high and the viscosity is low.

Finally, it is pertinent to note that the most important type of hydrodynamic instability in fine emulsion formation is that associated with the transition from laminar flow to turbulent flow. This occurs when the Reynolds number exceeds a certain critical value (about 10^3). At high Reynolds number, the viscous forces are overwhelmed by dynamic inertial forces, and so streamline flow is unstable with respect to infinitesimal perturbations (Gopal, 1968).

7.3.6 Role of the emulsifier

There are two main ways in which the efficiency of emulsification is affected by the presence of surface-active material (the emulsifier). Firstly, the emulsifier lowers the tension at the oil–water interface, which facilitates droplet deformation and disruption; secondly, once adsorbed at the interface, it protects the newly formed droplets against immediate recoalescence. In practice, it is found that the presence of rapidly adsorbing species during droplet disruption leads to smaller emulsion droplets in the resulting product. For instance, with protein-stabilized oil-in-water emulsions, the presence of additional oil- or water-soluble surfactants during homogenization leads to a substantial reduction in the average droplet size (Courthaudon *et al.*, 1991), and even a simple molecule like ethanol has a substantial effect on droplet size (Burgaud and Dickinson, 1990). We can attribute this improvement in the efficiency of emulsification to a lowering of the tension at the newly formed oil-water interface during the early stages of droplet break up, since theory predicts greater droplet deformation in viscous shear flow (see eqn (7.11)) and a smaller critical droplet diameter in turbulent flow (see eqn (7.6)) as the tension is reduced. In addition, however, the presence of more rapidly adsorbing species may also affect the dynamic surface rheological properties of the interface (Lucassen, 1981), and so an alternative explanation for the smaller average droplet size may be an improvement in the stability of freshly formed droplets against recoalescence. It is difficult to distinguish properly the droplet formation and stabilization processes, and in many emulsion systems it is the emulsifying agent itself which acts as the primary stabilizing agent against flocculation and coalescence.

During emulsification, the main stabilizing mechanism involving the emulsifier is the Gibbs–Marangoni effect (van den Tempel, 1960). The essence of the mechanism is illustrated schematically in Figure 7.10. The sketch shows two partially covered droplet surfaces coming together during emulsification. The region of bulk phase between the surfaces becomes depleted of surfactant as adsorption continues, and the amount of surfactant available for further adsorption is lowest where the film is at its thinnest. The non-uniform surface concentration leads to an interfacial tension gradient which causes streaming of liquid along the surface (the Marangoni effect) as surfactant molecules move in the plane of the interface towards the region of lowest surface concentration. The streaming liquid tends to drive the droplets apart, thereby stabilizing the thin film against rupture. The mechanism only operates if the surfactant is not present in the dispersed phase. If surfactant is available for adsorption from the droplet side of the interface, no significant tension gradient is set up, and so there is no liquid streaming. This means that, for the effective stabilization of oil droplets during emulsification by the Gibbs–Marangoni mechanism, we need to use a water-soluble emulsifier (Bancroft's rule).

The generation of a tension gradient implies that the area A of the oil–water interface is changing with time. The extent to which the local tension differs from the equilibrium value is normally expressed in terms of the surface dilational modulus E_s defined by

$$E_s = d\gamma/d\ln A \qquad (7.16)$$

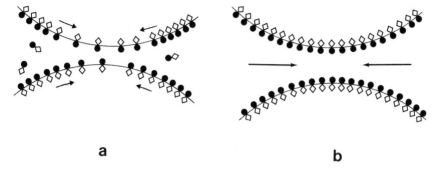

Figure 7.10 *Illustration of how droplets are stabilized against coalescence by the Gibbs–Marangoni effect: (a) adsorbed molecules move towards the region of lowest surfactant surface concentration where the film between the droplets is thinnest; (b) liquid streaming induced between the droplet surfaces drives them apart and so further film thinning is inhibited*

The instantaneous value of E_s depends on the chemical nature of the emulsifier, and on the rate of extension of the surface. In the restricted space between two closely approaching oil droplets, there is a finite amount of water-soluble emulsifier available for rapid adsorption at the two surfaces, and so the amount of available surfactant is the limiting factor determining E_s. The Gibbs elasticity of the aqueous film is simply twice the dilational modulus:

$$E_f = 2E_s \qquad\qquad (7.17)$$

In terms of the surfactant surface excess concentration Γ, Lucassen (1981) has shown that E_f is given by

$$E_f = -[2(d\gamma/d\ln\Gamma)]\,[1 + (h/2)(dc/d\Gamma)]^{-1} \qquad\qquad (7.18)$$

where h is the film thickness, and c is the (molar) concentration of surfactant in the continuous phase. Equation (7.18) shows that, when Γ is reasonably constant, the thinnest part of the film has the greatest Gibbs elasticity and hence the most resistance towards further stretching. This is the situation which occurs at high surfactant concentrations. When c is low, however, the stretching of the film causes a substantial reduction in Γ; this means that the Gibbs elasticity decreases as the film thins, which leads to spontaneous film rupture. (For a detailed thermodynamic description of Gibbs elasticity, the reader is referred to the paper by van den Tempel *et al.* (1965).)

This theoretical description of the role of the emulsifier in inhibiting droplet recoalescence by the Gibbs–Marangoni mechanism is strictly applicable only to an ideal system of small-molecule surfactants (Walstra, 1993). It is not possible to predict E_f reliably for polymeric emulsifiers or for mixtures of surfactants. This is, in part, due to deviations from ideal thermodynamic adsorption behaviour in such systems. More importantly, however, the surface dilational modulus depends on the rate of expansion of the surface, and values of E_s at the high deformation rates appropriate to emulsification are generally unknown. Estimates of $d\gamma/d\ln\Gamma$ determined in typical laboratory experiments are not relevant to the very short time-scales of droplet encounters during emulsion formation, especially for macromolecular emulsifiers which usually exhibit slow conformational rearrangement following initial adsorption. The most that can be said is that, for the emulsifier to be effective, it should be present in reasonably high concentration in the continuous phase and in negligible concentration in the dispersed phase.

Once the emulsion has been formed, and the oil–water interface has become fully saturated with emulsifier (stabilizer), the Gibbs–Marangoni mechanism is no longer operative. The main role of the emulsifier immediately following emulsification is the electrostatic or steric stabilization of the droplets against flocculation and coalescence. Most useful as steric stabilizers are polymeric emulsifiers (including proteins); the theory behind this type of stabilization can be found elsewhere (Dickinson and Stainsby, 1982; Tadros and Vincent, 1983). A common type of aggregation in freshly formed emulsions is bridging flocculation. This occurs when there is insufficient emulsifier present to cover fully the new interface without sharing of the surface-active material between adjacent droplet surfaces. In mixed emulsifier systems (e.g. two proteins of different surface activity), bridging flocculation may be induced by the competitive displacement of one species from the interface during or shortly after emulsification (Dickinson and Galazka, 1991).

7.4 Control of emulsion droplet formation

By control of emulsion droplet formation, what we usually mean is control of droplet-size distribution and the state of aggregation. These two physical attributes are interrelated – firstly, because the attractive forces between droplets are dependent on droplet size, and secondly, because any aggregation of droplets during testing may be interpreted in terms of larger 'effective particles' in the droplet-size distribution. The degree to which emulsion droplet formation may be controlled in practice depends in large part on the reliable assessment of average droplet size and the various types of aggregation phenomena in freshly made emulsions.

The most reassuringly direct way of assessing these two attributes is by optical microscopy. The routine, careful examination of diluted samples under the light microscope is a simple but invaluable procedure in the characterization of emulsion systems. Even a cursory examination tells us about the general effectiveness of the emulsification, the breadth of the size distribution, and the extent of clumping or flocculation. The freedom of movement of small droplets can be used to assess the strength of flocs and the reversibility of aggregation, and observations under polarized light may be used to detect liquid crystalline regions in the bulk phases or at the oil–water interface. As a quantitative probe of droplet size, however, traditional optical microscopic analysis is less than satisfactory. The two main disadvantages are the subjectivity of observations by eye which tends to give undue statistical bias to the largest droplets, and the general tediousness of making the measurements. Nowadays, by linking the microscope to an image analyser and data storage system, the tedium of particle counting and classification can be completely overcome, as can the subjectivity to a substantial degree, although there still remains a bias towards the clearly visible large droplets, with those below about 0.5 μm being missed altogether.

Various instrumental techniques are available for assessing average emulsion droplet size based on light scattering and turbidimetry (Farinato and Rowell, 1983). Turbidity measurement is often used because of its simplicity, but it is not to be recommended for emulsion droplets because the average size cannot be reliably determined from a single turbidity reading. The combination of turbidity measurement and centrifugal separation, however, does provide a satisfactory way of determining emulsion droplet diameters down to about 0.1 μm (Parkinson *et al.*, 1970). Total intensity time-averaged light scattering can also be used to determine the complete droplet-size distribution if measurements are made over a wide range of scattering angles and the data are properly analysed using Mie scattering theory (Kerker, 1969). Such analysis techniques are now built into modern commercial instruments (e.g. the Malvern Mastersizer), the only uncertainties being the values of the optical input parameters (real and imaginary parts of the refractive indices). For very fine emulsions, with droplets of diameter 0.5 μm or less, the technique of dynamic light scattering (photon correlation spectroscopy) is sometimes used, although the considerable poly-dispersity of most emulsions makes analysis difficult unless measurements are made over a wide range of scattering angles. The most common non-optical technique used for droplet-size distribution determination is the Coulter counter. An advantage of the Coulter counter is that droplets are counted individually; this leads to more reliable determination of the proportion of large droplets and the shape of the size distribution than with scattering techniques. Some dis-advantages of the Coulter counter are the necessity for dispersing droplets in electrolyte, the restriction to oil-in-water emulsions, the inability to detect droplets below about 0.6 μm, and the sensitivity of the orifice to blocking by large droplets and aggregates.

The size and strength of emulsion droplet aggregates can often be inferred during droplet-size determination. Weak flocs are broken down by gentle stirring or simply diluting the emulsion prior to analysis. Moderately strong aggregates may be resistant to stirring while being disrupted by ultrasonic agitation. In this way, any changes in size distribution following such treatments can be used as an indicator of the extent and strength of emulsion droplet aggregation. Less direct methods of assessing the state of aggregation involve measurements of creaming, sedimentation and various rheological parameters (Becher, 1965; Mulder and Walstra, 1974; Dickinson and Stainsby, 1982).

The formation of emulsions in which the dispersed phase is appreciably soluble in the continuous phase may lead to growth of larger droplets at the expense of smaller ones (Ostwald ripening). The rate of Ostwald ripening is given by

$$d\langle a\rangle^3/dt = 8V_1\gamma S_1 D_1/9RT \tag{7.19}$$

where $\langle a\rangle$ is the mean droplet radius, t is the time, D_1 and S_1 are the diffusion coefficient and solubility of component 1 (dispersed phase) in component 2 (continuous phase), V_1 is the molar volume of component 1, R is the gas constant, and T is the temperature. The main factor affecting the kinetics is the solubility S_1 of the dispersed phase in the continuous phase. In emulsions made with excess surfactant which forms micelles (or other association structures) in the continuous phase, mass transport of dispersed phase between droplets may be greatly enhanced. Control of Ostwald ripening may be achieved (Pertsov *et al.*, 1984) by addition to the dispersed phase of a small quantity of a soluble component that is poorly soluble in the dispersion medium. Macromolecular emulsifiers are particularly effective in inhibiting Ostwald ripening because of the resistance of their adsorbed films towards compression. Protein films are especially effective because of their substantial surface dilational elasticity (Dickinson and Stainsby, 1982).

By way of conclusion, then, let us recall some of the main factors required for the control of emulsion droplet formation. In the first place we must have an efficient emulsifier present in sufficient concentration to cover properly all the new interface created during emulsification. To control droplet size in the submicron range, we need a reliable high-performance homogenizer and a reproducible method for determining size distributions in the fresh emulsion and after storage. In emulsions formed by dispersion techniques, one should always expect a distribution of droplet sizes, although it is possible to make emulsions having a fairly narrow distribution using valve or jet homogenizers operating at high pressures. It is claimed that reasonably monodisperse emulsions can be produced from membrane emulsification (Nakashima *et al.*, 1991) and from aerosols formed by electric dispersion (Gopal, 1968), but probably the best way to make highly monodisperse emulsions is by fractional crystallization (Bibette *et al.*, 1990) or fractional centrifugation of existing polydisperse emulsions. In practice, the main requirement is not one of closeness to monodispersity *per se*, but the avoidance of a significant tail in the droplet-size distribution function, as it is the presence of particularly large droplets in the fresh emulsion, and the growth in the number of such droplets on storage, which generally has the predominant influence on the perceived instability. Apart from the emulsification equipment, the main factor under the control of the formulation technologist is the nature of the emulsifier. Ideally, the emulsifier should be insoluble in the dispersed phase, and the dispersed phase should be insoluble in the dispersion medium (including the dissolved emulsifier). Water-soluble polymers, as long as they are surface active, are especially suited to the making and stabilizing of fine oil droplets. Once formed, the state of aggregation of such droplets depends on the interaction between polymer layers on different droplets. Control of emulsion droplet flocculation is usually achieved by optimizing thermodynamic conditions (pH, ionic strength, temperature etc.) to avoid polymer aggregation or phase separation. Little can be done, however, to avoid flocculation of coarse emulsion droplets ($\geqslant 10\,\mu\text{m}$) – yet another reason for making the droplets small in the first place.

References

Ambruster, H. (1990) Ph.D. thesis, University of Karlsruhe

Becher, D. Z. (1985) Applications in agriculture. In *Encyclopedia of Emulsion Technology*, Vol. 2 (ed. Becher, P.) Marcel Dekker, New York, 239–320

Becher, P. (1965) *Emulsions: Theory and Practice*, 2nd edn, Reinhold, New York

Bibette, J., Roux, D. and Nallet, F. (1990) Depletion interactions and fluid–solid equilibrium in emulsions. *Phys. Rev. Lett.*, **65**, 2470–2473

Burgaud, I. and Dickinson, E. (1990) Emulsifying effects of food macromolecules in presence of ethanol. *J. Food Sci.*, **55**, 875–876

Burgaud, I., Dickinson, E. and Nelson, P. V. (1990) An improved high-pressure homogenizer for making fine emulsions on a small scale. *Int. J. Food Sci. Technol.*, **25**, 39–46

Castle, J., Dickinson, E., Murray, A. and Stainsby, G. (1988) In *Gums and Stabilisers for the Food Industry*, Vol. 4 (ed. Phillips, G. O., Wedlock, D. J. and Williams, P. A.) IRL Press, Oxford, 473–482

Courthaudon, J.-L., Dickinson, E. and Dalgleish, D. G. (1991) Competitive adsorption of β-casein and nonionic surfactants in oil-in-water emulsions. *J. Colloid Interface Sci.*, **145**, 390–395

Davies, J. T. (1972) *Turbulence Phenomena*, Academic Press, New York

Davies, J. T. (1985) Drop sizes of emulsions related to turbulent energy dissipation rates. *Chem. Eng. Sci.*, **40**, 839–842

Dickinson, E. and Galazka, V. B. (1991) Bridging flocculation induced by competitive adsorption: implications for emulsion stability. *J. Chem. Soc., Faraday Trans.*, **87**, 963–969

Dickinson, E. and Stainsby, G. (1982) *Colloids in Food*, Applied Science, London, Chap. 4

Dickinson, E., Murray, A., Murray, B. and Stainsby, G. (1987) Properties of adsorbed layers in emulsions containing a mixture of caseinate and gelatin. In *Food Emulsions and Foams* (ed. Dickinson, E.) Royal Society of Chemistry, London, 86–99

Evans, D. F., Mitchell, D. J. and Ninham, B. W. (1986) Oil, water and surfactant: properties and conjectured structure of simple microemulsions. *J. Phys. Chem.*, **90**, 2817–2825

Farinato, R. S. and Rowell, R. L. (1983) Optical properties of emulsions. In *Encyclopedia of Emulsion Technology*, Vol. 1 (ed. Becher, P.) Marcel Dekker, New York, 439–479

Gopal, E. S. R. (1968) Principles of emulsion formation. In *Emulsion Science* (ed. Sherman, P.) Academic Press, London, 1–75

Hinze, J. O. (1955) Fundamentals of the hydrodynamic mechanism of splitting in dispersion processes. *AIChE J.*, **1**, 289–295

Kerker, K. (1969) *The Scattering of Light and Other Electromagnetic Radiation*, Academic Press, New York

Kitchener, J. A. and Mussellwhite, P. R. (1968) The theory of stability of emulsions. In *Emulsion Science* (ed. Sherman, P.) Academic Press, London, 77–130

Kolmogorov, A. N. (1941) The local structure of turbulence in incompressible viscous fluid for very large Reynolds numbers. *Dokl. Akad. Nauk SSSR*, **30/32**, 16 (translations of these papers have recently been published in *Proc. R. Soc. (London)*, **434**, 9–18)

Korstvedt, H., Bates, R., King, J. and Siciliano, A. (1984) Microfluidization. *Drug Cosmet. Ind.*, **135**, 36–37, 40 (US Patent 4533254)

Levich, V. G. (1962) *Physicochemical Hydrodynamics*, Prentice Hall, Englewood Cliffs, NJ

Lucassen, J. (1981) Dynamic properties of free liquid films and foams. In *Anionic Surfactants: Physical Chemistry of Surfactant Action* (ed. Lucassen-Reynders, E. H.), Marcel Dekker, New York, 217–266

Moilliet, J. L. (1976) Emulsions in retrospect and prospect. In *Theory and Practice of Emulsion Technology* (ed. Smith, A. L.), Academic Press, London, 1–11

Mulder, H. and Walstra, P. (1974) *The Milk Fat Globule: Emulsion Science as Applied to Milk Products and Comparable Foods*, Pudoc, Wageningen

Nakashima, T., Shimizu, M. and Kukizaki, M. (1991) *Membrane Emulsification: Operation Manual*, Industrial Research Institute of Miyazaki Prefecture, Tsunehisa, Japan

Parkinson, C., Matsumoto, S. and Sherman, P. (1970) The influence of particle-size distribution on the apparent viscosity of non-Newtonian dispersed systems. *J. Colloid Interface Sci.*, **33**, 150–160

Pertsov, A. V., Kabal'nov, A. S. and Shchukin, E. D. (1984) Recondensation of the particles of a two-component disperse phase in the case of a large difference between the solubilities of compounds in a dispersion medium. *Kolloidn. Zh.*, **46**, 1172–1176

Phipps, L. W. (1975) The fragmentation of oil drops in emulsions by a high-pressure homogenizer *J. Phys. D*, **8**, 448–462

Phipps, L. W. (1985) *The High Pressure Dairy Homogenizer* (*Technical Bulletin 6*) National Institute for Research in Dairying, Reading, UK

Shinnar, R. (1961) On the behaviour of liquid dispersions in mixing vessels. *J. Fluid Mech.*, **10**, 259–275

Tadros, Th. F. and Vincent, B. (1983) Emulsion stability. In *Encyclopedia of Emulsion Technology*, Vol. 1 (ed. Becher, P.), Marcel Dekker, New York, 129–285

Taylor, G. I. (1934) The formation of emulsions in definable fields of flow. *Proc. R. Soc., London, Ser. A*, **146**, 501–523

van den Tempel, M. (1960) In *Proceedings of 3rd International Congress on Surface Activity* (Cologne) Vol. 2, 573

van den Tempel, M., Lucassen, J. and Lucassen-Reynders, E. H. (1965) Application of surface thermodynamics to Gibbs elasticity. *J. Phys. Chem.*, **69**, 1798–1804

Walstra, P. (1974) Influences of rheological properties of both phases on droplet size of oil-in-water emulsions obtained by homogenization and similar processes. *Dechema Mongr.*, **77**, 87–94

Walstra, P. (1975) Effect of homogenization on the fat globule size distribution in milk. *Neth. Milk Dairy J.*, **29**, 279–294

Walstra, P. (1983) Formation of emulsions. In *Encyclopedia of Emulsion Technology*, Vol. 1 (ed. Becher, P.), Marcel Dekker, New York, 57–127

Walstra, P. (1993) Principles of emulsion formation. *Chem. Eng. Sci.*, **48**, 333–349

Chapter 8

Particle size of emulsion polymers

K. R. Geddes

8.1 Introduction

Emulsion polymers are important articles of commerce. Those based on acrylic monomers alone enjoyed a world market of about 680 000 tonnes per year in 1990, while products containing vinyl acetate certainly now exceed 1 000 000 tonnes for northern Europe and the USA. In general, the term 'emulsion polymer' is understood to refer to dispersions of thermoplastic addition polymers in water. Polyvinyl acetate adhesives are included, but the wide range of synthetic rubber latices are rarely described in this way, nor products in processes where the emulsion polymer is an intermediate and is recovered as a solid as an integral part of the process. Thus emulsion polymers are sold and used in their dispersed form, where particle-size distribution has a profound effect on their viscosity, rheology and application properties.

Most applications lead to the isolation of the polymer by the removal of water. This brings into play the properties of the polymer, usually to act as a binder between particles of pigment; or as an adhesive for wood, paper or other surfaces; or perhaps as a decorative and protective film on a surface. The state of division of the polymer has a strong influence on its properties, even in the dry state. The key stage in many applications is the transition between wet, dispersed polymer and dry integrated film. Important properties include the ability to flow in the wet state and to coalesce smoothly without flocculation as the water evaporates. There is also a need to integrate further by deformation, flow and interparticle diffusion after the film has dried. This requires the polymer to have a glass transition temperature below the temperature of drying or to be plasticized by suitable solvent.

8.2 Chemistry of addition polymerization

8.2.1 Monomers

Suitable monomers for the process have at least one permanently activated double bond, or a bond capable of becoming active on the approach of a free radical. Major monomers of commerce include vinyl acetate, styrene, the lower alkyl esters of acrylic and methacrylic acids, and the vinyl ester of a mixture of tertiary aliphatic carboxylic acids of average formula $C_{12}H_{22}O_2$, known commercially as Veova 10 (Shell). Other monomers used in bulk include vinyl chloride and ethylene, but higher members of the olefin series are slow to react.

There is also a range of speciality difunctional and multifunctional acrylics and methacrylates which are mainly used for increasing molecular weight or introducing branched structures. Acid, hydroxy, amine, sulphonate and other groups may be added by the use of appropriate monomers. Of recent introduction are monomers for thickeners which develop interchain structures in solution.

Many monomers may be considered to be derivatives of ethylene.

$$\begin{array}{c}\text{H} \\ \\ \text{H}\end{array}\!\!\!\diagup\!\!\!\begin{array}{c}\end{array}\!\!\!\text{C}=\text{C}\!\!\!\begin{array}{c}\diagup\text{R}_1 \\ \\ \diagdown\text{R}_2\end{array}$$

Often, R_1 is simply hydrogen. R_2 may vary as follows:

Monomer	R_2
Vinyl acetate	$-\text{O}-\underset{\underset{\text{O}}{\|\|}}{\text{C}}-CH_3$
Vinyl chloride	$-Cl$
Sodium vinyl sulphonate	$-SO_3^-\ Na^+$
Styrene	$-C_6H_5$
Ethyl acrylate	$-\underset{\underset{\text{O}}{\|\|}}{\text{C}}-O-C_2H_5$

Methacrylate monomers are derivatives of methacrylic acid in which R_1 is a methyl group.

To illustrate the range of properties of common monomers, a selection of those sold commercially is given in Table 8.1.

8.2.2 Polymerization

The polymerization is an addition reaction:

$$\left(\begin{array}{c}CH_2=CH \\ \| \\ R\end{array}\right)_n \rightarrow -CH_2-\underset{R}{\underset{\|}{CH}}-CH_2-\underset{R}{\underset{\|}{CH}}-CH_2-\underset{R}{\underset{\|}{CH}}-$$

If a mixture of monomers is used, R may have two or more values. It can be seen that in this case the structure of the polymer can take many forms ranging from a random distribution of substituents, alternating monomer units or blocks of one type of polymer followed by blocks of another. For a single monomer the polymerized product is known as a 'homopolymer', for more than one monomer the product is known as a 'copolymer'.

Table 8.1 *The properties of some common monomers used in bulk*

Monomer	Molecular formula	Physical form	Melting point (°C)	Boiling point (°C)	Solubility in water (20°C) (%)	Special health problems	Special flammability
Acrylamide	$CH_2{=}CH{\cdot}CO{\cdot}NH_2$	Solid	–	–	100	Severe	No
Butyl acrylate	$CH_2{=}CH{\cdot}COOC_4H_9$	Liquid	–	149	0.2	Yes	Intermediate Flashpoint 49°C
Ethylene	$CH_2{=}CH_2$	Gas	–	-104	Low	No	Yes
Methacrylic acid	$CH_2{=}C{\cdot}CH_3{\cdot}COOH$	Liquid	14	160	100	Yes	No
Styrene	$CH_2{=}CH{\cdot}C_6H_5$	Liquid	-31	145	0.02	Yes	Intermediate
Veova 10 (Shell)	$CH_2{=}CH{\cdot}O{\cdot}COC_9H_{19}$	Liquid	-20	135	<0.1	No	No
Vinyl acetate	$CH_2{=}CH{\cdot}O{\cdot}COCH_3$	Liquid	-100	72.5	2.94	Yes	Yes. Flashpoint -8°C
Vinyl chloride	$CH_2{=}CH{\cdot}Cl$	Gas	-154	-14	0.11	Severe	Yes

8.2.3 Initiation and termination of the reaction

The reaction is a free radical process:

$$R\bullet + CH_2 = CHR_1 \rightarrow R - CH_2 - \overset{\displaystyle |}{\underset{\displaystyle R_1}{CH}}{}^{\bullet}$$

free radical   free radical

$$R - CH_2 - \overset{\displaystyle |}{\underset{\displaystyle R_1}{CH}}{}^{\bullet} + CH_2 = CHR_1 \rightarrow R - CH_2 - \overset{\displaystyle |}{\underset{\displaystyle R_1}{C}} - CH_2 - \overset{\displaystyle |}{\underset{\displaystyle R_1}{CH}}{}^{\bullet}$$

free radical   free radical

Free radical generation may be by heat (Hayashi, 1982), radiation or by shear. A more convenient way is to use thermally unstable compounds such as azo compounds, peroxides, persulphates, hydroperoxides or peroxydicarbonates. In emulsion polymerization, the mechanism requires water solubility which severely limits the choice of initiator. On heating, the persulphates (peroxodisulphates) break up as follows:

$$O=\overset{\displaystyle O}{\underset{\displaystyle O^-}{\overset{\displaystyle \|}{S}}}-O-O-\overset{\displaystyle O}{\underset{\displaystyle O^-}{\overset{\displaystyle \|}{S}}}=O \rightarrow 2O=\overset{\displaystyle O}{\underset{\displaystyle O^-}{\overset{\displaystyle \|}{S}}}-O^{\bullet}$$

The fragments can add to a monomer molecule:

$$O=\overset{\displaystyle O}{\underset{\displaystyle O^-}{\overset{\displaystyle |}{S}}}-O^{\bullet} + CH_2 = \overset{\displaystyle }{\underset{\displaystyle R_1}{CH}} \rightarrow O=\overset{\displaystyle O}{\underset{\displaystyle O^-}{\overset{\displaystyle |}{S}}}-O-CH_2-\overset{\displaystyle }{\underset{\displaystyle R_1}{CH}}{}^{\bullet}$$

Further monomer may add unless the chain is terminated by collision with a second free radical, resulting in addition and mutual elimination. Thorough agitation is required to separate the radical pairs and any growing chains, especially with low solubility monomers such as styrene. Alternatively, chain transfer may occur with any molecule containing a labile hydrogen group, e.g. 1-dodecanethiol ($C_{12}H_{25}SH$).

The $C_{12}H_{25}S^{\bullet}$ radical produced by hydrogen extraction is available for starting a new chain. The molecular weight is reduced but the number of growing chains is unchanged. Chain transfer with low-molecular-weight materials, including monomer, can result in the termination of the chain within the particle and radical desorption. Initiation of new particles by radicals derived from existing ones must be taken into account in any comprehensive theory.

Ammonium and sodium persulphates are very soluble in water, but the potassium salt less so. Comparisons given in Interox trade literature between the salts is given in Table 8.2.

Table 8.2

Persulphate Salt	Solubility (g l^{-1})		Decomposition half-life at 60°C (h)	
	20°C	40°C	5% solution	10% solution
Ammonium	620	700	–	30
Potassium	50	105	35	–
Sodium	545	605	–	30

It can be seen that there is little evolution of free radicals in aqueous solution below 60°C. Above 70°C rapid decomposition and rapid polymerization of monomers takes place. Alternatively, the activity of the persulphates, and t-butyl hydroperoxide, which has a half-life of 10 h at 170°C, may be enhanced by the use of reducing agents and a transition metal such as iron. The reaction under these circumstances may be illustrated with the very simple reaction of hydrogen peroxide and sulphur dioxide in the presence of iron:

$$Fe^{2+} + H_2O_2 \rightarrow Fe^{3+} + OH^- + OH^{\bullet}$$
$$2Fe^{3+} + SO_2 + 2H_2O \rightarrow 2Fe^{2+} + SO_4^{2-} + 4H^+$$

The energy of activation of a redox reaction is about 10 kcal mol^{-1} against 35 kcal mol^{-1} for a thermal decomposition of the type $S_2O_8^{2-} \rightarrow 2SO_4^{-\bullet}$. This leads to a faster evolution of radicals and a more rapid rate of reaction, especially at lower temperatures. At 70°C, 0.1 M persulphate solutions yield 3×10^{-4} mol l^{-1} of $SO_4^{-\bullet}$ per minute, while 1×10^{-4} M persulphate/l $\times 10^{-4}$ M Fe^{2+} solutions produce radicals at six times this rate, even at 10°C (Fordham and Williams, 1951; Kolthoff and Miller, 1951). A study of the kinetics of one particular system has been published (Badran *et al.*, 1990).

8.2.4 Copolymerization

Each monomer has its individual physical and chemical properties and the polymer derived from it is equally unique. By choosing a blend of monomers, the properties of the final copolymer may be determined. Particular technical requirements such as glass transition temperature, chemical resistance, polarity and durability influence this choice. Cost may be important, consistent with achieving acceptable technical standards. Modification of properties by the use of minor amounts of a speciality monomer may be an attractive alternative to more fundamental changes.

Most monomers polymerize together, although the rate at which they do so depends on both electronic and steric factors (Alfrey and Price, 1947; Ham, 1964; Young, 1975; Laurier *et al.*, 1985). There is a difference in the readiness of a monomer molecule to add to a free radical derived from one of its own kind to when the last unit of the chain is of a different type. These rates are defined as K_{11} and K_{12}, respectively. The ratio of the rates of homopolymerization to copolymerization k_{11}/k_{12} is defined as r_1 and for the second monomer $k_{22}/k_{21} = r_2$. The actual amount of monomer of each kind incorporated at any instant $(dm_1)/(dm_2)$ is also influenced by the relative concentration of the monomers $[M_1]$ and $[M_2]$. The relationship was first published by Mayo (1944):

$$\frac{dm_1}{dm_2} = \frac{[M_1](r_1[M_1] + [M_2])}{[M_2](r_2[M_2] + [M_1])}$$

Values of r_1 and r_2 have been published (Alfrey and Price, 1947; Kahn and Horowitz, 1961; Mark *et al.*, 1975.

Some monomers are difficult to copolymerize and traces of styrene inhibit vinyl acetate for electronic reasons. Conversely, some maleate esters will not homopolymerize because their bulk does not allow close enough approach of two molecules of the same kind. A particularly suitable combination is a mixture of vinyl acetate and Veova 10 (Shell) (a vinyl ester of a C_{10} mixed acid containing a tertiary carbon atom). Veova 10 brings flexibility and durability to the harder but lower cost vinyl acetate polymer. Other frequently used combinations are styrene and butyl acrylate and methyl methacrylate with ethyl acrylate. In these cases the styrene and methyl methacrylate give hard homopolymers with glass transition temperatures of 100°C and 105°C, respectively; while the homo-polymers of the acrylates are tacky, having glass transitions of –54°C and –22°C. To give an intermediate glass transition temperature of 15°C, it is necessary to use 44% butyl acrylate with styrene.

The transition from rubber to glass (glass transition temperature, T_g) is not sharp because the polymers are amorphous. In the case of copolymers, the randomness of the polymer units has an effect, but structural orientation (tacticity) is not a significant factor. T_g is often measured in dried films by differential scanning calorimetry, microindentometry, refractive index or specific volume changes. Apart from its obvious influence on dried film properties, it is the biggest factor in determining the minimum temperature at which an integrated film can be formed by drying. The T_g of a copolymer can be calculated approximately from the proportions of each monomer. Varying the monomer will also affect solubility and polarity, influencing both reaction mechanism and particle size.

8.2.5 Stabilizers for emulsion polymerization

Emulsion polymer particles are potentially unstable. They have a density 5–20% greater than the water in which they are dispersed, and for most applications are above their glass transition temperatures at ambient temperature. Their natural tendency is to settle and, in the absence of adequate stabilization, to coalesce.

The simplest form of stabilization is the incorporation of sulphate or sulphonate group on the surface of the particles. The use of peroxodisulphates incorporates a sulphate group as the first group of any chain. As this originates in the water phase, this group is well placed to remain at the interface of the polymer and water.

Additional sulphate may be introduced by the use of the speciality monomers 2-sulphoethyl methacrylate (Mills and Yocum, 1967) sodium vinyl sulphate (Hoechst, 1959) or sodium methallyl sulphonate (Geddes and Davis, 1974). Hydrogen peroxide, t-butyl hydroperoxide and most redox systems, however, give little or no sulphate substitution. An alternative method of giving stability is to use a surface active agent. Anionic types are generally favoured as reaction between oxidizing initiators and certain cationic groupings may result in poor reactions and the development of unwanted colours. Surface active agents are molecules with properties of oil and water solubility in different parts of the molecule. This behaviour is described as 'amphipathic'. Oil solubility is conferred by the presence of aliphatic or aromatic groupings, or a combination. Water solubility occurs through the presence of an ionizable group such as carboxyl, sulphate or phosphate; a quarternarized amine or similar nitrogen-containing grouping; or by the presence of a water soluble organic chain such as a polyethylene oxide condensate. The carboxyls, sulphates and phosphates give anionic properties; the quarternarized nitrogen groups are cationics, while the polyethylene oxide condensates are nonionic in their properties. Should both anionic and cationic groups be present, the predominant nature of the molecule may vary with pH. This type of surface activity is called 'amphoteric'.

The simplest form of anionic stabilizers are the soaps, e.g. $C_nH_{2n+1}COO^-Na^+$. These are available in a wide range of varieties derived from natural fats and synthetic fatty acids. They are of interest in the production of synthetic rubber because of their low cost and weak stabilizing action, allowing the rubber to be recovered as a solid by precipitation. In the general formula, n values from 1–9 are too water soluble and $n > 22$ is too oily to form micelles, a yardstick of surfactancy. Stronger stabilizing effects are needed for products which are stored, blended and processed in the wet form. One of the simplest of these is sodium dodecyl sulphate, $C_{12}H_{25}SO_4^-Na^+$. The most widely available non-ionic surfactants are the alkyl phenol ethoxylates, where the alkyl group is usually octyl or nonyl. Again, there are limits to the chain length. If n (the number of ethylene oxide units) is less than about 6, the value as a stabilizer diminishes. The low-chain-length variants are of more use as antifoams and hydrotropes. Values of n up to about 100 are available commercially, but higher members are more difficult to manufacture. At longer chain lengths the influence of the oil soluble alkyl phenol group diminishes.

On the addition of a surfactant to water, a small amount of true solubility first takes place. Additional material then concentrates in the surface, positioned with the water-soluble ends of the molecules in the water. At a certain concentration the surface becomes sufficiently saturated to force new material beneath the surface into clusters called 'micelles'. Micelles are usually spherical in form, although laminar and other shapes may be produced under some circumstances. Micelles consist of 50–100 molecules with simple compounds, but may have as few as 8–10 molecules with bulky, complex materials. The size of micelles is determined by a combination of steric factors and mutual repulsion of charged end-groups. These micelles are stable as the anionic or cationic groups give electrostatic stabilization by mutual repulsion. Nonionics stabilize by attracting loosely associated sheaths of water molecules.

Once micelles start to form, only a small decrease in surface tension occurs as very little of the added surfactant can pack into the surface. Graphs of surface tension against concentration reveal two distinct lines of different slope, the intersection of which marks the point at which micelles form. This concentration is called the 'critical micelle concentration' (CMC). A typical surface tension against concentration graph is given in Figure 8.1. Some interesting comparisons between the CMCs of octyl phenol and nonyl phenol ethoxylates of varying polyethylene oxide chain length are illustrated graphically in Figure 8.2, along with the surface tension at CMC. The influence of ethylene oxide chain length on the properties of ammonium salts of sulphated alkyl phenol ethoxylates is similarly shown in Figure 8.3 which was taken from GAF data. A linear relation of CMC against chain length is indicated for the simple nonionic ethoxylates, but there are insufficient points to be sure of the relationship with the sulphated versions. In each case the 'surface tension at the CMC' with 'moles of ethylene oxide' graph is best fitted by a line of the form $y = AX^B$. Lists of CMCs have been prepared (e.g. Gerrens, 1965) and most commercial manufacturers publish data for their own products.

Introduction of non-polar liquids such as monomers into a mixture of water and surfactant will result in the monomer being dispersed and migrating into the non-polar environment at the centres of micelles. As the proportion of monomer rises there may be insufficient surfactant to cover fully the surface of the drops

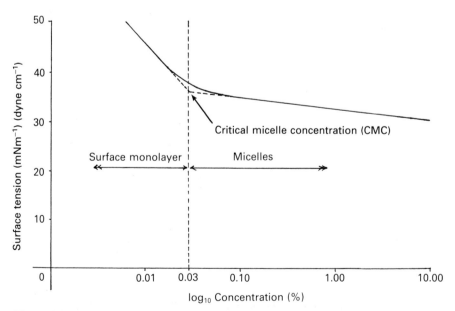

Figure 8.1   *Typical surface tension versus concentration graph*

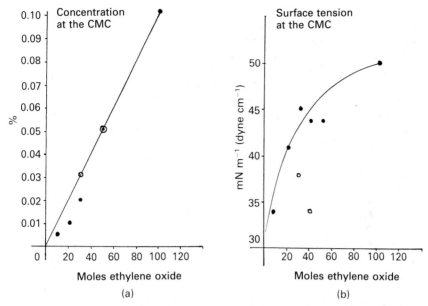

Figure 8.2   *Critical micelle concentration (CMC) (a) and the surface tension at the CMC (b) for nonionic ethoxylates: (○) octyl phenols; (●) nonyl phenols*

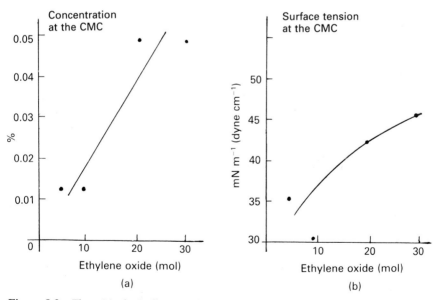

Figure 8.3   *The critical micelle concentration (CMC) (a) and the surface tension at the CMC (b) for ammonium salts of sulphated alkyl phenols*

of monomer which may then be only partially stabilized. The use of a mixture of surfactants of different molecular shapes may result in closer surface packing in micelles and give greater stability. Similarly, a mixture of ionic and non-ionic surfactants can reduce the mutual repulsion of ionic groups and lead to higher surface densities and more stability (Wang and Chu, 1990). Conversely, a mixture of anionic and cationic types can destroy each other's effectiveness.

An entirely separate class of stabilizers for emulsion polymerization is the protective colloids. These materials are water soluble polymers. They do not form micelles. A wholly synthetic example of the class is polyvinyl alcohol which is used extensively both as an adhesive in its own right and as a stabilizer for polyvinyl acetate adhesives. It is manufactured by the acid or alkaline hydrolysis or methanolysis of polyvinyl acetate. It is available in a range of molecular weights and in a range of degrees of hydrolysis. Higher molecular weights give greater solution viscosities, while higher degrees of hydrolysis (say 0–1.5% of residual acetate groups) are only dissolved by hot water, dried films giving a reasonable degree of water resistance. Once made, solutions are stable for several days. Higher levels of residual acetate have greater surface activity, especially where the polyvinyl acetate and polyvinyl alcohol units are arranged in blocks rather than randomly. Recently, modified polyvinyl alcohols have been claimed to give better stability against sedimentation (Baines *et al.*, 1990). Other synthetic colloids used are polyacrylic acid and polymethacrylic acid and their copolymers, and polyvinyl pyrrolidone.

Natural colloids and their modifications are also used. Exothylated starch and Gum Acacia had extensive use at one time. Dextrine is a key part of remoistenable envelope flap adhesives. Hydroxyethyl cellulose is also much used, especially for emulsion paint binders. A wide range of viscosities through different molecular weights can be obtained. Many colloid-stabilized emulsion polymers use an anionic or non-ionic surfactant (or both) in conjunction with the colloid.

Hence a range of stabilization mechanisms apply. Surface active agents are largely adsorbed onto the particle surfaces, although the non-polar chains in the molecules may diffuse into the polymer, giving firm attachment. Colloids appear also to adsorb onto polymer surfaces and may be grafted to the surface (Dong *et al.*, 1989; Geddes, 1990). Some monomer systems (notably styrene) may graft too readily, giving severe instability at the preparation stage, probably because of colloid chains bridging particles.

8.2.6 Process outline

The simplest form of the process for the manufacture of emulsion polymers is the true batch process. Water, monomer, initiator and a stabilizer are mixed together and heated under agitation until the initiator produces free radicals to start the polymerization. The stabilizer aids the dispersion of the monomer and prevents coagulation of the polymer particles.

The polymerization process is exothermic. Monomers liberate heat of polymerization of up to 22 kcal mol^{-1}. Once the reaction is underway, the problem is one of cooling. The need to avoid a runaway exotherm leading to uncontrollable boiling and consequent foaming or destabilization has led to the 'delayed addition' process, and to continuous processes.

The delayed addition process uses a stirred batch kettle (Figure 8.4) fitted with a reflux condensor, thermometer and a heating and cooling jacket. Most of the water, initiator and stabilizers are added, together with 0–10% of the monomer. On heating to a temperature which causes a breakdown in the initiator, the monomer polymerizes. The heat output assists the rise in temperature to the optimum for the process (say 75–85°C for thermally initiated preparations), at which point the balance of the monomer starts to be added over a period of 3–5 h. The rate at which monomer can be introduced depends on the rate of polymerization and the cooling available. The heat transfer not only depends on the geometry of the reactor, but also the viscosity and rheology of the emulsion polymer, the rate of agitation and the cleanliness of the cooling surfaces. Ultimately, the polymer content rises to 45–65% by weight and the conversion to better than 99%.

The continuous stirred tank reactor (CSTR) process uses one or more kettles similar to those used in the batch process, but with product overflowing to balance the feed of raw materials which must always include water to stabilize the polymer content.

The Loop reactor continuous process (Gulf Oil Canada, 1971; Geddes, 1983, 1989, 1993; Adams, 1985, 1990; Geddes and Khan, 1991; Khan, 1991) (Figure 8.5) approaches the removal of heat in a different way. Reaction mixture is recirculated by a pump through pipework fitted with cooling jackets or sprayed with variable water jets. Additional water phase and monomer is pumped into the circulating mixture and an equal volume of reaction mixture overflows through a pressure sustaining valve. The ratio of water phase to monomer and the total input volume can be controlled, as can the rate of recirculation. It is usual to employ a redox initiator system, one component dissolved in the monomer and the other in the water phase. The two streams are kept separate until they are mixed within the reactor.

In a batch kettle (whether or not a delayed addition process is used), the final volume is the working volume of the reactor. In commercial installations this may vary between about 2000 and 100 000 kg. The productivity of such plant varies. A production rate of 750–1000 kg h^{-1} is not unreasonable, but this can rise with upstream pre-mix and downstream cooling tanks; also with the use of external heat exchangers. In contrast, the Loop reactor uses a reaction volume of only 50–100 l. Depending on monomer type and sufficient chilled cooling water, production rates of 400–1000 kg h^{-1} are possible.

The temperature control of the Loop process is particularly simple. The flow of cooling water is varied by a standard temperature recorder/controller which gives an air pressure signal to a water valve. It is important that changes of cooling in the batch process act rapidly but with sensitivity as overcooling may retard the reaction, leading to an uncontrollable exotherm as the built-up monomer suddenly polymerizes. Conversely, overheating can deplete the initiator too rapidly, leading to loss of reaction later in the process or even destabilization of the polymer particles. Refluxing in a condensor is a very

Figure 8.4   *A batch process reaction kettle*

Figure 8.5 *A pilot scale Loop reactor*

effective way of cooling a batch process, but the amount of reflux varies greatly with availability of unpolymerized monomer, temperature and agitation. This may make the reaction temperature very difficult to stabilize. The Loop process, with such a small reaction volume and no reflux, is easier to control than a batch reactor in this respect.

8.3 Reaction mechanism and its relationship to particle size

8.3.1 Initiation theories

The theory and kinetics of emulsion polymerization are heavily dependent on the size and number of particles present. The composition of the growing particles is pure monomer and polymer, i.e. bulk polymerization. In the early stages when the particles are small and monomer is readily available, the polymerization is very rapid. Molecular weight will be very high as the reaction is free from the effects of chain transfer from solvents. In practice, the polymerization will terminate due to a mutual chain elimination, chain transfer or by a dis-proportionation reaction:

$$R - CH_2 - CH_2^{\bullet} + R_1 - CH_2 - CH_2^{\bullet} \rightarrow R - CH_2 - CH_3 + R_1CH = CH_2$$

Recent work (Giannetti, 1990) has shown that a mathematical model can be made to predict correctly the change in particle-size distribution with growth of styrene which excludes any contribution from bimolecular termination.

It is necessary in practice to provide a steady supply of free radicals to sustain the reaction. If the radical enters a 'dead' particle, polymerization will recommence with a new chain. This constant supply of radicals has the effect of switching the polymerization on and off. The average number of growing chains per particle can, it is argued, therefore approximate to 0.5.

Although not the first to express ideas in the field, the first comprehensive theory was published in 1948 by Smith and Ewart. Three cases were defined:

(1) where the average number of growing chains per polymer particle is much less than one
(2) where the average number of growing chains per polymer particle approximates to half
(3) where the average number of growing chains per polymer particle is large.

Case 1 can be seen to be applicable to systems starved of initiator. Case 2 is the situation with very small particles and an adequate supply of free radicals. Case 3 applies to large particles and beads where several chains can be growing simultaneously without mutual elimination.

Smith and Ewart, working with styrene, produced a rate of polymerization (R_p) proportional to the particle number (N) and also to the 0.6 power of the concentration of emulsifiers $[S]$ and the 0.4 power of the rate of initiation (R_i), i.e.

$$R_p \propto N \propto [S]^{0.6}[R_i]^{0.4}$$

This equation works well with styrene at low conversions and emphasizes that increasing surfactant levels increase the particle number, and hence the average particle size decreases. Greater initiator availability also decreases particle size by initiating more particles.

The exact mechanism of initiation of particles and the subsequent entry of new free radicals has been the subject of much study and debate. Original ideas were that the radicals formed by the breakdown of initiators simply found their way to micelles or particles. There would seem however to be very little driving force to make negatively charged inorganic radicals pass through a negatively charged surface of a micelle or particle into a predominantly organic and non-polar area of monomer or polymer. Polymerization in monomer droplets to form bead polymers can be achieved by the use of uncharged organic initiators but the rate of polymerization is fairly slow. Only by the introduction of special stabilizers to create ultrafine monomer droplets in the 'mini-emulsion' process can this mechanism compete with conventional emulsion polymerization. Here the monomer droplets and micelles act as reservoirs rather than as sites for polymerization. These ideas were discussed further (Casey *et al.*, 1990) on the basis of particle-size distribution at the early stages of the reaction. Dunn and Al-Shahib (1978) showed that the concentration of micellar emulsifier (rather than the total concentration) controlled the kinetics, but micelles were not essential for the formation of particles.

The alternative to 'micellar entry' is 'homogeneous nucleation', which has been strongly championed by Fitch and coworkers (Fitch and Tsai, 1971; Fitch, 1981a, b; Fitch *et al.*, 1985). The aqueous phase initiator decomposes to form soluble species which then add to solubilized monomer. At some stage, the diffusion-controlled solution polymerization produces a product of sufficient molecular weight to precipitate. This can be as high as about 50 units for the comparatively soluble vinyl acetate, but as few as 2–5 units for styrene. At the point of precipitation, the chains may migrate to particles with subsequent starting or termination of growing chains, or form primary particles (Alexander and Napper, 1970; Hansen and Ugelstad, 1977, 1982). The homogeneous nucleation theory has also been elaborated to postulate a two-stage process where the initial particles or precipitated chains (called 'precursor' particles) undergo coagulation with each other before they form true primary latex particles (Lichti *et al.*, 1977; Feeney *et al.*, 1984).

Homogeneous nucleation is the only mechanism that can take place in emulsifier-free systems, especially for more water soluble monomers. Interestingly, the two theories may both be correct, referring to different circumstances. A transition from homogeneous to micellar nucleation was found above a critical level of emulsifier (Song and Poehlein, 1988, 1989). In the Loop process formation of new particles was found only when surface area decreased to make free surfactant available (Geddes, 1989). Because of the sharpness of the appearance of new particles, it is tempting to assume a fundamental change in conditions such as the appearance of micelles, but no direct evidence for this has been presented.

The micellar nucleation mechanism still has adherents in the case of very insoluble monomers such as styrene, but as these monomers add so few units prior to precipitation from the water phase, there may be a unification between the two theories. Of significance is the observation that *p*-t-butyl styrene would not polymerize at all in the absence of emulsifier (Satpathy and Dunn, 1988).

The mainstay of the micellar nucleation theory according to Casey *et al.* (1990) has been the rapid change in particle number as the surfactant concentration is varied through the CMC. Nucleation in micelles cannot take place, by definition, below the CMC, but the main significance for other theories is that above this point, surfactant is readily available as a saturated solution, replenished by the micelles. In the presence of existing particles, the precipitating growing chains of the homogeneous theory have a choice of migrating to a particle or associating with the stabilizer dissolved in the water phase, thus forming a new particle. It is not inconsistent that such a one-chain precursor particle (of size similar to some complex surfactants and smaller than most colloids) would coalesce with others of a similar size to give the particles observed (Schmutzler, 1982; Zukaski and Saville, 1985). This collapsed mass of precursor particles would be large enough for the conventional view of stabilization of growing particles by adsorbed surfactant and colloid to apply. On a qualitative level there is some explanation of the observation that monomers such as styrene give finer particle size products compared to vinyl acetate. With vinyl acetate, soluble chains will grow to several times the molecular size of styrene before precipitation thus giving fewer, larger precursors. Also the precursor particles would have a greater tendency for self-stabilization and growth before collapse.

8.3.2 Application to the cycling of continuous reactors

What evidence is there for a two-stage particle formation process? In continuous emulsion polymerization by the continuous stirred tank reactor process (CSTR), the existence of start-up cycling in properties has been known for many years. There have been many attempts to understand the phenomenon by both practical trials and computer modelling (Rawlings *et al.*, 1986; Penlidis *et al.*, 1986; Dougherty, 1986; Brandolin and Garcia Rubio, 1990). The introduction of an externally generated feed of preformed particles (Penlidis *et al.*, 1986; Poehlien *et al.*, 1986; Mead and Poehlien, 1988, 1989) has often been suggested. More recently, cycles have been reported at start-up in the 'Loop reactor' process (Geddes, 1989) and in a recycling tube reactor more akin to the CSTR than the Loop (Lee *et al.*, 1990); and even in a delayed addition batch process using a pre-emulsified feed of styrene, surfactant and water (Snuparek, 1990).

In the case of the Loop reactor, examination of particle size distributions over about 35 residence times (280 min) from the start of a vinyl acetate-Veova 10, (Shell) copolymer run, indicated a single peak growing in size until 7.5 residence times. At this time a second peak was observed and in due course three, with traces of a fourth (Figure 8.6).

Concurrent tests for non-volatiles, unreacted vinyl acetate, viscosity, surface tension and pH, indicated cycles in these properties. Particle peaks faded into a broad single distribution and all other properties stabilized by about 20 or 25 residence times, confirming the possibility of production runs of 300–500 residence times and longer, which have been demonstrated repeatedly in practice. Plotting peak particle size against residence time produced a series of good, straight lines, indicating that particle diameter growth with time was linear (Figure 8.7). A similar result using styrene was also reported by Lee *et al.* (1990) on his recycling CSTR type reactor.

Linear growth has profound implications. To take place, the addition of polymer per unit surface area must be constant, and independent of particle diameter. This implies a steady arrival of material, and the high molecular weight of the product indicates that the majority of the material arriving at the particles must be monomer. As the particle grows, more than one active site must be present and therefore more than 0.5 free radicals per particle. Electron microscopy has confirmed circular particle shadows and the presence of at least two peaks in the later particle-size distributions.

Growth lines allow the identification of groups of particles at various times. Eventually the peaks become smaller due to elution from the reactor. It must be concluded that the series of peaks indicate a discontinuous creation of particles. This gives way gradually to a continuous creation where new particles balance those lost through elution and the average particle size remains constant.

Linear growth also allows the calculation of the total relative surface area of the particles at any time. In the formulae studied, diameter growth approximated

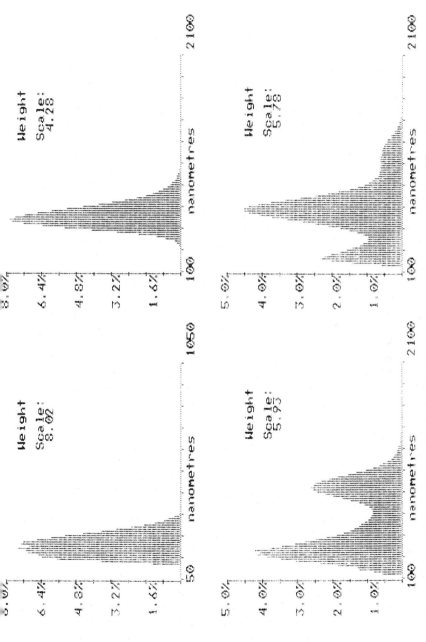

Figure 8.6 *Disc centrifuge particle-size distributions for the same Loop reactor run sampled at different times*

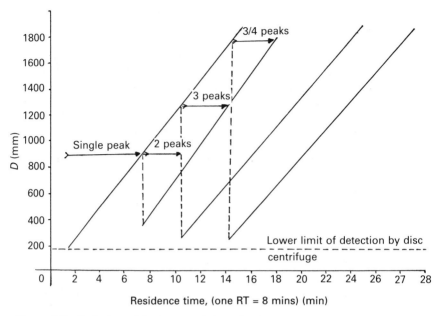

Figure 8.7 *Typical particle size growth lines for the Loop continuous reactor process*

to $15\,nm\,min^{-1}$, although there was some minor variation between different groups of particles. The interval for the creation of new crops of particles was about 31 min for three different formulae. This corresponds very well with the periodicity of other physical properties, strongly implying that such properties are secondary consequences of the particle size distribution.

These observations give no direct indication of the cause of the intermittent creation of new particles. At the start of these experiments, monomer, initiators and a water solution of stabilizers were fed at a constant rate into a Loop reactor prefilled with water phase. The initiators were active at ambient temperature and particles were formed in relation to the surfactant present. This is similar to the situation in the initial stage of a delayed addition batch process. In both processes the particle number is greater and the particle size smaller at higher levels of surfactant and this effect is used to control particle size. The available surfactant is largely scavenged by the freshly created surfaces, leaving insufficient available for later particle formation. In the batch process the monomer feed adds to existing particles, increasing them to their final size. Should surface area increase to the point where the surfactant is insufficient for stability, either steady particle coalescence occurs, leading to a constant total surface area and a skewed particle-size distribution, or total coagulation will take place.

The Loop reactor behaviour is different due to a further factor – that of particle elution. In this continuous process fresh monomer and water phase displaces the existing contents of the reactor, according to the exponential dilution law.

Surface areas were calculated against time (Geddes, 1989), using a series of growth rates from 10 to 20 nm min^{-1}, mean residence times in the reactor from 5 to 20 min and initial diameters from 25 to 500 nm. In each case the particles eluted were assumed to be of average surface area and the proportion remaining within the reactor was calculated from $f(t) = (1/\theta)e^{-t/\theta}$, where t is the particle age and θ is the reactor mean residence time. The results were of considerable relevance, showing in most cases the surface area to rise above its initial value due to growth and then to fall again because of elution. The sharpest rises were when the initial particles were small, the mean residence time was long and the growth rate was fast.

Growth rate is a factor of final polymer concentration and the number of particles, as well as the conversion. Mean residence time can be controlled, subject to satisfactory polymerization, while initial particle size can be varied by filling the reactor with various concentrations or types of surfactants or by using preformed particles. If 500 nm initial particles are assumed for the calculation at 8 min or less mean residence time, the surface area falls immediately. This makes free surfactant available for the immediate and continuous creation of new particles. This is the explanation for the self-equilibriation of the Loop process, once the average particle size grows to above some critical value. It also, by analogy, explains the value of the addition of preformed seed to a CSTR reactor. Furthermore, in the delayed addition batch process described by Snuparek *et al.* (1990), the continuously added surfactant was probably more than that required to balance exactly the increase in surface area of the particles due to growth. This excess resulted in new crops of particles giving the oscillation in the secondary properties observed. The hypothesis that precursor particles form constantly and either migrate to the particles where they turn the reaction on and off, or if surfactant is available, act as nuclei for a crop of new particles, is therefore consistent with the behaviour of all systems.

Although appearing a good explanation, some problems remain. In a continuous process the rate of generation of free radicals from the initiator must be reasonably constant. The initiation of polymer chains and hence formation of precursor particles will therefore be continuous but not necessarily constant as the monomer concentration oscillates. There are, however, no detectable discontinuities in the slope of the growth lines, even during the formation of primary particles at times of surfactant availability. Straight-line growth can be explained, however, if the rate of arrival of monomer in sufficient quantity is always sufficient even at troughs in the concentration cycles. This explanation is aided by the introduction of a 'sticking factor' concept or if a diffusion-controlled buffering mechanism regulates the transfer of monomer to the growing particles. The former does exist as monomer is in dynamic equilibrium between particles and water and will only join the particle permanently when it encounters a growing polymer chain. A steady attachment of radicals is also implied as the surface concentration in terms of radicals per unit area must remain constant to give linear diameter growth as the particle surface area increases. If the smallest particles are a condensation of precursors, the primary particles are already large enough to support several growing chains. The buffer idea is also possible as there are water, surfactant and colloid molecules associated with the particles.

These polar and viscous layers would slow the passage of monomer in either direction but whether they create a buffer supply of monomer by entrapment or semi-micelle formation is unknown. The linear growth rate has been found to be different between the first three seeding of particles, but to be constant even in the presence of each other. This implies a high degree of permanence of surface conditions.

It must be concluded however that there is always sufficient monomer available to maintain the rate of growth of the particles even at the troughs of the monomer concentration observed. As the total surface area cycles, the rate of uptake of monomer varies, hence the cycles in apparent reaction rate reported for so many systems.

8.4 Measurement of particle size

Particle size may be defined in a number of ways. The two most common are the weight average, given by many light-scattering methods:

$$\overline{D}_w = \frac{\Sigma W_i D_i}{\Sigma W_i}$$

and the number average:

$$\overline{D}_n = \frac{\Sigma N_i D_i}{\Sigma N_i}$$

where W_i is the weight of the particles and N_i is the number of particles with diameter D_i.

Number averages are given by counting micrographs or electron micrographs. A further value, the surface average, is sometimes encountered as it is given by soap titration.

Particle-size distributions in terms of weight or number can be calculated from distribution data. Number distributions are strongly influenced by fine particles which may total very little weight. Conversely, a 'tail' of a few large particles may make a very large difference to the weight distribution.

The majority of particles produced by emulsion polymerization are spherical, but aggregates occur, typically of size up to 50 000 nm. These may be found as strings or clusters. Often they can be dispersed by ultrasonic agitation, but if not, they are best treated as large single particles. Of special significance is the presence of aggregates which take the form of irregular clumps, needles, stars or sickles. Presence of such material can create dilatancy at high shear which may in turn cause shear instability. If suspected, a visible microscope may be used in their detection.

For simple quality control, average particle size is adequate and is best measured using a visible light spectrometer.

The turbidity (τ) of a sample is given by

$$\tau = k\lambda^{-n}$$

where λ is the wavelength of light (in the 500–1000 nm region) and n is related to the particle diameter D by a function of the type

$$\frac{dD}{dn} = kD$$

Loebel (1959) adjusted the optical density at 375 nm to 0.5 and plotted log (optical density) (OD)) against wavelength, the slope of which gives n. He showed that monodisperse particles gave straight lines, at least in the 375–550 nm range. Also that polydisperse emulsions, provided of the same kind of distribution, also gave straight lines but of different slope.

A simplification can be made to avoid the tedious plotting at 50-nm intervals used by Loebel. In this, the values of optical density at each end of the linear region are taken only. Loebel's log (OD) = n log (λ) equation can be rewritten as:

$$n' = \ln (OD_{375}/OD_{550})$$

Hence a plot of D against n' gives a straight line which is best calibrated against electron microscopy or some other technique such as disc centrifugy. For the calibration, the range of polymer types and particle sizes likely to be encountered should be used. This will introduce some scatter, but will give a more universal calibration, albeit at a lower standard of accuracy.

Experience has shown that this method gives consistent results of about 5–10% accuracy against disc centrifuge weight averages and about 3% discrepancy between operators. OD_{375} values from 0.4 to 0.6 are usable. Results seem acceptable from about 80 to 1500 nm, with some possible extension at either end. Skew and polydisperse distributions, and those containing aggregates are less easy to measure, but repeatable nominal values are produced. A more recent development of the turbidimetric method has been published (Zollars, 1981a). By the use of polarization sensitive light scattering and a double Fourier lens collection system, the entire range from 100 nm to 1 mm can be determined in a single measurement (Bott and Hart, 1990).

For particle size distributions, the Joyce–Loebl Disc Centrifuge 4 has given excellent service. Speeds up to 10 000 r.p.m. are available. Detection is by the attenuation of a narrow visible light beam, and the results are processed by an IBM personal computer or similar computer. Because of the reduced scattering by fine particles the raw data tends to underestimate the quantity of fine particle sizes. As the spinning disc separates the sample, a near monodisperse latex passes the detector at any time. It is therefore possible not only to calculate correction factors according to Rayleigh–Mie scattering theories, but also to calibrate by using monodisperse latices of known size and concentration.

The disc centrifuge works on a modified Stokes' law relationship. Thus, the time of detection is inversely proportional to the weight and hence diameter of the particle. Large particles are detected quickly, but fine particles take a considerable time. In practice, the rotational speed of the disc is adjusted to give an analysis time of 15–45 min. Particle density can be calculated from emulsion density for which a pressure density cup is advised. For accurate work, a more direct method for density using reprecipitated polymer would be better. With this type of measurement, the injection gives what is known as a 'line start'. 'Streaming' of large particles is avoided by using a small pre-injection of a buffer fluid. Interpretation of results have been discussed elsewhere (Oppenheimer, 1983).

Other instruments start from a homogeneous spin fluid. This avoids the streaming problem, but the analysis of the results is indirect, as the distribution is implied by the mode of reduction in optical density.

A combination of centrifugy with elements of chromatography is used by the DuPont Sedimentation Field Flow Fractionator (SFFF) where particles are first separated across a tube by spinning and then further separated by pumping in streamline flow. As the larger particles are closer to the walls, the velocity profile of the flow sweeps them along more slowly thus giving a double separation. The instrument is claimed to be easy to use, to be insensitive to particle shape and be capable of detection down to 10 nm. Particle sizes of a series of monodisperse emulsions were measured by using a disc centrifuge and SFFF and other methods. Disc centrifuge and SFFF methods gave similar results, while light scattering gave the highest results and electron microscopy the lowest. Small monodisperse samples gave the most consistent results between the methods (Koehler and Provder, 1987). An ultracentrifuge technique for the measurement of very broad distributions has also been described (Machtle, 1988).

Photon correlation spectrometry is a widely used technique, but relies on mathematical analysis of the change of frequency of the scattered light. Recent instruments are very good and are capable of separating close bimodal distributions, but care must be taken if the distributions overlap, as they may not be resolved. A Malvern instrument (the Autosizer Hi-C) is claimed to measure particles in the range 15–1000 nm and at concentrations of 0.01–50%. Three different light-scattering techniques including photon correlation spectroscopy were compared by Gulari *et al.* (1987).

Electrosensing systems such as the Elzone instruments and some of the Coulter range are very rapid to use, but have a resolution only down to 500 nm.

Micropore filters are not successful for particle sizing due to adsorption and blockage. Filtration by silane treated porous silica packings have been reported to give better results (Kato *et al.*, 1982). The problems of sizing particles with diffuse or extended boundary regions have been discussed (Zimehl *et al.*, 1990).

8.5 Wet emulsion polymer properties related to particle size

8.5.1 Viscosity

Emulsion polymers are generally supplied with non-volatiles between 45% and 65%. This figure includes contributions from salts, colloids and surfactants. The actual polymer content will probably be about 3% lower.

Consideration of equal-diameter spheres, small in comparison with the total space, allows a packing factor of over 70%. If the particle-size distribution is widened, higher packing factors may be achieved.

Theoretical packing factors are not easy to achieve in practice. To achieve stability polymer particles have to be separated by a greater distance than the range of electrostatic forces. The preparation process results in a mixture of particles controlled by factors other than ideal distribution (Kazanskaya, 1982; Chen and Wu, 1990). On shearing, the structure may take up a different configuration, resulting in time-dependent changes in apparent viscosity. Time-dependent reduction in shear is known as 'thixotropy' while an increase is called 'rheopecticity' (see Figure 8.8).

Four other viscosity-related terms have been defined: yield point, Newtonian rheology, pseudoplasticity and dilatancy. Yield point indicates the initial force that must be applied to cause the emulsion polymer to flow. It is related to the network of charged particles or hydrogen-bonded colloid molecules which must be disturbed to allow flow. Once the shear is removed, these networks slowly re-establish.

Most stable emulsions are pseudoplastic in nature. Pseudoplasticity is a reduction of viscosity with shear. Near-Newtonian systems are often fairly narrow in particle-size distribution. They may contain a fairly low polymer content but can also have relatively high levels of a colloid such as polyvinyl alcohol. In this case the viscosity and rheology is largely controlled by the colloid.

Dilatant emulsions are frequently unstable as any attempt to force them to flow merely serves to tear the stabilizer from the surfaces, causing coagulation or grits. Some products may exhibit a Newtonian or a slightly dilatant property over only part of the shear range, while many dilatant emulsions are shear thinning at the time and temperature of manufacture, only to develop dilatancy on cooling.

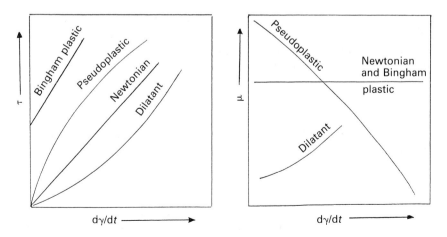

Figure 8.8   dγ/dt, *shear strain rate;* τ, *shear stress;* μ, *viscosity*

Dilatant products may result if irregular fused aggregates are present. These are often the result of excessive grafting of the polymer onto the colloid. Some grafting may make tougher films, but the degree of grafting has to be carefully controlled as the results are unpredictable. High temperatures and the use of oxidizing free-radical generators may increase grafting (Geddes, 1990), but also cleave the colloid.

There is a general rule that at constant non-volatiles content, finer particle-size emulsions have higher viscosities. Also, for any constant average particle size, wider distributions are more pseudoplastic than narrower distributions. Addition of water to any emulsion polymer sharply reduces its viscosity and rheological characteristics. It is claimed that 25% solids is the maximum for Newtonian properties (Candau *et al.*, 1990). Larger particle-size products, especially those exhibiting dilatant behaviour, usually give poor stability through rapid gravitational settling if diluted, although modified polyvinyl alcohols may help (Baines, 1990).

8.5.2 Colour

Emulsion polymers of average particle size 1000 nm or greater are white in colour due to back scattering of the incident light. As the average particle size is reduced they go through the stage of appearing to have a blue edge to the liquid, and then the bulk becomes blue-grey in colour and ultimately translucent. These effects may be accentuated by the examination of thin wet films, drawn down on glass plates. Emulsion colours produced by light scattering should not be confused with colours introduced by the incorporation of dyes (Winnik and Ober, 1987) which are retained in the dry state.

Some emulsions, particularly of the styrene–acrylic type, may include a proportion of larger particles with a narrow distribution. This can produce a pink or green colouration by reflected light. On standing in a transparent container, such particles form an array on the inside walls due to separation from the bulk of the emulsion. Causes of such separation are density and incompatibility with the stabilizers. In polystyrene containers a strong electrostatic charge may accumulate on the walls, giving a further force for separation of inhomogeneous particles. Composition analysis has indicated a higher styrene content for the fluorescent array than for the bulk of the copolymer.

8.5.3 Stability

Emulsion polymers may be unstable on storage by settling, phase separation (syneresis), or by gellation. Most of these effects are only partly dependent on particle size, although in low viscosity systems large heavy particles settle rapidly under gravity.

Some grafting between polymer and colloid is essential for stability in polyvinyl alcohol or hydroxyethyl cellulose stabilized emulsions, but excess surfactant (Yoshimura *et al.*, 1978) or colloid (Sperry *et al.*, 1981) introduced to reduce particle size may be disadvantageous. There is always competition between the particle surfaces and the water phase for ungrafted colloid and surfactant, and conversely post-additions of strong colloid solutions may cause destabilization by stripping protective water layers away from particle surfaces. This effect is called 'colloid shock'. The interaction of salts and particles leading to instability has recently been studied (Elaissari and Pefferkorn, 1991). Electrolyte added to increase particle size or buffer pH may cause destabilization by reduction of the electrical double layer on particle surfaces (Hergeth *et al.*, 1980; Pefferkorn and Stoll, 1990). Sulphonate containing monomers can be used to improve stability (Geddes and Davis, 1974; Fischer and Noelken, 1988).

Another desirable property is the ability to resist shear when pumped and during application. A degree of grafting to colloid and the copolymerization with unsaturated carboxylic acids or sulphonates has a positive effect, as has the avoidance of undersaturated surfaces, indicated by high surface tension. Similar improvements are noticed in the high-temperature and freeze–thaw stabilities. In the former, care should be taken not to rely on stabilization by surfactants with low cloud points or ready decomposition at high temperatures. Freeze–thaw is also improved by copolymerized acids, but fully hydrolysed polyvinyl alcohols may irreversibly precipitate out of solution when frozen.

8.6 Film formation in relation to particle size

Film forming emulsions have a multitude of applications. The largest volumes are for water thinnable paints, adhesives and textiles, including non-wovens. To give the required film, the temperature of drying must be above the 'minimum film-forming temperature' (MFT). The MFT is largely controlled by the polymer or copolymer glass transition temperature (see Section 8.2.4), but is also influenced by a number of other factors. These include particle size through capillary forces in the semi-dried polymer film. Vanderhoff and coworkers have published a series of papers on both practical and theoretical aspects from the late 1950s (Vanderhoff and Bradford, 1963; Vanderhoff *et al.*, 1966; Vanderhoff, 1970). Recent analyses of the forces acting have also been published (Guerro, 1989; Leonard, 1990). The process is a two-stage one. First is the drying process as the water evaporates, then the coalescence of the deposited polymer particles. In practice the two stages overlap.

The size and size distribution of the particles, as well as the state of the particle surfaces, have influence over the forces acting. A recent ingenious study (Zhao *et al.*, 1990) using the reaction of dissimilar pendant groups on polybutyl methacrylate particles, confirmed diffusion, which was slow to start, but increased with increasing time above the glass transition of the polymer. As expected, the diffusion rate then decreased with time, but was generally greater at higher temperatures. The diffusion data were consistent with the Williams–Landel–Ferry equation. A method for the non-destructive measurement of surface levelling by use of a two-beam optical interference method was reported (Klarskov, 1990). As coalescence is by interdiffusion of the material of different latex particles (Hahn *et al.*, 1986), the behaviour of core-and-shell systems is of obvious importance (Devon *et al.*, 1990). For practical applications, studies of film formation in the presence of pigment particles are of great interest (Lepoutre and Alince, 1981; Nicholson, 1989).

8.7 Influence of particle size on dry film properties of emulsion polymers

If the polymer surfaces were uncontaminated with stabilizers, fully coalesced unpigmented film properties should be independent of the particle size of the emulsion from which they were formed.

In practice, unpigmented films retain voids and pathways filled with surfactant and colloid, along with residual salts from buffers and initiator residues. Unless soluble in the polymer or grafted, these materials partially migrate to the surface during the drying process, but the slower processes of film consolidation (driven especially by surface tension) continue to eject the water soluble material for a considerable time. In general, fine particle-size products contain more stabilizers due to their greater surface area, but dry faster due to stronger capillary forces. When used as a coating, exuded solubles often wash off, but may leave residual pathways. A quantitative study of porosity has recently been published (Balik *et al.*, 1989). The positive effect of volatile plasticizing solvents, including water, should not be overlooked.

In the presence of pigments or fillers, other effects occur. Surfaces of pigments and fillers may adsorb stabilizers, giving less tendancy to exude. More particularly, the concept of pigment volume concentration (PVC) arises. At some intermediate point between 100% film forming polymer and 100% pigment is a composition identified as the critical pigment volume concentration (CPVC). Below the CPVC, the properties of such films are essentially the properties of the polymer. Above the CPVC, the pigment dominates and the strength of the film is the strength of pigment–polymer–pigment links. The effect of PVC on the properties of pigmented systems is shown qualitatively in Figure 8.9. Because the packing as well as volume of the polymer and pigment determines the density of the final film and the presence or absence of voids and pathways, the particle size and particle-size distribution of the polymer are highly significant. CPVC is not a sharp transition and is measured practically by permeation determinations. Finer particle-size emulsions may be used at higher pigment volumes before they reach the CPVC.

Quantitative studies of pigmented polymers in practical cases are difficult. One study (Hearn *et al.*, 1990) avoided the variability by employing surfactant-free, specially cleaned polybutyl acrylate latices. Permeability decreased with increasing drying time and temperature, factors which would be expected to consolidate the film more rapidly. The effect on permeability of post-added surfactants and salts was also reported. As in non-pigmented films, incompatible surfactants result in surface exudations which reduce reflection. To give a sharp, clear gloss, even to semi-gloss paints, requires a reasonably narrow distribution with no skew or tail above 2000 nm.

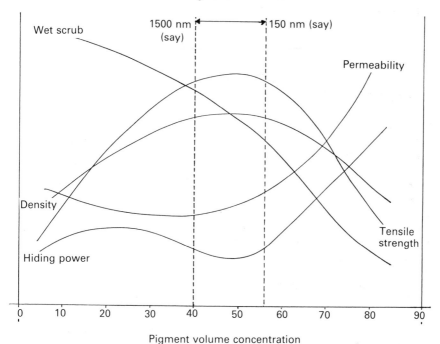

Figure 8.9   *Qualitative effect of pigment volume concentration (PVC) on paint properties*

Even in films used as clear overvarnishes or glossy top coatings, particle size is important. Finer particles, especially those with diameters less than the wavelength of light (say of 200 nm and below) give films of high gloss. Average particle sizes of 1500 nm and upwards give low glosses. In each case, the residual surface irregularities are the basic control, although temperature of drying is also important (Figure 8.10).

One further effect of incompatible surfactants or a mixture of incompatible polymers may be the production of stress-release patterns. If severe, these may give an appearance of frosting. The onset of the problem may be delayed for hours or days after the film has dried.

An interesting reverse phenomenon is the need to have particles of sufficiently large size to remain on the surface of porous substrates, otherwise gloss will not be achieved without application of a sealer prior to the main coating. Higher surface tension also helps to prevent penetration. These requirements were found in the manufacture of glossy washable wallpapers (Figure 8.11), although a straight line relationship should not be assumed.

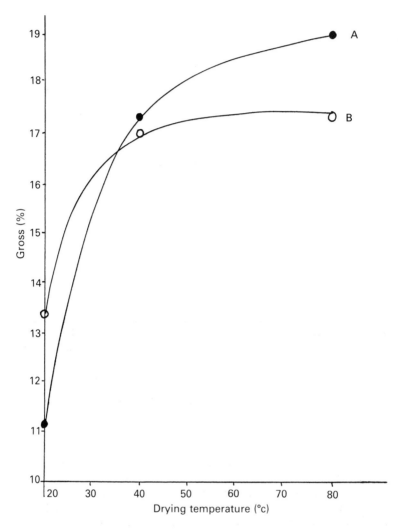

Figure 8.10   *Plot of gloss versus drying temperature for polyvinyl acetate homopolymers 'A' (●) and 'B' (○)*

8.8 Core-and-shell particles

Suppression of new particle formation during the process is necessary to produce core-and-shell particles. This type of product has layers of different compositions caused by changes in monomer feed composition during monomer addition in a delayed addition batch process. The monomer change is introduced when the level of free surfactant is low to ensure that no new particles are formed but the

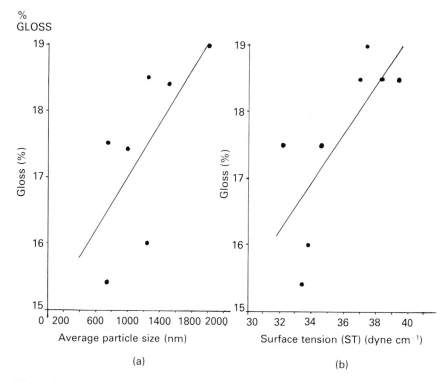

Figure 8.11   *Plots of* (a) *gloss versus particle size and* (b) *gloss versus surface tension*

growth of the existing particles continues. An interesting discussion on the conditions required is given by Erickson and Seidewand (1981).

An early application of this technique was for a glossy wallpaper coating, exhibiting exceptional resistance to blocking under hot and humid conditions. The product had an ethyl acrylate–styrene core, soft enough to film form readily, but with a pure styrene shell. On drying, the integration forces were sufficient to burst the particles, giving a continuous film with hard partial shells on the surface. A disadvantage was the porosity of the film due to the impossibility of the soft core to perform more than a 'spot weld' between particles. Interestingly, the reverse situation, i.e. with a hard core and a film-forming shell, has also been claimed to give blocking resistance (Nippon Shokubai Chemical Industry Co., 1988), but the nature of some systems may not be what is assumed. Hard hydrophobic monomers such as styrene have been found to migrate to the core of such particles (Lee and Ishikawa, 1983). These techniques are of importance in the paint field where non-blocking properties are essential in glossy emulsion paints. Care has to be taken not to create too great an incompatibility between core and shell, as the former may burst through the shell because of the strains created. This effect was studied by Ugelstad and illustrated in a remarkable series of electron micrographs (Ugelstad *et al.*, 1983). The stress was assumed to be caused by differential thermal properties and could be minimized by grafting between the components (Kumiawan and Mullen, 1989). Differential scanning calorimetry showed two distinct glass transition regions and the interfacial layer was estimated to be 2–7 nm (Hergeth *et al.*, 1990a). Multishell particles have also been reported (Kowalski *et al.*, 1988; Chen and Lee, 1990). The technique seems of special interest where functionality is required only on the surface of particles (Nitto Electric Ind. KK, 1982; Bayer, 1988).

Another application in the paint field is the coating of pigment or extender particles. For the most efficient hiding power the size of titanium dioxide particles must be only a little greater than the wavelength of light and be dispersed evenly throughout the continuous or polymer phase. Simple mixtures of pigment and polymer exhibit various degrees of flocculation on drying which, if severe, lead to the creation of unobstructed light paths through the coating. Polymer coated pigment, whether used as the sole component of the film, or in conjunction with added conventional emulsion polymer may be effective at preventing flocculation. To keep the pigment particles sufficiently spaced, it would be desirable to have a polymer layer greater than the wavelength of visible light. The total size of such a particle (of at least 750 nm diameter) would then be undesirable for the creation of glossy films, so severe compromises have to be made. The technique of building the first layers of polymer coating, after which the particle behaves as all-polymer, is difficult. Various approaches have been claimed in the literature (Caris *et al.*, 1989; Technische Universiteit Eindhoven, 1990; Hoshino, 1990), including the use of ultrasonics (Templeton-Knight, 1990).

An even more extreme form of core-and-shell polymer is the encapsulation of voids for use as a pigment. In one process (Kowalski and Vogel, 1985; Tioxide Group plc, 1989; Rohm and Haas, 1985) the particles have been made hollow by the use of a core or intermediate acidic layer which is swollen by neutralization with an alkali and which on drying collapses to leave an air void. In another patent (Hoshino, 1990) a core of polyvinyl acetate which is subsequently hydrolysed is used. These hollow particles have a tough outer shell consisting of polymer which is below its glass transition temperature at the temperature of use. Hence shape is retained, protecting the void. This type of hollow particle scatters light internally and acts as a pigment, especially in dry films. This enhances the dry hiding power of a paint at the expense of wet hiding (Andrew *et al.*, 1986). It also improves the printability of paper coatings (Haskins and Lunde, 1989). Alternative processes have been claimed, e.g. the use of volatile solvents (Rohm and Haas, 1988; Dow, 1990), blowing or foaming agents (Asahi Chemical Industry Co., 1990; Pierce and Stevens Corp., 1990), or the simple employment of core-and-shell polymers with polymers of different solubility parameters (Dainippon Ink and Chemicals, 1986).

Characterization of core-and-shell latices of any description is difficult, but some methods have been put forward (Hergeth *et al.*, 1990) and the special difficulty of using the disc centrifuge for hollow particles has been discussed (Cooper *et al.*, 1989).

8.9 Non-spherical particle formation

It was noted earlier that fused aggregates may have beneficial effects on both rheology and binding power, provided the aggregates are in limited numbers. A process was claimed (Berger Jenson and Nicholson Ltd, 1981) for the production of such aggregates for use as opacifiers for paints.

Non-spherical particles also result if core-and-shell polymers are made with the two stages widely different in polarity. In this case the second monomer tends to concentrate in domains at or near the surface of the core because of the incompatibility and grows outward non-uniformly (Shen *et al.*, 1989; Rohm and Haas, 1989; Daicell Chemical Industries, 1990).

Disc-like particles with a diameter/thickness ratio in the range of 1.2–5 have been prepared by a two-stage emulsion polymer process in which an organic solvent such as heptane is added to the second stage. Uses in the fields of printing, paper coating and paints are claimed (Mitsui Toatsu Chemicals Co., 1990). A patent application (Kao Corporation, 1985) with interesting implications for the theory of emulsion polymerization was published in 1985. This claimed rod-shaped particles by polymerizing rod-shaped micelles. A high degree of thixotrophy was claimed for the product.

A number of attempts have been made to develop a process which creates even more exaggerated non-spherical particles in a controlled manner. One possibility is to create controlled aggregation by mixing dissimilar particles in appropriate ratios. By careful choice of particle sizes and polarities, many fine particles can be precipitated onto the surface of large particles. These are obvious nuclei for the further growth of the particles in a non-spherical manner. Particles with lobes have been described where the lobes are grown directly on seed particles rather than on precipitated secondaries (Schaller *et al.*, 1987; Chou *et al.*, 1987; Kowalski *et al.*, 1988; Strauss, 1988). Because of polarity differences between the first- and second-stage monomer and the need to establish the presence of radicles prior to the introduction of the second-stage monomer, the process employed may be an extreme development of non-uniformity mentioned previously. The products are claimed to increase brushing viscosity in paints without decrease of pigment binding ability.

8.10 True spherical particle formation

Batch, continuous or single monomer feed delayed addition processes give nominally spherical particles. While Brownian movement and general agitation cause particles to turn during growth, perfect symmetry cannot be obtained in a gravitational field.

Vanderhoff arranged for the production of large particle-size monodisperse emulsion polymers under zero gravity in a space shuttle. It was found that perfect spheres were formed in space and that particle sizes narrowed. On the continuation of polymerization on Earth, the particle-size distribution broadened again and more coagulum was formed. The work and results have been reported in a series of papers (NASA and Vanderhoff, 1984; Vanderhoff *et al.*, 1984, 1986, 1987).

8.11 Particle-size distribution

Particle-size distribution is often the most vital factor in the application properties of emulsion polymers. The subject has been discussed from a theoretical standpoint by a number of authors (Fitch, 1981b; Lichti, 1981; Lichti *et al.*, 1982; Cauley and Thompson, 1982; Poehlein *et al.*, 1985; Dunn, 1985; Napper and Gilbert, 1989).

8.11.1 Monodisperse particles

In a true monodisperse distribution all the particles have exactly equal diameter. The term 'monodisperse' is used loosely to describe a narrow, single-peak distribution and should ideally be qualified by the standard deviation of diameter. A number of processes have been claimed (Laaksonen and Stenium, 1980; BASF, 1982; Vanderhoff *et al.*, 1984, 1986, 1987; Kawaguchi *et al.*, 1985; Ugelstad *et al.*, 1986; Ober and Lok, 1987; Fuji Xerox, 1988, Eckersley *et al.*, 1989), including the use of surface active azo initiators used without other surfactants (Feike *et al.*, 1990). The primary application is for the calibration of particle-sizing instruments, but the materials have interesting properties in their own right (Furusawa and Yamashita, 1982).

8.11.2 Skewed and multi-peak distributions

The shape of single-peak distributions can vary from symmetrical to strongly skewed in either direction. If aggregation has taken place, particles may be grouped in twos, threes or larger clusters. These aggregates may be broken by the addition of a little extra emulsifier and ultrasonics; or, if formed during the preparation, may have coalesced and taken spherical shape through surface tension forces. If formed early in the reaction from a wide distribution, growth will often merge them into a large particle size 'tail'. Coalescence from narrow distributions can sometimes be detected through multiple peaks representing the volume of two, three or more particles. Colloid-stabilized preparations appear particularly successful at allowing smooth coalescence to take place to give a large particle-size 'tail' (Donescu *et al.*, 1989), but the straight-line particle growth in the Loop reactor (Geddes, 1989) shows that no general coalescence has taken place under the reaction conditions employed. In this case large particle-size tails are due to a wide range of reactor residence times.

Sharper multiple peaks may be introduced by the availability of free surfactant at some intermediate stage during the course of polymerization. This may occur due to some facet of the process (Geddes, 1989; Snuparek *et al.*, 1990) or by the deliberate addition of extra surfactant. This may be to restore stability to systems (like methyl methacrylate or styrene) which are below their glass transition temperature during preparation and grow without coalescence. Other reasons for the deliberate creation of new particles is to increase the rate of monomer incorporation to ensure good overall conversion, and broaden distributions to produce high solids emulsions at reasonable viscosities (SCM Corp., 1981; Polysar Ltd, 1984). Bimodal distributions were reported when ABA non-ionic block copolymers were used as stabilizers. This was due to cloudpoint properties of the surfactant which also led to abnormal dependence of reaction rate on temperature (Hergeth *et al.*, 1990b).

8.11.3 Practical aspects

The development of new emulsion polymers involves the development of formulations and processes with improved application or performance, or to achieve properties similar to those existing, but at a lower cost. Sometimes more efficient production is sought, especially with regard to a faster rate of production with less waste in the form of polymerization grit or reactor surface fouling. However, at some stage in every development average particle size and particle-size distribution must be considered.

The major formulation decision is the choice of emulsifier (Dunn, 1982; Kine and Redlich, 1988) and colloid. Type, concentration and mode of addition have to be considered. The level of salts, either added as initiators or as pH control agents, has also to be assessed (Zollars, 1981; Dunn and Said, 1982). Process factors may help considerably (Dunn and Hassan, 1986) especially in continuous processes (Adams, 1985, 1990; Gugliotta and Meira, 1986). By the use of a cosolvent to stabilize monomer droplets of size similar to micelles, the reaction mechanism and kinetics can be radically altered from classical emulsion polymerization. Higher steady-state conversion and an absence of transient oscillations is claimed for this 'mini-emulsion' system in a continuous process (Barnette and Schork, 1986). Some processes may be intermediate, with both mechanisms competing.

Samples may be taken at intervals during any process for particle-size analysis, but speciality on-line techniques have been developed (Kourti *et al.*, 1987; Van Gilder and Langhorst, 1987) for following the changes.

An example of the effect of formulation factors on a styrene acrylic copolymer is given below.

Basic formulation

Initial water phase	Nonyl phenol 10 mol ethylene oxide	*
	Nonyl phenol 40 mol ethylene oxide	*
	Deionized water	210.0
Pre-emulsion	Styrene	262.0
	Butyl acrylate	204.0
	Methacrylic acid	19.0
	Nonyl phenol 20 mol ethylene oxide	*
	Nonyl phenol 40 mol ethylene oxide	*
	Ammonium persulphate	0.5
	Deionised water	211.0
Feed initiator	Sodium metabisulphite	0.5
	Deionized water	1.0

* A total of 1.75% on total formulation weight, and average HLB of 16. HLB is defined as the ratio of the molecular weight of the hydrophilic portion divided by the molecular weight of the whole molecule, multiplied by 20.

The basic formulation was varied in 12 ways as follows:

Variation from basic	Factor	High level	Low level
Non-ionic concentration (%)	a	2.5	1.0
Reaction temperature (°C)	b	60.0	40.0
Initial (seed) monomer (%)	c	20.0	5.0
HLB	d	17.0	15.0
Ratio of surfactant between initial water phase and pre-emulsion	e	1:1	1:3

A full factorial plan was then run where (l) indicates all factors at a low level, a indicates all low except a etc. The following results were obtained for particle size (nm):

Treatment	Result	Treatment	Result	Treatment	Result	Treatment	Result
(l)	180	abcd	122	ce	195	bde	194
b	200	ad	166	bce	223	acde	116
ac	155	abd	133	e	195	abcde	120

Overall average particle size was 167 nm. Individual changes gave the following average results:

Process or formulation change	Effect on particle size
a Increased total surfactant from 1.0% to 2.5%	Decreased by 79 nm
b Increased reaction temperature from 40°C to 60°C	Unchanged
c Increased initial monomer from 5% to 20%	Decreased by 22 nm
d Increased HLB of the surfactant blend from 15 to 17	Decreased by 50 nm
e Increased proportion of surfactant initially from 25% to 50%	Increased by 15 nm

The two most significant factors were shown to be the total concentration of surfactant used, and the length of the ethylene oxide chain on the non-ionic (the longer chains giving smaller particle sizes than shorter chains).

With the Loop reactor continuous process, process factors have been found to be especially powerful in determining particle size. The reaction mixture is pumped around the reactor by a circulation pump and new raw material fed in by a metering pump. Several parameters can be calculated; of significance are the recycle rate and the feed rate. From these can be calculated the recycle-to-feed ratio and the mean residence time. With a measured or estimated viscosity under reaction conditions and the internal diameter of the reaction piping, the shear on the reaction mixture may be calculated.

Comparisons were made (Adams, 1985, 1990) between two Loop reactors of internal volume 2300 ml, the first conventional and the second comprising a recycling section of 500 ml and a plug of 1800 ml to complete the reaction. The second format is known as a' P-loop' reactor. The recycle rate and the feed rate were the same in each case, only the residence time in the recycling section varied. In the conventional reactor it was 7 min, while in the other it was 1.5 min in the recycling section and 5.5 min in the plug.

Using two vinyl acetate–Veova 10 (Shell) copolymer formulae, one containing colloid (A) and the other colloid free (B), the values listed in Table 8.3 were obtained.

Table 8.3

Formulation Reactor	A		B	
	Conventional	*P-Loop*	*Conventional*	*P-Loop*
Average particle size (nm)	0.74	0.63	0.49	0.28
Particle-size range (nm)	0.15–2.5	0.15–1.95	0.1–1.1	0.1–0.55

The reduction in both average particle size and range is striking.

Similar results were quoted in a patent (Geddes and Khan, 1991) for multipassageway Loop reactors of the same total cross-section and volume as a single pipe conventional Loop. In this case, as all other factors were equal, the shear on the reaction mixture was the only variable, being greater with the multipassageway variant. For a number of monomer types and formulations, the multipassageway reactor gave finer particle sizes. The claim was for shear rates of above 800 s^{-1} while sustaining laminar flow.

8.12 Conclusions

Every year the consumption of water-based emulsion polymers increases and steady growth has been given an additional boost by restrictions on the use of solvent-based paints and adhesives. In step with the increased consumption is a greater realization of the importance of particle-size distribution and greater skill in its control under production conditions.

Despite intensive investigation of the mechanism of reaction and the incentive of commercial producers to understand better the processes they use, many problems remain. It has become clear in recent years that electronic and steric factors, and solubility in water, result in the mechanism being different for every monomer. Certainly, variations in initiator type and electrolyte level are important. Sufficient agitation to separate radical pairs and radical oligomers is essential. Earlier theories tended to oversimplify and more recent ones to overelaborate through the use of computer modelling. A useful summary of some of the remaining problems is given by Dunn (1991), but Alexander and Napper's comment made in 1970 that 'emulsion polymerization is more complicated and rather more subtle than first envisaged' is just as true over 20 years later.

References

Adams, D. A. (1985) Reactor for making emulsion polymerised products. *European Patent 0145325-A2*

Adams, D. A. (1990) Developments in tubular reactors for emulsion polymerisation. In *Proceedings of the 16th International Conference on Organic Science and Technology* (Athens), in press

Alexander, A. E. and Napper, D. H. (1970) Emulsion polymerisation. In *Progress in Polymer Science*, Vol. 3 (ed. Jenkins), Chap. 3, 146–197

Alfrey, T. and Price, C. C. (1947) *J. Polym. Sci.*, **2**, 101 ff.

Andrew, R. W. *et al.* (1986) The formulation of lower cost decorative coatings using opaque polymer. *Polym. Paint Colour J.*, **176**(4172), 567, 568, 570, 572

Asahi Chemical Industry Co. (1990) *Japanese Unexamined Patent JP01/301730*

Badran, A. S. *et al.* (1990) Study of the parameters affecting the polymerisation of vinyl acetate. *Acta Polymerica*, **41**(3), 187–192

Baines, S. J. *et al.* (1990) *European Patent 0389268*

Balik, C. M. *et al.* (1989) Residual porosity in polymeric latex films. *J. Appl. Polym. Sci.*, **38**(3), 557–569

Barnette, D. T. and Schork, F. J. (1986) Continuous polymerisation in mini-emulsions. In *Polymer Reaction Engineering* (ed. Reichert, K.-H. and Geiseler, W.) Huthig and Wepf, Basel, 71–76

BASF (1982) *European Patent 0043464*

Bayer A. G. (1988) *European Patent 0279260*

Berger, Jenson and Nicholson Ltd (1981) *British Patent 1591924*

Bott, S. E. and Hart, W. H. (1990) *Polym. Mater. Eng. Sci.*, **62**, 381–389

Brandolin, A. and Garcia-Rubio, L. H. (1990) On-line particle size distribution measurements for latex reactors. *Polym. Mater. Eng. Sci.*, **62**, 312–316

Candau, F. *et al.* (1990) *J. Colloid Interface Sci.*, **140**(2), 466–473

Caris, C. H. M. *et al.* (1989) Copolymerisation at the surface of inorganic sub-micron particles in emulsion-like systems. In *Proceedings of the 2nd CNRS International Conference* (Lyon), 128–130

Casey, B. S., Maxwell, I. A., Morrison, B. R. and Gilbert, R. G. (1990) Establishing mechanisms for emulsion polymerisation. *Makromol. Chem., Macromol. Symp.*, **31**, 1–10

Cauley, D. A. and Thompson, R. W. (1982) Particle size distribution during latex growth. *J. Appl. Polym. Sci.*, **27**(2), 363–379

Chen, S.-A. and Lee, S.-T. (1990) Shell region polymerisation characteristic of large emulsion particles. *Makromol. Chem., Rapid Commun.*, **11**(9), 443–450

Chen, S.-A. and Wu, K.-W. (1990) Emulsion polymerisation: determination of the average number of free radicals per particle by the use of number-average volume of the particles. *J. Polym. Sci. Polym. Chem.*, **28**(10), 2857–2866

Chou, C.-S. *et al.* (1987) Non-spherical acrylic latices. *J. Coatings Technol.*, **59**(755), 93–102

Cooper, A. A. *et al.* (1989) Use of a disk centrifuge to characterise void-containing latex particles. *J Coatings Technol.*, **61**(769), 25–29

Daicell Chemical Industries (1990) *Japanese Unexamined Patent JP02/123113*

Dainippon Ink and Chemicals (1986) *European Patent Application EP0173789A*

Devon, M. J. *et al.* (1990) Effects of Core/Shell latex morphology on film-forming behaviour. *J. Appl. Polym. Sci.*, **39**(10), 2119–2128

Donescu, D. *et al.* (1989) Emulsion polymerisation of vinyl acetate in a semi-continuous system. *Mat. Plast. (Bucharest)*, **26**(3), 144–146

Dong, H. S., Gilbert, R. D. and Fornes, R. E. (1989) Grafting of vinyl acetate on hydroxyethyl cellulose during emulsion polymerisation. *ACS Polym. Preprints*, **30**(1), 101–102

Dougherty, E. P. (1986) 'Scope' dynamic model for emulsion polymerisation. *J. Appl. Polym. Sci.*, **32**(1), 3051–3095

Dow Chemical Co. (1990) *US Patent 4973670-A*

Dunn, A. S. (1982) Effects of the choice of emulsifier in emulsion polymerisation. In *Emulsion Polymerisation* (ed. Piirma, I.) Academic Press, New York, 221–245

Dunn, A. S. (1985) Factors determining the breadth of particle size distributions in the emulsion polymerisation of styrene. In *Proceedings of the 2nd International Conference on Polymer Latex* (London), 1/1–1/6

Dunn, A. S. (1991) Problems of emulsion polymerisation. *JOCCA*, **74**(2), 50–54, 65

Dunn, A. S. and Al-Shahib, W. (1978) Effect of size of initial micelles on nucleation of latex particles in emulsion polymerisation of styrene. In *American Chemical Society Division of Colloid and Surface Chemistry, 176th Meeting* (Miami Beach, FL) Abs. 131

Dunn, A. S. and Hassan, S. A. (1986) The effect of reaction variables on particle size distributions in the emulsion polymerisation of styrene. *American Chemical Society. Polym. Mater. Sci. Eng.*, **54**, 439–443

Dunn, A. S. and Said, Z. F. M. (1982) Effect of electrolyte concentration in emulsion polymerisation of styrene. *Polymer*, **23**(8), 1172–1176

Eckersley, S. T., Vandezande, G. and Rudin, A. (1989) *JOCCA*, **72**(7), 273–275

Elaissari, A. and Pefferkorn, E. (1991) Aggregation modes of colloids in the presence of block copolymer micelles. *J Colloid Interface Sci.*, **143**(2), 343–355

Erickson, J. R. and Seidewand, R. J. (1981) Latex seed particle growth at high surfactant surface coverage. *American Chemical Society Symp. Ser.*, **168**, 483–504

Feeney, P. J., Napper, D. H. and Gilbert, R. G. (1984) Coagulative nucleation and particle size distributions in emulsion polymerisation. *Macromolecules*, **17**(12), 2520–2529

Feike, H. *et al.* (1990) *East German Patent 0277690*

Fischer, J. P. and Noelken, E. (1988) Correlation between latex stability data determined by practical and colloid chemistry based methods. *Prog. Colloid Polym. Sci.*, **77**, 180–194

Fitch, R. M. (1981a) Latex particle nucleation and growth. *American Chemical Society Symp. Ser.*, **165**, 1029

Fitch, R. M. (1981b) Growth of particle size and size distributions in latices and emulsions. In *Proceedings of the 8th Water-borne and High Solids Coatings Symposium* (New Orleans), Vol. 2, 25–43

Fitch, R. M. and Tsai, C. H. (1971) In *Polymer Colloids* (ed. Fitch, R. M.) Plenum Press, New York

Fitch, R. M. *et al.* (1985) Kinetics of particle nucleation and growth in the polymerisation of acrylic monomers. *J Polym. Sci., Polym. Symp.*, **72**, 221–224

Fordham, J. W. L. and Williams, H. L. (1951) *J. Am. Chem. Soc.*, **73**, 4855 ff.

Fuji-Xerox KK (1988) *Japanese Patent 63254104-A*

Furusawa, K. and Yamashita, S. (1982) Phase transition behaviour of monodisperse latices under various conditions. *J. Colloid Interface Sci.*, **89**(2), 574–576

Geddes, K. R. (1983) The Loop process. *Chem. Ind.*, **21 Mar.**, 223–227

Geddes, K. R. (1989) Start-up and growth mechanisms in the Loop continuous reactor. *Br. Polym. J.*, **21**, 433–441

Geddes, K. R. (1990) The use of hydroxyethyl cellulose in emulsion polymerisation. In *Cellulose Sources and Exploitation* (ed. Kennedy, J. F., Phillips, G. O. and Williams, P. A.) Ellis Horwood, Chichester, 287–294

Geddes, K. R. (1993) The Loop reactor process. *Surface Coatings Int.*, **8**, 330–339

Geddes, K. R. and Davis, J. W. (1974) Improvements in or related to emulsion polymers. *British Patent 1350282*

Geddes, K. R. and Khan, M. B. (1991) *UK Patent: Polymerisation process and reactors. GB 2235692-B*

Gerrens, H. (1965) In *Polymer Handbook* (ed. Brandup, J. and Immergut, E.) Wiley-Interscience, New York, II-399–II417

Giannetti, E. (1990) Comprehensive theory of particle growth in the Smith-Ewart interval II of emulsion polymerisation systems. *Macromolecules*, **23**(22), 4748–4759

Guerro, L. R. (1989) Film formation: aspects of the coalescence of latex particles. *Proceedings of the 2nd CNRS International Conference* (Lyon), 96–98

Gugliotta, L. M. and Meira, G. R. (1986) Control of particle size distributions in latices through periodic operation of continuous polymerisation reactors. *Makromol. Chem. Macromol. Symp.*, **2**, 209–218

Gulari, E. *et al.* (1987) Determination of particle size distribution using light scattering techniques. *American Chemical Society Symp. Ser.*, **332**, 133–145

Gulf Oil Canada (1971) *British Patent 1220777*

Hahn, K. *et al.* (1986) Particle coalescence in latex films. *Colloid Polym. Sci.*, **264**(12), 1092–1096

Ham, G. E. (1964) In *High Polymers*, Vol. VIII, *Copolymerization* (ed. Ham, G. E.) Interscience, New York

Hansen, F. K. and Ugelstad, J. (1977) Particle nucleation in emulsion polymerisation. I: A theory for homogeneous nucleation. *J. Polym. Sci., Polym. Chem. Edition.*, **16**, 1953–1979

Hansen, F. K. and Ugelstad, J. (1982). In *Emulsion Polymerisation* (ed. Piirma, I.) Academic Press, New York

Haskins, W. J. and Lunde, D. I. (1989) Hollow-sphere pigment improves gloss, printability of paper. *Pulp Paper*, **63**(5), 53–54

Hayashi, K. *et al.* (1982) Formation of fine particle size emulsions by high-dose polymerisation. *J. Polym. Sci., Polym. Lett.*, **20**(12), 643–645

Hearn, J., Roulstone, B. J. and Wilkinson, M. C. (1990) Transport through polymer latex films in relation to morphology. *JOCCA*, **73**(11), 467–470

Hergeth, W.-D. *et al.* (1980) Secondary aggregation in vinyl acetate emulsion polymerisation. *Acta Polym.*, **31**(11), 704–708

Hergeth, W.-D. *et al.* (1990a) Differential scanning calorimetry study of polyvinyl acetate/polymethyl methacrylate two-stage emulsion polymers. *Colloid Polym. Sci.*, **268**(11), 991–994

Hergeth, W.-D. *et al.* (1990b) Makromol. Chem. *Macromol.Chem. Phys.*, **191**(12), 2949–55

Hoechst (1959) *German Patent 1060600*

Hoshino, F. (1990) *European Patent 0376684*

Hoshino, F. *et al.* (1990) *European Patent 0349319*

Kahn, D. J. and Horowitz, H. H. (1961) *J. Polym. Sci.*, **54**, 363 ff.

Kao Corporation (1985) *European Patent 0154831-A2*

Kato, T. *et al.* (1982) *Colloid Polym. Sci.*, **268**(10), 924–933

Kawaguchi, H. *et al.* (1985) Preparation of large and monodisperse copolymer latex particles. *Makromol. Chem., Rapid Commun.*, **6**(5), 315–319

Kazanskaya, V. F. *et al.* (1982) Influence of the rate of stirring on the particle size of a suspension polymer. *Int. Polymer. Sci. Technol.*, **9**(9), 33–34

Khan, M. B. (1991) The Loop continuous process. *Eur. Coatings J.*, **5/91**, 307–319

Kine, B. B. and Redlich, G. H. (1988) Role of surfactants in emulsion polymerisation. *Surfactant Sci. Ser.*, **28**, 263–314

Klarskov, M. (1990) New non-destructive test method for determination of rheological properties of paints during film formation. *Farg Lack*, **36**(3), 53–58

Koehler, M. E. and Provder, T. (1987) Comparative particle size analysis. *American Chemical Society Symp. Series*, **332**, 231–239

Koltoff, I. M. and Miller, I. K. (1951) *J. Am. Chem. Soc.*, **73**, 3055 ff.

Kourti, T. *et al.* (1987) Measuring particle size distributions of latex particles in the sub-micron range using size-exclusion chromatography and turbidity spectra *American Chemical Society Symp. Ser.*, **332**, 242–255

Kowalski, A. and Vogel, M. (1985) Process for making multistage polymer particles and their use. *European Patent Application 0022633*

Kowalski, A. *et al.* (1988) Sequentially-produced polymer particles, processes and uses. *European Patent Application 0254467*

Krumrine, P. R. and Vanderhoff, J. W. (1982) Applicability of DLVO theory to the formation of ordered arrays of monodisperse latex particles. In *Polymer Colloids* (ed. Fitch, R. M.) Plenum Press, New York, 289–312

Kumiawan, A. S. and Mullen, R. L. (1989) Induced stresses in core/shell latexes. *J Colloid Interface Sci.*, **128**(1), 201–207

Laaksonen, J. and Stenium, P. (1980) Mechanism of emulsifier-free emulsion polymerisation of vinyl chloride. *Plastics Rubber: Mater. Appl.*, **5**(1), 21–24

Laurier, G. C., O'Driscoll, K. F. and Reilly, P. M. (1985) Estimating reactivity in free-radical copolymerisations. *J. Polym. Sci., Polym. Symp.*, **72**, 17 ff.

Lee, D. I. and Ishikawa, T. (1983) Formation of 'inverted' core/shell latices. *J. Polym. Sci., Polym. Chem.*, **21**(1), 147–154

Lee, D.-Y., Kuo, J.-F., Wang, J.-H. and Chen C.-Y. (1990) Study of a continuous loop tubular reactor for emulsion polymerisation of styrene. *Polym. Eng. Sci.*, **30**(3), 187–192

Leonard, R. G. (1990) *Makromol. Chem., Macromol. Symp.*, **35/36**, 389 ff.

Lepoutre, P. and Alince, B. (1981) Dry sintering of latex particles in pigmented coatings: influence of coating structure and properties. *J. Appl. Polym. Sci.*, **26**(3), 791–798

Lichti, G. (1981) Styrene emulsion polymerisations: particle size distributions. *J. Polym. Sci., Polym. Chem.*, **19**(4), 925–938

Lichti, G., Gilbert, R. G. and Napper, D. H. (1977) *J. Polym. Sci., Polym. Chem.*, **15**, 269 ff.

Lichti, G. *et al.* (1982) In *Emulsion Polymerisation* (ed. Piirma, I.) Academic Press, New York, 93 ff.

Loebel, A. B. (1959) Determination of the average particle size of synthetic latices by turbidity measurement. *Off. Digest*, **Feb.**, 200–212

Machtle, W. (1988) Coupling particle size distribution technique: a new ultracentrifuge technique for determination of the particle size distribution of extremely broad distributed distributions. *Makromol. Chem.*, **162**, 35–52

Mark, H. *et al.* (1975) Reactivity ratios. In *Polymer Handbook* (ed. Brandup, J. and Immergut, E.) Wiley-Interscience, New York, II291–II336

Mead, R. N. and Poehlein, G. W. (1988) Emulsion copolymerisation of styrene/methyl acrylate and styrene/acrylonitrile in a continuous stirred tank reactor, I. *Ind. Eng. Chem. Res.*, **27**(12), 2283–2293

Mead, R. N. and Poehlein, G. W. (1989) Emulsion copolymerisation of styrene/methyl acrylate and styrene/acrylonitrile in continuous stirred tank reactors, II: aqueous phase. *Ind. Eng. Chem. Res.*, **28**(1), 51–57

Mills, J. A. and Yocum, R. H. (1967). *J. Paint Technol.*, **39**, 532 ff.

Mitsui Toatsu Chemicals Co. (1990) *European Patent 0349319*

Napper, D. H. and Gilbert, R. G. (1989) Polymerisation in emulsions. In *Comprehensive Polymer Science*, Vol. 4, Part 2 (ed. Allan, G. and Bevington, J. C.) Pergamon Press, London, 171–218

NASA and Vanderhoff, J. W. (1984) Polymeric particles and their preparation. *Chem. Week*, **135**(16), 52

Nicholson, J. W. (1989) Film formation by emulsion paints. *JOCCA*, **72**(12), 475–477

Nippon Shokubai Chemical Industry Co. (1988) *Japanese Unexamined Patent JP63/022812*

Nitto Electric Ind. KK (1982) *Japanese Patent Application J 01030510-A*

Ober, C. K. and Lok, K. P. (1987) Formation of large monodisperse copolymer particles. *Macromolecules*, **20**(2), 268–273

Oppenheimer, R. E. (1983) Interpretation of disc centrifuge data. *J. Colloid Interface Sci.*, **92**(2), 350–357

Pefferkorn, E. and Stoll, S. (1990) Aggregation/fragmentation processes in unstable latex suspensions. *J. Colloid Interface Sci.*, **138**(1), 261–272

Penlidis, A., MacGregor, J. F. and Hamielec, A. E. (1986) Mathematical modeling of emulsion polymer reactors: a population balance approach. *J. Coatings Technol.*, **58**, 49–60

Pierce and Stevens Corp. (1990) Preparation of opacifier intermediates. *US Patent 4912139*

Poehlein, G. W. *et al.* (1985) Latex particle size distributions from steady-state continuous stirred-tank reactor systems. *J. Polym. Sci., Polym. Symp.*, **72**, 207–220

Poehlein, G. W, Lee, H.-C. and Chern, C.-S. (1986) Free radical transport and reactions in emulsion polymerisation. In *Polymer Reaction Engineering* (ed. Reichert, K.-H. and Geiseler, W.) Huthig and Wepf, Basel, 59–70

Polysar Ltd (1984) High Solids latex. *British Patent Application 2138830*

Rawlings, J. B., Prindle, J. C. and Ray, W. H. (1986) A better understanding of continuous emulsion polymerisation reactor behaviour through mathematical modeling. In *Polymer Reaction Engineering* (ed. Reichert, K.-H. and Geiseler, W.) Huthig and Wepf, Basel, 1–18

Rohm and Haas (1985) Multistage polymer particles. *European Patent 0022633-B1*

Rohm and Haas (1988) Process for preparing core/shell particle dispersions and core/shell particles. *European Patent Application 027726*

Rohm and Haas (1989) Modified latex composition. *US Patent 4814373*

Rohm and Haas (1990) Multistage polymer particles. *European Patent 0341944-A1*

Satpathy, Y. S. and Dunn, A. S. (1988) Emulsion polymerisation of highly water-soluble monomers. *Br. Polym. J.*, **20**(6), 521–524

Schaller, E. J. *et al.* (1987) Multi-lobed acrylic latices. *Proceedings of the Paint RA 7th International Conference* (London), 107–114

Schmutzler, K. (1982) Mechanism of particle formation in the emulsion polymerisation of vinyl acetate. *Acta Polym.*, **33**(8), 454–458

SCM Corp. (1981) *US Patent 4254004*

Shen, H. R., El Asser, M. S. and Vanderhoff, J. W. (1989) Uniform, nonspherical latex particles as model interpenetrating networks. *American Chemical Society Div. PMSE Papers*, **59**, 1185–1189

Smith, W. V. and Ewart, R. H. (1948) Kinetics of emulsion polymerisation. *J. Chem. Phys.*, **16**, 592–599

Snuparek, J., Skoupil, J., Podzimek, S. and Kastanek, A. (1990) Non-seeded semi-continuous emulsion polymerisation. *Makromol. Chem., Macromol Symp.*, **31**, 89–105

Song, Z. and Poehlein, G. W. (1988) Particle formation in emulsion polymerisation: particle number at steady state. *J. Macromol. Sci.*, **A25**(12), 1587–1632

Song, Z. and Poehlein, G. W. (1989) Particle nucleation in emulsifier-free aqueous-phase polymerisation: stage 1. *J. Colloid Interface Sci.*, **128**(2), 486–500

Sperry, P. R. *et al.* (1981) Flocculation of latex by water soluble polymers: experimental confirmation of a non-bridging, nonabsorptive, volume restriction mechanism. *J. Colloid Interface Sci.*, **82**(1), 62–76

Strauss, J. (1988) Multilobed acrylic latexes. *Surface Coatings Austr.*, **25**(11), 6–12

Technische Universiteit Eindhoven (1990) Process of preparation of polymers in emulsion, pigment particles and a process for the modification of pigment particles. *European Patent 0328219*

Templeton-Knight, R. L. (1990) A process for encapsulating inorganic pigment with polymeric materials. *JOCCA*, **11**, 459–464

Tioxide Group plc (1989) Polymeric particles and their preparation *British Patent Application GB 2207680-A1*

Ugelstad, J. *et al.* (1983) Preparation and application of monodisperse polymer particles. In *Science and Technology of Polymer Colloids* (ed. Poehlein, G. W., Otterwill, R. H. and Goodwin, J. W.) Martinus Nijhoff, The Hague, 51–99

Ugelstad, J. *et al.* (1986) New developments in production and application of monosized polymer particles. *American Chemical Society Polym. Mater. Sci. Eng.*, **54**, 521–523

Vanderhoff, J. W. (1970) Mechanism of film formation of latices. *Br. Polym. J.*, **2**, 161–173

Vanderhoff, J. W. and Bradford, E. B. (1963) *TAPPI*, **46**, 215 ff.

Vanderhoff, J. W. *et al.* (1966) *J. Macromol. Chem.*, **1**, 361 ff.

Vanderhoff, J. W. *et al.* (1984) Preparation of large-particle size monodisperse latexes in space: polymerisation kinetics and process development. *J. Dispers. Sci. Technol.*, **5**(3/4), 231–246

Vanderhoff, J. W. *et al.* (1986) Preparation of large-particle-size monodisperse latexes in space. *American Chemical Society Polym. Mater. Sci. Eng.*, **54**, 587–592

Vanderhoff, J. W. *et al.* (1987) The first products made in space: monodisperse latex particles. *American Chemical Society Polym. Preprints*, **28**(2), 455–456

Van Gilder, R. L. and Langhorst, M. A. (1987) Application of highspeed, integrated, computerised hydrodynamic chromatography for monitoring particle growth during latex polymerisation. *American Chemical Society Symp. Ser.*, **332**, 272–286

Wang, H.-H. and Chu, H.-H. (1990) Stabilisation effect of mixed surface active agents in the emulsion polymerisation of methyl methacrylate. *Polym. Bull.*, **24**(2), 207–214

Winnik, F. M. and Ober, C. K. (1987) Coloured particles by dispersion polymerisation. *Eur. Polym. J.*, **23**(8), 617–622

Yoshimura, S. *et al.* (1978) Attraction between latex particles in the presence of excess surfactant. In *Polymer Colloids*, Vol. II (ed. Fitch, R. M.) Plenum Press, New York, 139–151

Young, L. J. (1975) In *Polymer Handbook* (ed. Brandup, J. and Immergut, E.) Wiley-Interscience, New York, 291ff., 341ff.

Zhao, C. L., Wang, Y., Hruska, Z. and Winnik, M. A. (1990) Molecular aspects of latex film formation: an energy transfer study. *Macromolecules*, **23**(18), 4082–4087

Zimehl, R. *et al.* (1990) *Colloid Polym. Sci.*, **268**(10), 924–933

Zollars, R. L. (1981a) Turbimetric method for determination of latex particle size distribution. *J. Dispersion Sci. Technol.*, **2**(2/3), 331–344

Zollars, R. L. (1981b) Effects of particle number and initiator level on the kinetics of vinyl acetate polymerisation. In *Emulsion Polymerisation of Vinyl Acetate* (ed. El-Asser, M. S. and Vanderhoff, J. W.) Applied Science Publishers, London, 31–48

Zukaski, C. F. and Saville, D. A. (1985) Formation of small scale granularities in latex particles. *J. Colloid Interface Sci.*, **104**(2), 583–586

Chapter 9

Microencapsulation: techniques of formation and characterization

P. F. Luckham

9.1 Introduction

It has been said that the process of microencapsulation started with Nature's creation of the first living cell. Man, on the other hand, did not consciously attempt to copy this ingenious process of nature until about 50 years ago. The first practical use of microencapsulation was the National Cash Register Company's 'no carbon required' copying paper, used as an improvement over carbon paper copying. In this process, two dyes were coated with a clay and when ruptured the dyes formed a coloured imprint. Subsequently, many researchers have adopted this and similar microencapsulation techniques to specific industrial applications such as agrochemicals pharmaceuticals, bio-technology, the oil industries and the food industry, to name the more common ones.

'Microencapsulation' is a term which encompasses many different techniques where a solid or liquid is enclosed with some other material such as a polymer. The oldest and perhaps most widely used microencapsulation technique has been coacervation – phase separation (Bungenberg de Jong, 1949). This involves the deposition of polymer around the material to be coated by, for example, a temperature change. Another approach has been emulsion polymerization (Deasey, 1984), where microspheres are formed by polymerization of a monomer dispersed as droplets throughout a continuous phase in which the reaction is initiated. This technique produces small microspheres and has the potential for producing colloidal-sized delivery systems required for certain drug targeting applications (Couvreur et al., 1979; El-Samaligy and Kohdewald, 1982; Tomlinson, 1983). A more recently developed technique is an emulsification/solvent evaporation procedure, where a polymer solution containing the material to be encapsulated is emulsified into a second immiscible phase and the solvent subsequently removed (Beck et al., 1979; Watts et al., 1990). A further method involves interfacial polymerization where polymerization occurs at the interface between two immiscible phases.

In addition to these chemical methods, microcapsules may also be prepared mechanically, using specialized equipment such as spray drying, extrusion or spray coating. Electrostatic methods may also be used to form microcapsules. The term 'microencapsulation' has also been used to cover materials entrapped within vesicles. As the scope of this book is the preparation of colloidal formulations, the reader is referred to the chapter on vesicles (Chapter 10) for further information on this procedure. In this chapter we shall not refer any further to vesicle microcapsules.

Microencapsulation has a number of applications, these include:

(a) protection of reactive species from their environment
(b) safe handling of otherwise toxic or noxious materials
(c) as a means of handling liquids as a solid
(d) taste or smell masking of unpleasant materials
(e) as a means of providing controlled release of materials following application.

It is the latter which is perhaps the most important industrially and the most stimulating intellectually. The essential point is that the encapsulated active ingredient needs to diffuse through the encapsulating material in order to escape from the capsule and perform the task for which it was synthesized. For example, if a drug is encapsulated, the drug can slowly diffuse from the capsule, into the patient, ensuring a continuous, fairly constant supply of the drug for a long time period (many months) into the body. Similarly perfumes may be encapsulated and incorporated into paper products to ensure a slow release of the fragrance. Clearly the rate of release of the active ingredient is an important property of the microcapsule. It will depend on the material used to encapsulate, the thickness of the material and also the structure of the microsphere. These factors are discussed further below.

Somewhat surprisingly, perhaps, microencapsulation can sometimes make the active ingredient more effective. For example, 'Knox Out' is a micro-encapsulated formulation of the pesticide diazinon, with a polyamide–polyurea wall. This microcapsule has relatively low oral and dermal mammalian toxicity, but has outstanding cockroach control properties and, in addition, the volatility of the active ingredient is reduced. Furthermore, field tests have shown that this microencapsulated form of diazinon controls normally diazinon-resistant cockroaches as well. Apparently cockroaches rub their legs together and comb their legs and bodies with their antennae to remove foreign substances from their bodies. In turn they clean their antennae by passing them through their mandibles. This indicates that the enhanced activity of Knox Out is due to pick up of the microcapsules and subsequent ingestion, followed by stomach poison activity. Diazinon usually controls cockroaches by acting as a contact poison. The enhanced activity against resistant cockroaches is an additional benefit of the microencapsulated pesticide (Lowell *et al.*, 1980).

9.2 Microsphere structure

The structure of a loaded microsphere varies greatly and in turn will have a large influence on the properties of the particle. Thus it is important, if possible, to characterize the structure of the microcapsule.

Figure 9.1 contains four possible structures. Structure (1) has a continuous core of active ingredient surrounded by a continuous sheath. This is the so-called 'core shell structure'. The second case is where the encapsulated material is

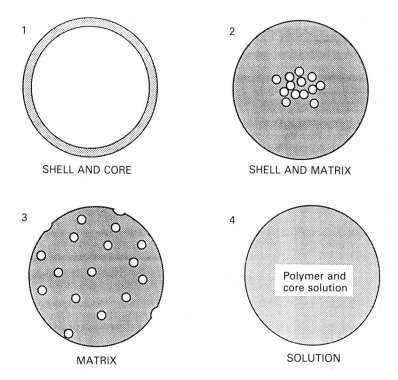

Figure 9.1   *Schematic representation of four hypothetical polymer microspheres.* (1) *Continuous core of active ingredient surrounded by a continuous shell.* (2) *Discrete core of active ingredient surrounded by a continuous shell.* (3) *Discrete regions of active ingredient dispersed in polymer matrix.* (4) *Active ingredient molecularly dispersed in polymer matrix*

scattered uniformly through the interior of the microsphere. The third structure represents a matrix through which particles of the encapsulated material are dispersed. The final case is where the encapsulated material is either dissolved, or molecularly dispersed in the carrier material from which the microsphere is dispersed.

Which structure is present will determine the release characteristics of the microsphere. In the first two cases the microsphere acts as a reservoir and provides a constant rate of release of material, while in the following two cases the particles do not act as reservoir devices and neither will give constant release characteristics over a prolonged time period. This may have detrimental effects in certain applications such as drug delivery or pesticide leaching for example; in other cases this may not be important.

The type of microsphere formed depends to some extent on the method of preparation. These are outlined below, although in many instances it is not known how the encapsulated material is actually dispersed in the microsphere.

9.3 Encapsulant materials

Although other materials may be used, for example clays in the 'no carbon required' paper described in the introduction, polymers are by far the most common materials used to form microcapsules. The choice of material will depend, to some extent, on the method of microsphere preparation of course, but other factors also are important. Various polymers have been used for microencapsulation, they should satisfy the following conditions in general (Kydonious, 1981):

(a) the polymer should have the correct molecular weight, glass transition temperature and molecular structure to ensure the desired diffusion and release rates of the encapsulated active ingredient
(b) the polymer should not react with the active ingredient
(c) the polymer should be stable during storage and general usage
(d) the polymer should be easily manufactured and fabricated into the desired product
(e) in biological and environmental applications, the polymer must be bio-compatible, biodegradable and the degradation products must not be toxic
(f) the polymer should not greatly add to the expense of the final product.

In large volume industries, such as agrochemicals, cheap, fairly crude encapsulants may suffice. For this purpose several derivatives from natural polymers, such as cellulose, starch, sawdust and bark have been developed for microcapsule formation (Kydonious, 1981). While in the preparation of progesterone microcapsules for contraception by intracervical injection, a more expensive and specifically 'tailored' coating of poly D, L-lactic acid has proven to be a commercially viable proposition (Mason *et al.*, 1984). It is really a question of 'horses for courses' and in a chapter of this length it is not possible to describe the specific merits, or otherwise, of the whole range of polymers used in microencapsulation.

9.4 Release from microspheres

9.4.1 Theoretical considerations

Diffusion of materials occurs through the walls of the capsules and the normal starting point for most diffusion-based release studies is Fick's first law of diffusion:

$$\frac{dc}{dt} = -DA \frac{dx}{dl} \tag{9.1}$$

where dc/dt is the mass of solute which diffuses in unit time through a cross-sectional area A in terms a concentration gradient dx/dl, in the direction l, and D is the diffusion coefficient.

Assuming the particles are spherical, of uniform diameter and do not change in size during the dissolution, and that the medium is an infinite sink, eqn (9.1) may be rewritten as

$$\frac{dc}{dt} \approx -ADC \qquad (9.2)$$

where C is the concentration of the encapsulated material, which on integration yields

$$\ln (C - C_0) = -DAt + \ln C \qquad (9.3)$$

where C_0 is the medium concentration. Provided that the area can be estimated, a plot of $\ln (C - C_0)$ versus t will allow the diffusion coefficient D to be estimated.

When applying Fick's law to diffusion from microcapsules, one problem is of how the thickness of the microcapsule wall affects the rate of dissolution. The first law may be rewritten as

$$\frac{dc}{dt} = -DA \frac{(C - C_0)}{h} \qquad (9.4)$$

where h is the thickness of the barrier. Under sink conditions this becomes

$$\frac{dc}{dt} = -KC \qquad (9.5)$$

where $K = AD/Vh$ and V is the volume of the microcapsules. Integration of eqn (9.5) gives a first-order kinetic equation

$$\ln C = \ln C^t - Kt \qquad (9.6)$$

where C^t is the concentration in the microcapsules at time t. This relationship shows how increasing the wall thickness of the microcapsule will decrease the rate of release of 'active'. It is also clear that the rate of release will decrease with increasing particle size as a result of the surface area present decreasing. In point of fact it is found experimentally that there is often a linear relationship between mean size and the time for 50% of the active to be released (Pongpaibul et al., 1984; Suzuki and Price, 1985). The rate of active release also tends to increase with increased active loading (Suzuki and Price, 1985; Tsai et al., 1986; Wada et al., 1988; Spenlehauer et al., 1988; Huang and Ghebre-Sellassie, 1989). Theoretically this should only apply to systems where there is some active present in particulate form because, as particles dissolve, pores will be created and a higher number of particles thereby creates a more porous structure enabling more rapid release (Nixon, 1984; Traisnel, 1984).

9.4.2 Experimental aspects

Release from microspheres is generally assessed by dispersion in an agitated release medium with samples being periodically withdrawn for assay (Cha and Pitt, 1988, 1989). An alternative technique for the assessment of release from microspheres is dual wavelength ultraviolet (UV) spectroscopy (Shively and Simonelli, 1989). Samples of the microsphere suspensions are taken from the release flask and placed directly into a cuvette of the spectrometer. By simultaneous measurement of the absorbance at λ_{max} for the active wavelength and at another wavelength at which the active does not adsorb, it is possible to remove the contribution of λ_{max} due to the scattering by the microspheres, hence enabling the active concentration to be monitored. This technique allows a more continuous measurement of the appearance of the active material in the release medium.

9.5 Methods for microsphere preparation

9.5.1 Simple and complex coacervation

The term 'coacervation' has been used by chemists to describe the 'salting out', or phase separation, of lyophilic solids into liquid droplets rather than into solid aggregates. According to Bungenberg de Jong (1949) coacervation can be considered to be either 'simple' or 'complex'.

Simple coacervation is a process involving the addition of a lyophilic substance to a polymer solution, e.g. an electrolyte to gelatin. This added substance causes two phases to be formed, one which is rich in the polymer and the other poor. The polymer-rich phase, under the correct conditions, forms microcapsules. The process depends primarily on the degree of hydration (solvation) produced and its principal requirement is the lack of sufficient water to hydrate fully the whole system. The procedure for producing microspheres in this way is illustrated schematically in Figure 9.2.

Complex coacervation is primarily a pH-dependent effect. The acidic, or basic, nature of the system is manipulated to give microcapsules. Above (or below), depending on the acidic/basic nature of the constituents, a critical pH value the system may produce microcapsules. Usually complex coacervation deals with systems containing more than one hydrophilic polymer. The classical explanation for complex coacervation was offered by Bungenberg de Jong (1949). It is fundamentally a result of salt bond formation. For example-complex coacervation of gelatin with acacia gum is only possible at a pH below the isoelectric point of gelatin, under these conditions the gelatin, which contains both acidic and basic amino acid groups, becomes positively charged, while the acacia gum, which contains only acidic groups, remains negatively charged. In complex coacervation, the two polymers essentially salt each other out, and under the correct conditions encapsulate the active material. A schematic representation of microencapsulation by a complex coacervation process is given in Figure 9.3.

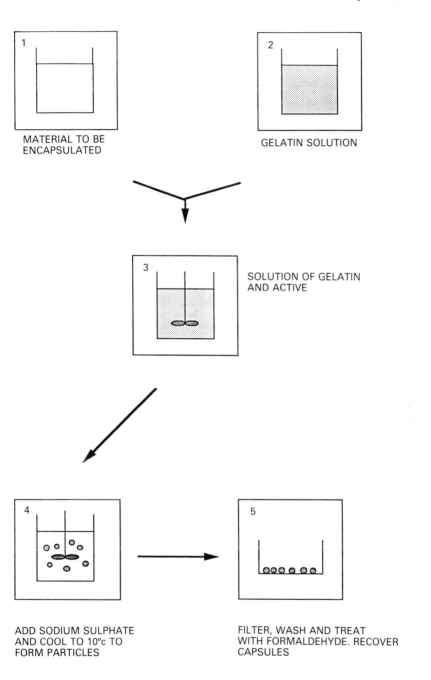

Figure 9.2   *Schematic diagram of microsphere formation by the simple coacervation technique*

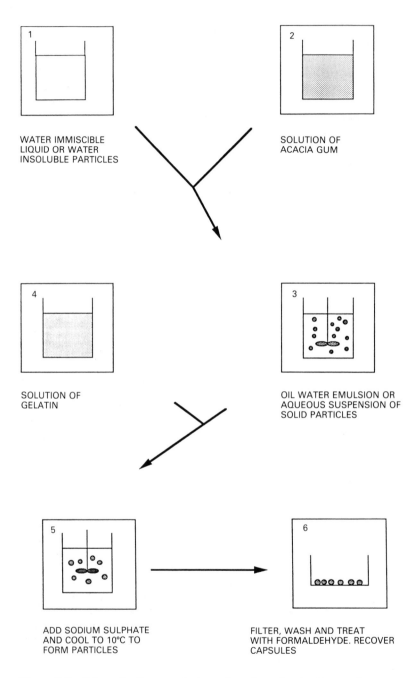

Figure 9.3 *Schematic diagram of microsphere formation by the complex coacervation technique*

The structure of microcapsules produced by coacervation is essentially capsular in nature, i.e. structure 1 or 2 in Figure 9.1. This means that coating integrity is essential to preserve the microsphere properties, especially in release. Therefore the coating walls need to be relatively thick and elastic to cope with the loss of material from the core of the particle into the environment. As gelatin, which is frequently used as the encapsulant in coacervation processes, is structurally weak, it is necessary to have thick films. Thus release rates are of necessity slow, and this limits the usage of this procedure.

Many materials, particularly drugs (Palmieri, 1977; Madan, 1980), have been encapsulated via coacervation processes. However, these systems frequently are difficult to scale up to manufacturing quantities, are difficult to reproduce accurately and are expensive to prepare. As a consequence, coacervate microcapsules are seldom commercially viable.

9.5.2 Emulsification/solvent evaporation

Microencapsulation by emulsification/solvent evaporation is conceptually a simple procedure (Watts *et al.*, 1990). It involves firstly the emulsification of a polymer solution containing the material to be encapsulated, the active, which may either be dispersed or dissolved, into a second, immiscible liquid phase. On the addition of an emulsifier, droplets containing polymer, active ingredient and solvent, dispersed in the immiscible liquid (generally a polar liquid such as water), are formed. In the next step the solvent is removed from the droplets by the application of heat, vacuum, or simply by allowing evaporation at room temperature to occur, leaving a suspension of active containing polymer microspheres that can be readily separated by filtration, or centrifugation, washed and dried. A schematic representation of the preparation of microcapsules by this route is given in Figure 9.4. This method can be modified to produce microspheres over a wide size range (<200 nm to 500 μm) (Pongpaibul *et al.*, 1988; Koosha *et al.*, 1989).

The method outlined above and in Figure 9.4 can be used for oil soluble materials. If, on the other hand, the material to be encapsulated is water soluble, emulsification into an aqueous phase is not generally successful. This can be overcome by the formation of an oil-in-oil emulsion, e.g. acetone in mineral oil, but there is always the problem of oil removal (Wada *et al.*, 1988). However, this may be overcome by the formation of a multiple emulsion, i.e. a water-in-oil-in-water multiple emulsion. Here a water-in-oil emulsion is formed, with the active dissolved in the water, and this emulsion, in the presence of an oil-in-water emulsifier, is formed into an oil-in-water emulsion, giving an aqueous continuous phase in which the oil droplets contain within them the active-ingredient-containing water droplets (Ogawa *et al.*, 1988).

The principal means of controlling particle size in the emulsification/solvent evaporation microencapsulation route are the mixing conditions. Essentially, the higher the shear which can be generated, the smaller the particles. Indeed, it is

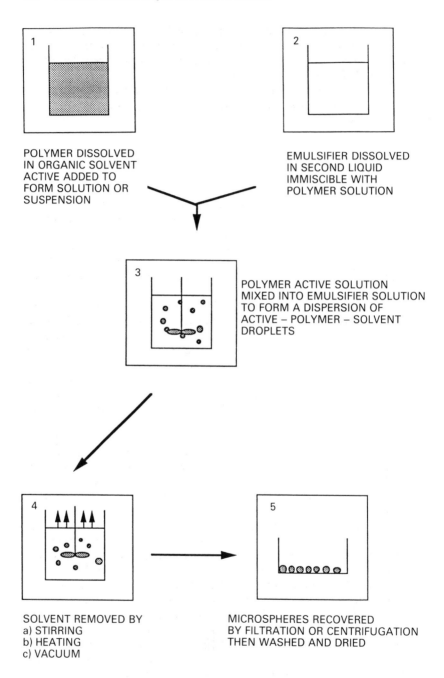

Figure 9.4   *Schematic diagram of microsphere formation by the emulsification/solvent evaporation technique*

observed that the particle size decreases exponentially with mixer speed (Benita *et al.*, 1984). Impeller and mixer design are also important. Vortex formation needs to be avoided to prevent aggregate formation, and this may be achieved by the presence of baffles in the mixing tank.

The structure of microspheres prepared in these ways are essentially that of material dispersed through a matrix of the encapsulating polymer either as a particulate or as a molecular dispersion.

9.5.3 Interfacial polymerization

Microencapsulation by interfacial polymerization is a process whereby a monomer is made to polymerize at the interface of two immiscible substances. The two immiscible liquids are then emulsified until the desired particle size is reached, whereupon a cross-linking agent is added to the external phase of the emulsion (see Figure 9.5). Nylon has been used extensively to form microspheres by interfacial polymerization. This results in polymerization at the interface between the droplets and the external phase and is an excellent means of preparing liquid-containing microcapsules. For example, Luzzi *et al.* (1970) have prepared nylon capsules of sodium phenobarbital. This procedure used 25 ml of a 2% aqueous solution of carboxymethylcellulose, 25 ml of a 2% aqueous solution of sodium phenobarbital, and 25 ml of a 6.7% aqueous solution of hexamethylene diamine. This mixture was emulsified in 165 ml of an organic solvent (4:1 (v/v) cyclohexane/chloroform). A further 165 ml of the mixed organic solvent containing 0.2% sebacyl chloride was added. Nylon micro-capsules were formed as a result of the interfacial polymerization of hexamethylene diamine and sebacyl chloride.

This procedure generally leads to the formation of core shell microcapsules (structure 1 in Figure 9.1). One great advantage of interfacial polymerization over coacervate encapsulation is that the film formed is thin and usually very durable. This is principally a consequence of the bulk properties of the common polymers used – gelatin for coacervation and nylon for interfacial polymerization.

A variation of interfacial polymerization, *in situ* polymerization, was patented in 1961 (Brynko and Scarpelli, 1961a, b). Monomers were contained only in the dispersed oil phase and polymerized *in situ* without any coreactive monomers in the water phase. A second shell was then formed by coacervation of the polymers used to disperse the oil phase. *In situ* polymerization with all monomers in the aqueous phase and no coreactive monomers in the dispersed oil phase has also been reported (Foris *et al.*, 1978a, b, c). Another adaption of *in situ* polymerization utilized isocyanate monomers in the dispersed oil phase and no coreactive monomers in the aqueous phase (Scher, 1977, 1979). Hydrolysis of isocyanate to the corresponding amine groups made possible reaction of the amine groups with remaining isocyanate groups to form polyurea shell walls. The major limitation of *in situ* polymerization from an industrial point of view is the low concentration of capsules formed and the slow reaction conditions.

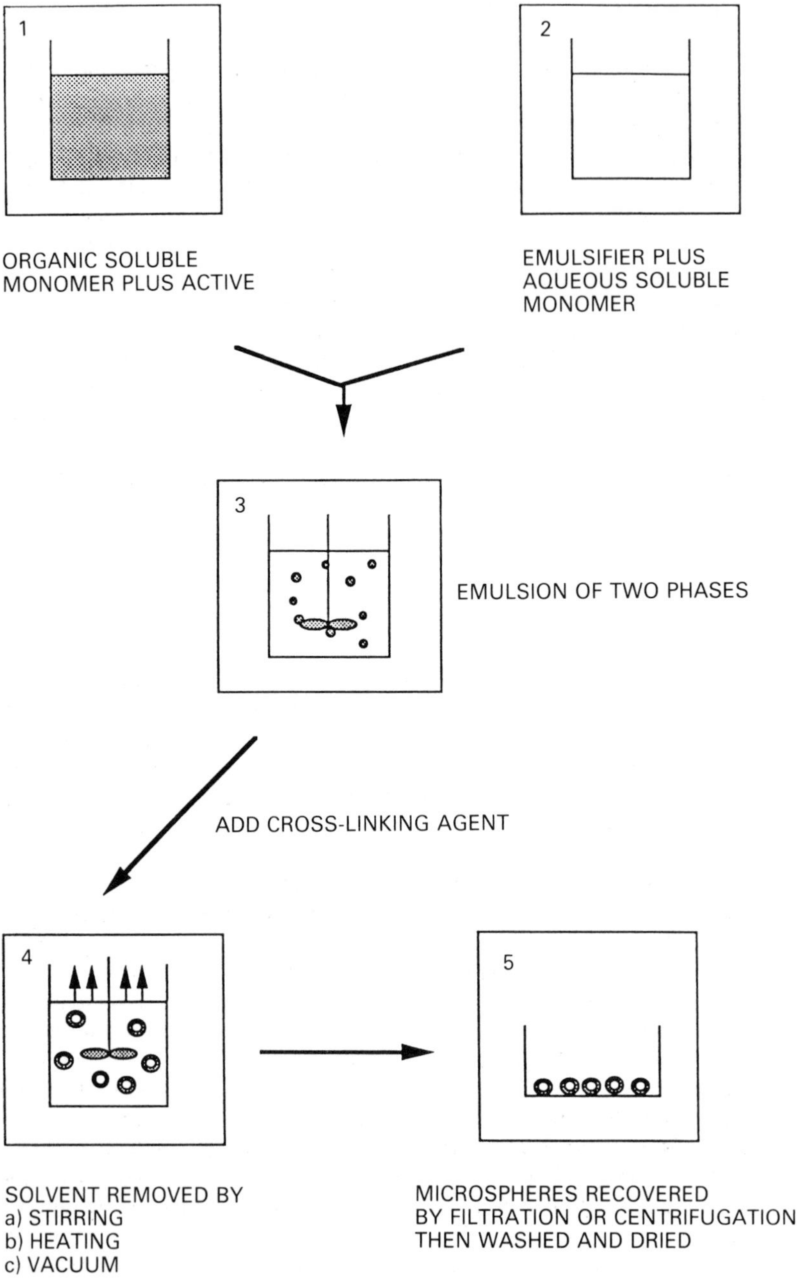

Figure 9.5 *Schematic diagram of microsphere formation by the interfacial polymerization technique*

9.5.4 Electrostatic methods

Preparation of these microcapsules involves the bringing together of coating material and the material to be encapsulated when both are aerosols (Luuzi, 1970). The coating material must be liquid during the encapsulation stage and must be capable of wetting the core material. Should the internal phase be liquid during the process then the two liquids must be immiscible and the coating liquid must have a higher interfacial tension to ensure spreading and hence encapsulation. The two aerosols must be oppositely charged, this can be accomplished prior to mixing of the two. Three chambers are used in this process: two for atomization and one for mixing. Oppositely charged ions are generated into the two atomization chambers and are deposited onto the aerosols while they are being atomized. The droplets are then mixed in the third chamber, the oppositely charged droplets attract each other and collide and one droplet spreads over the surface of the other. The droplets are cooled to solidify the coating and collected. The microspheres formed in this way tend to be of the type 1 structure shown in Figure 9.1, although if aggregation also occurs structure types 2 and 3 may also form.

9.5.5 Mechanical methods

A mechanical method is frequently the most commercially successful system for preparing microcapsules since the process often adapts well to large-scale manufacturing and tends to be more reproducible than the other encapsulation technique. A variety of patented microencapsulation equipment is now commercially available and for further information the reader is referred to the recent review by Li *et al.* (1988). Here we shall describe one of the best-known methods for mechanically encapsulating materials, namely the Wurster fluidized bed apparatus (Wurster, 1953, 1965; Hall and Pondell, 1980).

A schematic representation of the Wurster coating chamber is given in Figure 9.6. The particles to be coated are pneumatically conveyed upward through the central tube of the coating chamber. As the particles pass the atomizer they are wetted by the coating fluid, and then immediately subjected to a drying process, due to the heated support air moving upwardly in the column. Partially coated particles move downward in a near-weightless environment along the periphery of the tube (C), where further drying occurs. When the particles reach the lower end of the column they are directed back into the upwardly moving bed and the entire process is repeated. The amount of coating material applied (and hence the wall thickness) is proportional to the atomizing time. The microspheres formed in this way tend to be of the type 1 structure (Figure 9.1), although if aggregation also occurs structure types 2 and 3 may also form.

The Wurster process has found applications in many industrial areas such as the food and feed industries, pharmaceuticals, agrochemicals as well as the general chemical industry. Excellent drying conditions and the ability of the process to handle almost any shaped particle, as well as being able to encapsulate

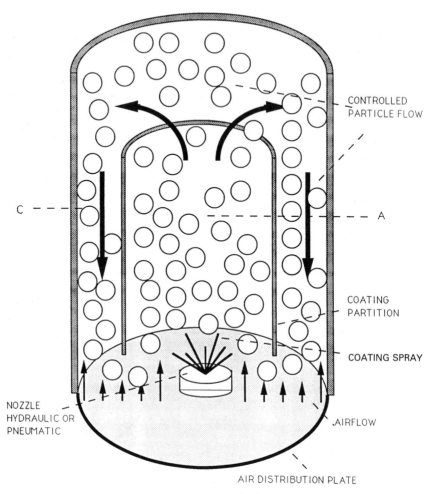

CONTROLLED
PARTICLE FLOW

C

A

COATING
PARTITION

COATING SPRAY

NOZZLE
HYDRAULIC OR
PNEUMATIC

AIRFLOW

AIR DISTRIBUTION PLATE

Figure 9.6   *Schematic diagram of a Wurster coating chamber*

hydroscopic particles, are the main advantages of the system. The principal disadvantages are the inability to form capsules smaller than about 50 mm and the inability to encapsulate volatile substances, including of course liquids.

9.6 Characterization of microspheres

9.6.1 Size and morphology

The most commonly used method for assessing particle size and size distribution is sieving. Occasionally, light or electron microscopy have been used for particle sizing, although microscopy is generally used to study the morphology of microspheres (Tsai *et al.*, 1986). Optical microscopy has generally only been used for monitoring microsphere shape and the presence of aggregates, while scanning electron microscopy (SEM) has been used to assess surface details such as pores, pitting and the presence of surface crystals of encapsulated material. Confocal microscopy should also be able to detect these structures. In addition, light-scattering techniques such as photon correlation spectroscopy (PCS) for submicrometre particles and Mie scattering for larger particles have occasionally been used to determine the sizes of microcapsules (Koosha *et al.*, 1989).

In order to obtain information of the internal structure of microspheres, freeze fracture followed by transmission electron microscopy needs to be employed, although care needs to be used in interpreting the results because artefacts are easily introduced in sample preparation (Suzuki and Price, 1985).

9.6.2 Surface analysis

Surface analytical techniques such as X-ray photoelectron spectroscopy (XPS) or static secondary ion mass spectroscopy (SSIMS) may be used to determine the chemical nature of microspheres (Koosha *et al.*, 1989). In particular, the presence of any encapsulated material or emulsifiers in the surface of particles may be identified by using these techniques. As these techniques are both vacuum techniques, they cannot be reliably used when volatile substances are present.

References

Beck, L. R., Cowsar, D. R., Lewis, D. H., Cosgrove, R. J., Riddle, C. T., Lowry, S. L. and Epperly, T. (1979) A new long acting injectable microcapsule for the administration of progesterone. *Fertility Sterility,* **31**, 545–561

Benita, S., Benoit, J. P., Puisieux, F. and Thies, C. (1984) Characterisation of drug loaded poly(D,L-lactide) microspheres. *J. Pharm. Sci.,* **73**, 1721–1726

Brynko, C. and Scarpelli, J. A. (1961a) *U.S. Patent No. 2,969,330*

Brynko, C. and Scarpelli, J. A. (1961b) *U.S. Patent No. 2,969,331*

Bungenberg de Jong, H. G. (1949) Coacervation. In *Colloid Science*, Vol. II (ed. Kruyt, H. R.) Elsevier, Amsterdam, 144–163

Cha., Y. and Pitt, C. G. (1988) A one week sub-dermal delivery system for L-methadone based on biodegradable microcapsules. *J. Control. Rel.,* **7**, 69–78

Cha., Y. and Pitt, C. G. (1989) The acceleration of degredation controlled drug delivery from polyester microspheres. *J. Control. Rel.,* **8**, 259–265

Couvreur, P., Kante, B., Roland, M. and Speiser, P. (1979) Adsorption of antineoplastic drugs to polyalkylcyanoacrylate nanoparticles and their release in calf serum. *J. Pharm. Sci.,* **68**, 1521–1524

Deasy, P. B. (1984) *Microencapsulation and Related Drug Processes*, Marcel Dekker, New York

El-Samaligy, M. and Kohdewald, P. (1982) Triamcinolone diacetate nanoparticles, a sustained release drug delivery system suitable for parenteral administration. *Pharm. Acta Helv.,* **57**, 201–208

Foris, P. L., Brown, R. W. and Phillips, P. S. (1978a) *U.S. Patent No. 4,087,376*

Foris, P. L., Brown, R. W. and Phillips, P. S. (1978b) *U.S. Patent No. 4,089,802*

Foris, P. L., Brown, R. W. and Phillips, P. S. (1978c) *U.S. Patent No. 4,100,103*

Hall, H. S. and Pondell, R. E. (1980) The Wurster process. In *Controlled Release Technologies: Methods Theory and Applications*, Vol. II (ed. Kydonieus, A. F.) CRC Press, Boca Raton, FL, Chap. 7

Huang, H.-P. and Ghebre-Sellassie, I. (1989) Preparation of microspheres of water soluble pharmaceuticals. *J. Microencapsulation,* **6**, 219–225

Koosha, F., Muller, R. H., Davis, S. S. and Davies, M. C. (1989) The surface chemical structure of poly (β-hydroxybutyrate) microparticles produced by the solvent evaporation process. *J. Control. Rel.,* **9**, 149–157

Kydonoius, A. F. (1981) *Controlled Release Technology: Methods, Theory and Applications*, Vols I and II, CRC Press, Boca Raton, FL

Li, S. P., Kowarski, G. R., Feld, K. M. and Grim, W. M. (1988) Recent advances in microencapsulation technology and equipment. *Drug Dev. Ind. Pharm.,* **14**, 353–376

Lowell, J. R., DeSavignu, C. B. and Curl, G. D. (1980) *Polymer Preprints,* **21**, 118–120

Luzzi, L. A. (1970) Microencapsulation. *J. Pharm. Sci.,* **59**, 1367–1376

Luzzi, L. A., Zoglio, M. A. and Maulding, H. V. (1970) Preparation and evaluation of the prolonged release properties of nylon microcapsules. *J. Pharm. Sci.,* **59**, 338–341

Madan, P. L. (1980) Clofibrate capsules, the mechanism of release. *Drug Dev. Ind. Pharm.,* **6**, 629–643

Mason, N. S., Gupta, D. V. S., Keller, D. W., Youngquist, R. S. and Sparks, R. E. (1984) Microencapsulation of progesterone for contraception by intracervical injection. In *Biomedical Applications of Microencapsulation* (ed. Lim, F.) CRC Press, Boca Raton, FL, 75–85

Nixon, J. R. (1984) Release characteristics of microcapsules. In *Biomedical Applications of Microencapsulation* (ed. Lim, F.) CRC Press, Boca Raton, FL, 19–51

Ogawa, Y., Yamamoto, M., Okada, H., Yashiki, T. and Shimamoto, T. (1988) A new technique to efficiently entrap leuprolide acetate into microcapsules of polylactic acid or copoly(lactic/glycolic) acid. *Chem. Pharm. Bull.,* **36**, 1095–1103

Palmieri, A. (1977) Dissolution of prednisone microcapsules in conditions simulating the pH of the gastro-intestinal tract. *Can. J. Pharm. Sci.,* **12**, 88–101

Pongpaibul, Y., Price, J. C. and Whitworth, C. W. (1984) Preparation and evaluation of controlled release indomethacin microspheres. *Drug Dev. Ind. Pharm.,* **10**, 1597–1616

Pongpaibul, Y., Maruyama, K. and Iwatsuru, M. (1988) Formation and *in vitro* evaluation of theophylline loaded poly (methylmethacrylate) microspheres. *J. Pharm. Pharmacol.,* **40**, 530–533

Scher, H. B. (1977) *U.S. Patent No. 4,046,741*

Scher, H. B. (1979) *U.S. Patent No. 4,140,516*

Shively, M. L. and Simonelli, A. P. (1989) The investigation of dual wavelength spectroscopy for the analysis of dissolved drug in microcapsule suspensions. *Int. J. Pharm.,* **50**, 39–43

Spenlehauer, G., Vert, M., Benoit, J.-P., Chabot, F. and Veillard, M. (1988) Biodegradable cisplatin microspheres prepared by the solvent evaporation method; morphology and release characteristics. *J. Controlled Release,* **7**, 217–229

Suzuki, K. and Price, J. C. (1985) Microencapsulation and dissolution properties of a neuroleptic in a biodegradable polymer, poly(D,L-lactide). *J. Pharm. Sci.,* **74**, 21–24

Tomlinson, E. (1983) Microsphere delivery systems for drug targeting and controlled release. *Int. J. Pharm. Technol. Prod. Manufacturer,* **4**, 49–56

Traisnel, M. M. (1984) Controlled release from microspheres. *Bol. Chim. Faem.,* **123**, 303–312

Tsai, D. C., Howard, S. A., Hogan, T. F., Malanga, C. J., Kandzari, S. J. and Ma, J. K. H. (1986) Preparation and *in vitro* evaluation of polylactic acid – mitomycin C microcapsules. *J. Microencapsulation,* **3**, 181–187

Wada, R., Hyon, S.-H., Ike, O., Watenabe, S., Shimizu, Y. and Ikada, Y. (1988) Preparation of lactic acid oligomer microspheres containing anticancer drugs by o/o type solvent evaporation process. *Polymer Materials, Sci. Eng.,* **59**, 803–810

Watts, P. J., Davies, M. C. and Melia, C. D. (1990) Microencapsulation using emulsification/solvent evaporation: an overview of techniques and applications. *Crit. Rev. Therup. Drug Carrier Systems*, **7**, 235–259

Wurster, D. E. (1953) *U.S. Patent No. 2,648,609*

Wurster, D. E. (1965) *U.S. Patent No. 3,207,824*

Chapter 10

Production of vesicles of defined size

M. J. Lawrence

10.1 Introduction

Vesicles are a family of closed-membrane capsules consisting of single or multiple bilayers of non-covalently assembled amphiphilic molecules (Figure 10.1). In a vesicle each bilayer is separated from its neighbour by an aqueous compartment. Although phospholipids have been known for a long time to form the bilayer matrix of biological membranes, it was only 30 years ago that Bangham and coworkers discovered that, under certain conditions, naturally occurring phospholipids could aggregate in water to form vesicles (Bangham *et*

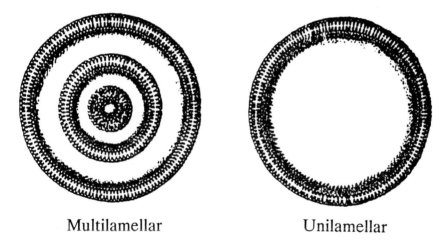

Multilamellar   Unilamellar

Figure 10.1 *Diagrammatic representation of multilamellar and unilamellar vesicles (from Ostro (1987))*

al., 1965, 1967a, b). When composed of phospholipids, vesicles are called 'liposomes'. Vesicles have now been produced from a whole variety of surfactants. Vesicles formed from completely synthetic surfactants are designated as surfactant vesicles (Fendler, 1980).

Since Bangham's original observations, the number of reported investigations into vesicles has mushroomed from hundreds to thousands. Many of the early studies examined the potential of liposomes as models for the study of biological membrane structure and function. More recently, in addition to their membrane mimetic properties, vesicles formed from both natural and synthetic surfactants have been investigated for their potential use in drug delivery and targeting (Gregoriadis, 1988), vaccination (Six *et al.*, 1988), diagnostics (Ho and Huang, 1988), medical imaging (Seltzer, 1988), as a carrier for haemoglobin (Kondo, 1988), catalysis (Fendler, 1982), energy conservation (Fendler, 1982).

Depending upon the method of preparation, a range of uni- or multilayer vesicles can be produced which vary in size from 25 nm up to 50 μm. The choice of what type of vesicle to use for a given application depends partly on what one wishes to put inside the vesicle (if anything at all) and partly on what one wants to do with the vesicle once it is made. For example, giant empty vesicles of tens of micrometres are used in the study of membrane mechanical and adhesion properties in order to allow the insertion of the necessary microprobes (Evans and Needham, 1986), whereas multilamellar vesicles are ideally suited to determine details of bilayer structure (Hope *et al.*, 1986). Consequently, considerable effort has focused on preparation methods which permit vesicle properties to be tailored for specific applications. An ideal preparation method would allow the investigator to produce vesicles with predetermined properties from a variety of amphiphile(s) over a broad range of amphiphile concentrations. The vesicle size would be under experimental control, and size distribution would be relatively homogeneous. The method would allow a predictable and reproducible entrapped volume. Finally, the method would require minimal time input and would not damage or contaminate the amphiphile(s). As yet, no single preparation method encompasses all of these objectives, but there are several general methods now available, one of which will often provide a satisfactory approximation to the desired characteristics.

10.2 Background to vesicle formation

It is generally considered that the prerequisite of vesicle formation is that an amphiphile produces lamellar phases, although this in itself is insufficient to guarantee that it will form vesicles (Israelachvili *et al.*, 1980). At a first approximation it is reasonable to say that vesicles are formed from double-chain amphiphiles; although it must be noted that double-chain phospholipids containing chains less than eight carbon atoms form micelles (Schmidt, 1981), while some long double-chained phospholipids produce inverted hexagonal phases in preference to bilayers, for example phosphatidylethanolamine at physiological temperatures (Cullis and Kruijff, 1979). Vesicles, however, have been formed by a wide range of surfactants including single-chain, double-chain and triple-chain amphiphiles having a wide range of cationic, anionic, non-ionic and zwitterionic head groups. Table 10.1 lists some of the surfactants known to form vesicles.

When preparing vesicles it is important to remember that unlike other surfactant–water phases (e.g. micelles) vesicles are rarely formed as equilibrium surfactant systems (Safran *et al.*, 1991). Generally non-equilibrium methods such as sonication of lamellar liquid crystalline phases are usually necessary to obtain

Table 10.1   *Some vesicle forming amphiphiles*

Single-chain amphiphiles

Ionic

Oleic and linoleic acids	Gebicki and Hicks (1973, 1976)
C_8–C_{18} monoalkyl amphiphiles	Hargreaves and Deamer (1978)
Fatty acids (and fatty acid esters)	Bittman and Blau (1986)
1-Alkyl-4-dodecylpyridinium iodide	Nusselder and Engbert (1989)
Diphenylazomethine containing quaternary ammonium amphiphiles	Kunitake and Okahata (1980)
Naphthyloxy containing quaternary ammonium amphiphiles	Nagamura *et al.* (1985)
Monoalykyl amphiphiles containing large rigid segments	Kunitake *et al.* (1981)
Hemisuccinate esters of cholesterol and tocopherol	Lai *et al.* (1985)
Cholesterol sulphate	Brockerhoff and Ramsammy (1982)

Non-ionic

Monoalkyl glycosides	Kiwada *et al.* (1985)
Monoalkyl polyglycerols	Vanlerberghe *et al.* (1978)
Monoalkyl polyoxyethylene glycols	Ribier *et al.* (1984)
Monoalkyl amphiphiles containing a rigid hydrophilic portion	Hundscheid and Engberts (1984)
Monoalkyl amphiphiles containing a rigid segment	Kunitake *et al.* (1981)
Triethoxycholesterol	Patel *et al.* (1984)

Zwitterionic

Diphenyl and azobenzene containing phosphocholine amphiphiles	Kunitake *et al.* (1985)
Monoalkyl amphiphiles containing a rigid segment	Kunitake *et al.* (1981)

Double-chain amphiphiles

Ionic

Didodecyldimethylammonium bromide	Kunitake and Okahata (1977)
Didodecyldimethylammonium salts	Brady *et al.* (1984)
Dialkyldimethylammonium bromides	Kunitake *et al.* (1977)
Didodecyldimethylammoinium hydroxide	Ninham *et al.* (1983)
Dialkyl amphiphiles containing phosphate, sulphate and carboxylate groups	Kunitake and Okahata (1978)
Cationic amphiphiles containing amino acid residues	Murakami *et al.* (1982)

Non-ionic

Dialkyl polyoxyethylene glycols	Okahata *et al.* (1981)
Dialkyl polyoxyethylene glycols	Chauhan and Lawrence (1989)
Dialkyl polyoxyethylene glycols	Lawrence *et al.* (1991)
Long chain dialkyl polyglycerols	Vanlerberghe *et al.* (1978)

Zwitterionic

Synthetic phospholipids	Kunitake *et al.* (1985)

Triple-chain amphiphiles

Ionic

Triple-chain ammonium amphiphiles containing 1–3 fluorocarbon chains	Kunitake and Higashi (1985)

Non-ionic

Hydrogenated castor oil	Tanaka *et al.* (1990)

a metastable phase of vesicles, which may re-equilibrate back to the multilamellar, liquid crystalline phase from which they originated.

Recent work has shown that it is possible to produce equilibrium phases of vesicles of controlled size. In all cases, however (with one notable exception, Cantu *et al.* (1991)), formation of spontaneous vesicles has involved mixing of two simple ionic surfactants with oppositely charged head groups (Kaler *et al.*, 1989). In systems composed of a single surfactant, the curvature energy of a bilayer dictates that the energy of a phase of spherical vesicles is never lower than that of a multilamellar liquid crystalline phase (Safran *et al.*, 1990). This is because the bilayer is composed of two surfactant monolayers which, in the single surfactant case, have the same spontaneous curvature (Helfrich, 1973). Since in a vesicle the two layers have curvature of opposite sign (i.e. the inner one being concave with respect to the water, the outer one being convex), the system is frustrated. Single surfactants are thus not generally considered to form vesicles spontaneously, although Israelachvili *et al.* (1976, 1977, 1980) argue that, provided the total lipid concentration is low, small unilamellar vesicles are thermodynamically favoured over extended bilayers.

10.3 Terminology

In the present chapter the term 'vesicle' is used to describe spherical or ellipsoidal single- or multicompartment closed bilayer structures, regardless of their chemical composition. It is an important point to note that while vesicles are generally considered to be spherical or onion shaped, depending on the method of preparation, vesicles may vary both in morphology (ranging from dics to long tubules) and in size (from several hundred angstroms to fractions of a millimetre). Consequently, it is not unusual to find a whole range of sizes and shapes of vesicles coexisting within one vesicle preparation.

Three main parameters are usually sufficient to characterize a vesicle preparation: namely its lamellarity, size distribution and trapped aqueous volume. Partly because it turns out that liposomes of different sizes often require completely different methods of manufacture, and partly because different applications often demand the use of liposomes within particular size ranges, classification of liposomes according to size is the most common index of characterization in current use. Within a particular size range the vesicles are further subdivided according to the number of lamellae.

When vesicles are classified according to the number of bilayers, vesicles containing more than one bilayer are referred to as *multilamellar* (MLV) while those consisting of a single bilayer are called *unilamellar*. Unilamellar vesicles can be further distinguished on the basis of their size. Vesicles with a diameter of approximately 100 nm or more are usually considered to be *large unilamellar vesicles* (LUV), whereas those that are smaller are referred to as *small unilamellar vesicles* (SUV). In the former category the vesicles are generally considered to be of a sufficiently large size that the curvature is small enough for them to be indistinguishable from a planar lamellar phase. In the SUV, the vesicle radius is of the order of the surfactant size, consequently the bilayer is highly curved and its constituent molecules highly constrained. A third class of unilamellar vesicle are the *giant vesicles* (GV). GV generally have diameters between 1000 and 50 000 nm (Hub *et al.*, 1982). In cases where a few vesicles are concentrically entrapped in LUV or GV the term *large oligolamellar vesicles* (LOV) is used (Lasic, 1988). Other terms, such as reverse-phase evaporation vesicle (REV) and dehydrated–rehydrated vesicle (DRV) are used in the literature. However, these relate to the method of manufacture rather than vesicle type (see below).

10.4 Methods of vesicle preparation

Table 10.2 gives the main methods of vesicle production and summarizes the advantages and disadvantages of the various preparations. Note that while one technique may give vesicles of a certain size range, small modifications to the procedure may very often alter the resulting nature and size distribution of the vesicles considerably. For a comphrehensive practical review of vesicle preparation the reader should consult New (1990). It is important to realize that, in practice, most methods of vesicle preparation give a fairly heterogeneous population of vesicles with a wide distribution of sizes and lamellarity. Furthermore, it should be noted that since the lipid-to-aqueous entrapped-volume ratio varies markedly with liposome size and lamellarity, the observed vesicle size distribution does not correlate with the entrapped volume. Indeed, even vesicles that are superficially identical with respect to size and lamellarity can exhibit considerable differences in trapped aqueous volume and solute distribution (Gruner *et al.*, 1985).

Surprisingly, in spite of the amount of work that has been performed on vesicle preparation there has been very little systematic work examining how size distribution and lamellarity are affected by the multitude of experimental variables involved in vesicle preparation: pH, temperature, ionic strength, buffer type, concentration of lipid, presence of cholesterol, occupancy by a probe in the enclosed aqueous space, vesicle age etc. (Menger *et al.*, 1989). In addition, it is important to point out that size may be affected by experimental parameters, e.g. the power of the sonicator used, so that the exact size may conceivably vary from laboratory to laboratory (Menger *et al.*, 1989).

It is also pertinent to mention that although surfactants are known to form vesicles the majority of the work examining vesicle preparation has utilized phospholipids as their main component. Most of the examples referred to in this section therefore use phospholipid as the main vesicular constituent. Although most studies examining the formation of synthetic surfactant vesicles have utilized sonication as their method of vesicle preparation, it appears that surfactant vesicles can be produced by the same techniques as those used for liposome production (Ballie *et al.*, 1985; Chauhan and Lawrence, 1989). Furthermore, while most work examining the formation of vesicles involves the dispersion of the vesicles in water, it has been shown that vesicles can be prepared in other non-aqueous polar solvents such as glycerol (McDaniel *et al.*, 1983). Recent papers have also reported the formation of reverse vesicles in non-

Table 10.2 *Preparation of vesicles*

Type of vesicle	Method of preparation	Advantages	Disadvantages
Multilamellar	Mechanical dispersion (Bangham et al., 1965)	Ease of preparation	Very heterogenous low entrapment (typically 2–10%)
	Sequential extrusion (Olson et al., 1979)	Defined size, high homogeneity, high entrapment (typically 30%), high lipid concentration	Special equipment required
	Dehydration–rehydration (Kirby and Gregoriadis, 1984a)	High entrapment (typically >40%), high lipid concentration	Difficult to control hydration rate
	Freeze–thaw (Mayer et al., 1985)	High entrapment, high lipid concentration	Require charged lipids
	Formation from templates (Payne et al., 1986a)	Ease of preparation, can be stored dry	Low entrapment, special equipment required
	Reverse-emulsions (Gruner et al., 1985; Pidgeon et al., 1986)	High entrapment (typically >35%), high stability, good for basic studies	Heterogeneous exposure to organic solvents
	Two-step emulsification (Batelle Memorial Institute, 1979)	High entrapment	Heterogeneous exposure to organic solvents
	Dilution from a gel (Perrett et al., 1991; Kaneko and Sagitani, 1992)	No solvents, readily scaled up, high entrapment	High concentration of ethanol or polyol
	Simple dispersion (Handjani-Vila et al., 1979)	No solvents, readily scaled up	Degradation of lipids possible

Table 10.2 *(Continued)*

Type of vesicle	Method of preparation	Advantages	Disadvantages
Small unilamellar	Sonication (Huang, 1969)	Miminal size vesicles, relative ease of preparation	Degradation of lipid, low entrapment, contamination with titanium
	Infusion (Batzri and Korn, 1973; Kremer *et al.*, 1977)	Fairly good homogeneity, no degradation of lipid	Difficult to control injection rate, ethanol present
	Detergent dilution (Brunner *et al.*, 1976; Rhoden and Goldin, 1979)	High homogeneity, high entrapment, no degradation, high lipid concentration	Residual detergent, need to control rate of dialysis, special equipment required
	French press (Barenholz *et al.*, 1977, 1979; Hamilton *et al.*, 1980)	High homogeneity, high entrapment, no degradation	Special equipment required
	Formation from templates (Lasic *et al.*, 1987)	Ease of preparation	Removal of support, require charged lipids

Table 10.2 *(Continued)*

Type of vesicle	Method of preparation	Advantages	Disadvantages
Large unilamellar	Reverse-phase evaporation (Szoka and Papahadjopoulos, 1978)	Entrapment of large macromolecules	Exposure to organic solvents, heterogeneous
	Extrusion (Hope *et al.*, 1985)	Defined size, high homogeneity, high entrapment, high lipid concentration	Special equipment required, narrow size range
	Infusion methods (Deamer, 1984)	Entrapment of large molecules	Difficult to control injection rate
	Formation from templates (Lasic, 1988)	Ease of preparation	Removal of support
Giant vesicles	Slow hydration (Reeves and Dowben, 1969; Hub *et al.*, 1982)	Entrapment of large molecules	Difficult to control hydration rate
	Reverse-phase evaporation (Kim and Martin, 1981)	High encapsulation	Heterogeneous, exposure to organic solvents
	Freeze–thawing (Oku and MacDonald, 1983a, b)	High encapsulation	Complicated procedure

polar solvents such as decane (Kunieda *et al.*, 1991a, b). In one instance the production of reverse vesicles has been reported to be spontaneous (Kunieda *et al.*, 1991b).

10.4.1 Multilamellar vesicles

Multilamellar vesicles (MLV) usually consist of a population of vesicles covering a wide range of sizes (typically 100–1000 nm); each vesicle generally consists of five or more concentric lamellae. Vesicles composed of just a few concentric lamellae are sometimes called 'oligolamellar liposomes', or 'pauci-lamellar vesicles'. MLV are the simplest type of vesicle to prepare and their manufacture can be readily scaled up. They are also mechanically stable for long periods of time as they can be considered to be planar on a molecular scale.

Thin-film method
MLV are typically formed by the hydration of a dry lipid film. The standard method for producing MLV is as follows and is based on the original method of Bangham *et al.* (1965). The lipid (or lipid mixture) is dissolved in an organic solvent, such as diethyl ether or chloroform. The solvent is removed, typically using a rotary film evaporator, to leave behind a thin film of dry lipid on the surface of the flask. An alternative method of dispersing the lipid in a finely divided form, prior to the addition of the aqueous media, involves freeze drying the lipid dissolved in an appropriate organic solvent (ICI, 1978). The choice of solvent is determined by its freezing point, and its inertness; tertiary butanol is frequently considered to be the most suitable solvent (New, 1990).

Vesicles are formed by adding to the thin film, the aqueous solution and shaking, or vortexing, until the dry film is removed from the wall of the flask. Dispersion of the amphiphile can be aided by the addition of glass beads. It is important to note that vesicles formed by gentle hand shaking have an appreciably larger diameter. The temperature of the aqueous phase should be above the phase transition temperature (T_c) of the amphiphile, or in the case of an amphiphile mixture, above the T_c of the higher melting amphiphile.

When vesicles are prepared from a mixture of lipids, it is important that the lipids are homogeneously dissolved in the organic solvent system used in order to obtain bilayers with evenly distributed lipids after hydration. If the lipids exhibit different solubilities in the chosen organic solvent(s) an inhomogeneous lipid film is formed on the glass wall and upon hydration vesicle dispersions containing vesicles with different compositions will be formed (Estep *et al.*, 1978, 1979).

The vesicles produced by the thin-film technique are very heterogeneous, both with respect to size and lamellarity. Thus, for many purposes the MLV formed by this method are too large or too heterogeneous a population to work with, and so although this method of preparation is quick and simple, its drawbacks mean that it is not often used *per se* as a method of preparation, rather as a preliminary stage in the preparation of the vesicles. Alternative methods have therefore been devised: (a) to reduce size and, in particular, to convert vesicles in the larger size range into smaller vesicles and/or reduce their lamellarity; and (b) to increase the aqueous entrapped volume. As the thin-film method is frequently used as a preliminary step it is worthwhile listing the problems encountered when using the method.

One potentially serious drawback with the method is the difficulty of removing all traces of organic solvent(s); frequently the thin film is left overnight under vacuum in an attempt to remove all of the solvent(s), or alternately the final vesicle preparation may be dialysed. As solvent contamination can affect experimental results it is usually worthwhile reducing the organic solvent concentrations to trace levels. In practice, one cannot be certain that all traces of solvent have been removed. However, it is possible to add small amounts of the potential contaminant to determine whether it affects the parameter being measured. If there is no effect, one can conclude that traces of that contaminant will not produce artifactual data. In the event of MLV being used for drug delivery, traces of organic solvent present a toxicity hazard.

Another problem is the heterogeneity of the resulting vesicle suspension. Attempts to obtain a more homogeneous MLV preparation by high-speed centrifugation have largely been unsuccessful; the vesicles in the pellet generally have the same distribution as the uncentrifuged mixture (Olson *et al.*, 1979). The use of high-speed centrifugation (typically $100\,000\,g$ for $1\,h$) as a method of increasing homogeneity is only feasible if two very distinct size populations of vesicles are present; namely large multilamellar and small unilamellar vesicles (Litman, 1974) and is useful only for making gross cuts between the small and large vesicles, but not for generating narrow size distributions (SUV generally remain suspended). Size exclusion chromotagraphy has also been used to produce vesicles of fairly homogeneous sizes. As with centrifugation, this method is only suitable for separating small vesicles from larger structures. Both methods are quite limited in terms of volume and throughput, and must be carried out in batches. Chromatography also suffers from the added disadvantage that it results in a significant dilution of the product, making it necessary to concentrate the product, usually by high-speed centrifugation.

One of the main problems with MLV produced by this method is the low entrapped volume and even lower solute entrapment. Solute to be entrapped is added to the hydrating solution. Upon addition of the aqueous solution the dry film will first adopt a bilayer structure. Water can permeate readily through such bilayers, but ions and other solutes permeate much more slowly. It may therefore be expected that the outermost bilayers act as a molecular sieve, where water permeates through to achieve equilibrium hydration of interior bulk liquid but where solutes are excluded. This means that the interbilayer solute concentrations are lower than exterior concentrations for MLV systems produced by this method. MLV produced by this method are known as solute deficient. One consequence of this is that osmotic differences arise between adjacent compartments giving rise to membranes in different parts of the vesicle which are 'stressed' – either under compression or expansion. This stress gives rise to the instability experienced with this type of MLV. The trapped volume can be enhanced by including charged molecules in the bilayer; these produce electrostatic repulsion between successive lamellae of the vesicles, thus swelling the bilayers.

Modifications of the thin-film method

Sequential extrusion
It is possible to achieve a reduction in the size and polydispersity of MLV by sequential extrusion through polycarbonate filters. The pressure required is fairly low, i.e. less than 25 N/m^2 (Olson *et al.*, 1979). Jousma *et al.* (1987) found that both multiple extrusion through the same filter and sequential extrusion reduced the number of bilayers per vesicle. They also observed that multiple extrusion ($\times5$) through membranes with a pore diameter of 0.2 μm yielded a dispersion of mainly unilamellar vesicles. They stated, however, that the number of bilayers after extrusion will depend not only on the diameter of the vesicle used, but also on the lamellarity of the vesicles in the initial preparation. In the case of vesicles of the same size but with differing lamellarity after the same number of extrusions, the vesicle with the most bilayers will initially still retain the greatest lamellarity. The method does however produce MLV with a reasonably homogeneous size distribution centred on the pore size of the filter. This technique thus provides a reproducible protocol for the formation of MLV of definite size distribution while maintaining high efficiency of entrapped aqueous volumes. This technique can give an apparent increase in encapsulation efficiency when the extrusions are carried out in the original 'mother liquor' (Olson *et al.*, 1979). Extrusion is convenient, easily reproducible, does not introduce impurities into the vesicles, and does not induce breakdown of sensitive molecules such as phospholipids. Vesicles extruded through a 0.22 μm or smaller filter allow the production of a sterile preparation for *in vivo* injection.

In an obvious extension of this work, Hope *et al.* (1985) produced large unilamellar vesicles by repeated extrusion through polycarbonate filters (100 nm pore size). They reasoned that extrusion through smaller pores may result in the production of vesicles of sufficiently small size that the presence of inner lamellae is unlikely. (It is generally assumed that, due to geometric constraints, vesicles of diameter less than 100 nm can contain only one bilayer.)

Dehydration–rehydration
This technique involves the drying of either MLV (or unilamellar vesicles) in aqueous medium followed by controlled rehydration above the phase transition temperature of the lipid (Kirby and Gregoriadis, 1984a, b; Shew and Deamer, 1985). Both freeze drying (lyophilization) and direct drying by vacuum or convection are effective. Before freeze drying, the solute to be encapsulated is added to the empty vesicle preparation. During the drying process, vesicles are concentrated concomitantly with solute, and at some point fuse into large aggregates. If rehydration is performed gradually, MLV are produced. As solute is added before the lyophilization stage it is already present in many of the interbilayer spaces, and so efficient trapping occurs. Kirby and Gregoriadis (1984a, b) observed that sonicating the original MLV preparation leads to higher entrapment levels.

Recently, Gregoriadis *et al.* (1990) reported that microfluidization of DRV produces liposomes down to 100 nm in diameter which retain 10–100% of the originally entrapped solute. Solute retention depends on the number of microfluidization cycles, the medium in which microfluidization is carried out, and on whether or not, before processing, DRV are previously separated from the unentrapped solute. Compared with procedures (Mayhew *et al.*, 1984) that employ much larger amounts of lipid to achieve efficient entrapment, the present approach provides preparations with augmented solute to lipid mass ratios and should therefore be more economical.

The DRV method is a variant of an earlier one developed by Pick (1981), after Kasahara and Hinckle (1977), to produce LUV. In Pick's method the material to be entrapped is also introduced into vesicles after they have been formed. In this case, a freezing and thawing process is used to rupture and re-fuse the SUV, during which time the solute equilibrates between the inside and outside, and the vesicles themselves fuse and markedly increase in size.

Freeze–thaw

Repeated freezing and thawing of MLV preparations results in the formation of MLV with increased trapped volume and enhanced solute entrapment (Mayer *et al.*, 1985). Frozen-and-thawed MLV (FATMLV) are obtained by freezing MLV in liquid nitrogen and thawing the sample in a water bath at the same temperature used for hydration. The freeze–thaw cycle is usually repeated several times. The mechanism by which freezing and thawing improves the swelling of MLV and induces solute equilibrium is not clear.

Formation from templates (pro-liposomes)

Recently a method has been devised in which the lipids are dried down on to a finely divided particulate support, such as powdered sodium chloride, or sorbitol or other polysaccharide (Payne *et al.*, 1986a, b). Upon adding water, with mixing on a whirlimixer, to the dried lipid-coated powder (known as 'pro-liposomes') the lipids swell, while the support rapidly dissolves to give a suspension of MLV in aqueous solution. As with other methods of preparation, efficient dispersion is brought about when carried out above the phase transition temperature of the lipids. The particulate size of the carrier influences the size and heterogeneity of the liposomes finally produced, but in general the size and lamellarity seem to be smaller than for MLV produced by the conventional thin-film method.

One of the advantages of this technique is that, because the molecular area for deposition is considerably greater than that calculated on the basis of the geometric surface of the granule (and obviously that of a round-bottomed flask) the bilayer built up on the sorbitol is equivalent to a continuous coating of just a few bilayers over the whole surface. Because water has much greater access to lipid in the form of pro-liposomes than when dried on to a glass wall, vesicles are formed much more rapidly by this method, and a higher proportion of smaller vesicles is obtained.

This method of liposome preparation is a good one for commerical applications where large quantities of pro-liposome can be prepared and stored dry before use in sealed vials, then resuspended when required to give batches of liposomes of reproducible size over a long period of time. This method also overcomes problems encountered when storing liposomes in either liquid, dry or frozen form, and is clearly best suited for preparations where the material to be entrapped is incorporated in the lipid bilayer. It is of much less use in the entrapment of hydrophilic material.

In a related procedure, Schneider and Lamy (1979) reported that the addition of hydrophilic polymeric materials, such as dextran, to a preformed liposomal dispersion prior to lyophilization produced a free-flowing powder which reconstituted readily in aqueous media.

Reverse-phase evaporation

Recently two new methods of producing MLV based on emulsion technology have been published: the stable plurilamellar (SPLV) method of Gruner *et al.* (1985) and the reverse-phase evaporation (MLV-REV) process of Pidgeon *et al.* (1986, 1987). Although both processes are claimed to give MLV, a recent paper by Talsma *et al.* (1987) suggests that the vesicles formed by these processes are not actually multilamellar but rather multivesicular, i.e. LOV. These methods are modifications of the reverse-phase-evaporation (REV) technique of Szoka and Papahadjopoulos (1978) generally used to make large uni- or oligolamellar vesicles. By slightly altering several 'critical' parameters in the REV procedure these workers were able to produce MLV with higher entrapment than those produced by the thin-film method.

When compared to MLV made via conventional methods, these newer preparations have different (higher) stabilities, better entrapment efficiencies, and improved biological effects, even when made from the same materials and appearing the same under the electron microscope. Depending upon the lipid and solute, the SPLV process gives entrapment values of 35–40%, while the MLV-REV procedure frequently gives values greater than 65%. These values obviously depend upon the container used to prepare the vesicles, the solute and its concentration, and the preparation conditions, e.g. temperature (Pidgeon *et al.*, 1987).

The major difference between the REV method and the newer methods is the difference in the ratio of organic solvent to initial aqueous phase used to make the water-in-oil emulsion. In the SPLV method of Gruner *et al.* (1985) typically 3 ml of solvent and 0.3 ml of aqueous phase are emulsified, whereas in the MLV-REV method of Pidgeon *et al.* (1987) 10 ml of solvent is emulsified with 0.3 or 0.5 ml of aqueous phase. In both cases, because of the lower initial aqueous phase/ organic solvent ratio, the internal structure of the final vesicle differs from REV in that they lack a large aqueous core. Furthermore, unlike MLV formed by the thin-film technique, any solute these MLV contain is evenly distributed throughout the different aqueous compartments. Consequently these vesicles are not under stress, thus accounting for their greater stability compared with other types of MLV.

Two-step emulsification

In this method, an aqueous solution is emulsified in an organic solvent. This primary emulsion is then re-emulsified in a second aqueous solution to produce a 'water-in-oil-in-water' system. The phospholipid contained in the system is oriented at each of the oil–water interfaces in such a way that multicompartment 'vesicles' are obtained. The inner monolayer is obtained in the primary emulsification step, and the outer layer is created as a consequence of the second emulsification procedure. In the 'vesicle' the two phospholipid layers are separated from each other by a thin film of organic solvent. Removal of this solvent results in intermediate-sized unilamellar vesicles. The theoretical entrapment yield is 100% and can approach this value on occasions, depending upon the nature and the concentration of the material to be entrapped (Batelle Memorial Institute, 1979).

Dilution from a gel

This new method for the preparation of liposomes avoids the use of pharmaceutically unacceptable solvents and energy intensive procedures such as sonication which cannot readily be scaled up (Perrett *et al.*, 1991). The method is based on the initial formation of a pro-liposome mixture containing lipid, ethanol and water, which is converted to liposomes by a simple dilution step. Water-soluble drugs or markers included in the pro-liposome mixture are trapped within the liposomes formed by this technique with very high efficiency. The liposomes produced by the pro-liposome technique are multilamellar vesicles with diameters centred around 0.5 μm, although the average diameter of liposomes formed from mixtures containing charged lipids is found to be smaller than that obtained with neutral lipids. The main disadvantage of this technique is the relatively high concentrations of ethanol present in the pro-liposome mixture, but this can be used to advantage as it allows the storage of the pro-liposome mixture in a sterile form suitable for the subsequent *in situ* formation of liposomes.

Simple dispersion

To date this technique has only been used for non-ionic surfactants (Handjani-Vila *et al.*, 1979). An equivalent mixture of surfactant and aqueous phase is stirred to form a homogeneous lamellar phase which is then diluted and homogenized by temperature-controlled agitation or centrifugation. This method has been used to produce large (kilogram) scale, solvent-free vesicle preparations with diameters of several micrometres, suitable for use in drug delivery and cosmetics.

10.4.2 Small unilamellar vesicles

These are those vesicles at the lowest limit of size. A bilayer vesicle composed of lipid has a size limit, below which the radius of curvature at the inner monolayer becomes so acute that lipid packing constraints prevent smaller diameters being achieved. For egg phosphatidylcholine this diameter is about 25 nm; vesicles of this diameter exhibit a 2:1 excess of lipid in the outer layer compared with the inner monolayer (Huang, 1969). Since, according to the definition, SUV are at, or close to, the lower size limit, they will form a relatively homogeneous population in terms of size. However, because of the highly constrained nature of the lipid molecules, SUV are generally thermodynamically unstable entities exhibiting a marked tendency to increase in size with time.

Sonication

Sonic dispersal of surfactant is the most general procedure for vesicle formation. In fact most surfactant vesicles have been made by this method. It was the earliest method used for forming single-compartment small layer vesicles (Huang, 1969). Usually sonication is performed on an MLV preparation of the lipid, athough certain workers have sonicated a dispersal of the lipid in water. However, since any vesicles initially present are broken down in the process, it is not necessary to be too concerned about their initial size.

Either a probe or a bath sonicator can be used. In the former case, a metal probe is directly inserted into the sample. In the latter the sample is sealed in a glass vial and suspended in an ultrasonic cleaning bath. Inserting the probe directly into the lipid dispersion results in greater energy input and generally smaller vesicle size. Although the probe sonicator gives fast and reliable results and is therefore often used, it has a number of drawbacks, including a limit on the size of sample that can be handled, and the production of aerosols and heat during the sonication process. In addition, contaminating fragments of titanium probe are released during sonication and these obviously need to be removed from the samples by either centrifugation, ultrafiltration or gel permeation chromatography (New, 1990). These drawbacks are overcome by the use of bath sonication. However, there are also several drawbacks associated with the use of bath sonicators: the process is more time consuming, the sample size, container, and location in the bath are critical (so that reproducibility can be problematic) and the residual concentration of large particles is greater (Szoka and Papahadjopoulos, 1980).

During sonication the MLV structure is broken down and SUV with a high radius of curvature are formed. Sonication has to be carried out above the T_c of the amphiphile or else the vesicles are formed containing structural defects in the bilayer (Lawaczeck *et al.*, 1976). Such vesicles are unstable and highly permeable. Structural defects can be annealed simply by incubating the vesicles above the T_c of the lipid. The length of time required to ensure complete annealing is dependent on the T_c of the lipid. A few minutes is sufficient when the temperature is substantially above (+10°C) and an hour when only slightly (+3°C) above the T_c. If the lipids are prone to oxidation the sonication needs to be performed under an inert atmosphere of nitrogen or argon. As already stated the type of sonicator, the applied power and the time and temperature of the sonication need to be specified as these parameters can affect the final vesicle. The minimal diameter of SUV prepared by sonication varies depending on lipid composition and time of sonication (Menger *et al.*, 1989) but generally sizes less than 50 nm are obtained. Inclusion of charged phospholipids in the bilayer facilitates a reduction in vesicle size, whereas the presence of cholesterol produces relatively large vesicles (Forge *et al.*, 1978).

Generally, increasing the sonication for a given lipid dispersion results in an exponential decrease in size. At the plateau region of sonication liposomes are optically transparent and are relatively homodisperse. Vesicles of more uniform size distribution can be obtained using gel filtration on Sepharose 4B (Huang, 1969) or ultracentrifugation (Cornell *et al.*, 1980).

Infusion methods

Hydration of lipids from a polar organic solvent can result in the formation of SUV. Batzri and Korn (1973) found that injection of an ethanolic solution of egg phospholipid through a small-bore Hamilton syringe into a well-stirred aqueous (salt) solution above the phase transition of the lipid produces a reasonably homogeneous preparation of unilamellar vesicles, diameter about 25 nm. In this procedure dilution of the ethanolic lipid solution is performed rapidly. Vesicle formation, which occurs upon solvent dilution, presumably occurs locally near the point of injection. As the lipid is used in ethanolic solution the method has no degrading effect on the lipid and, although the presence of ethanol is a disadvantage, it can easily be removed by dialysis.

By varying injection conditions, Kremer *et al.* (1977) modified the technique to allow the formation of SUV of variable but controlled diameter (15–50 nm). Although very high injection rates combined with rapid stirring of the aqueous solution produced vesicles with smaller polydispersity, only the lipid concentration in the injected ethanol influenced the size of the vesicles markedly (Kremer *et al.*, 1977). The main problem with the technique is that the resulting preparations are relatively dilute so that the encapsulation efficiency is low. The procedure has the advantage that it is applicable to a wide variety of lipids and can be completed in less than an hour.

Detergent dilution

Detergent dilution has provided a means to form both SUV and LUV depending on the surfactant used. Lipids are cosolubilized with a surfactant to form mixed micelles and the surfactant is subsequently removed by gel filtration (Brunner *et al.*, 1976), dialysis (Rhoden and Goldin, 1979) or dilution, whereupon the unilamellar vesicles form spontaneously. This is a gentle method where no strong mechanical forces and no high temperatures are applied. It is important to minimize/remove all detergent. However, as the removal of the detergent is performed using techniques which inevitably remove other small soluble water molecules (e.g. dialysis), this method is not very efficient in terms of the resulting percentage entrapment. Furthermore, many of the techniques used to remove detergent involve a considerable dilution of the vesicle preparation, so that techniques such as ultrafiltration may have to be used to concentrate the vesicles. The technique does, however, provide the ability to vary the size of the vesicles by allowing precise control of the conditions of detergent removal (giving vesicles of very high size homogeneity).

Vesicle size depends upon experimental conditions and in particular the type of detergent used. For example, sodium cholate produces unilamellar vesicles in the range 20–100 nm, while N-octylglucoside gives unilamellar vesicles of the order of 100–250 nm and n-octyltetraoxyethylene unilamellar or oligolamellar vesicles greater than 500 nm. The most frequently used detergents are sodium cholate, alkyl glycoside and Triton-X 100. The use of the alkyl glycosides enables a wide variety of vesicles sizes to be obtained by simply lengthening the alkyl chain of the detergent. For example, the substitution of octyl glycoside by heptyl glycoside reduces the diameter from 180 to 80 nm, with an intermediate value of 140 nm being obtained with a 1:1 mixture of the two detergents (New, 1990).

For a particular detergent, under conditions of constant pH, ionic strength and temperature, the factors affecting size and homogeneity are: the rate of removal of the detergent, the initial lipid/detergent ratio, and the presence of cholesterol and/or charged lipids. It is essential for a homogeneous preparation that the detergent is removed at a constant rate.

French press

This process was developed to overcome the problems inherent in ultrasonic irradiation. In a French press the lipid suspension is forced through an orifice and subjected to very high hydraulic pressures of up to $13\,000\,N/m^2$. These forces are sufficient to cause mechanical disruption of MLV; the higher the force the smaller the vesicular fragments. From the theory of hydrophobic interactions, these fragments will form vesicles spontaneously. The technique yields rather homogeneous uni- or oligolamellar vesicles of intermediate sizes (30–80 nm depending upon the pressure used) from an MLV preparation (Barenholz *et al.*, 1977, 1979; Hamilton *et al.*, 1980). The size of the vesicles formed is determined by the original size of the membrane fragments created by the shear forces across the orifice. The homogeneity of the resulting vesicle suspension is inversely proportional to the flow rate through the orifice. The slower the flow rate, the better the chances of disrupting the vesicles as they pass through the orifice: although it is difficult to remove the larger vesicles completely.

Vesicles prepared by this technique, although small, are somewhat larger than those prepared by sonication, and are thus less likely to suffer from structural defects and instabilities. Advantages of this technique include the fact that high lipid concentrations can be used.

Formation from templates

Lasic and coworkers (1987) have described a quick and simple technique to produce SUV. The process involves coating an inert support with lipid doped with a small amount of charged surfactant, typically cetrimide. The size and distribution of the vesicles produced depends upon the size and homogeneity of the support used. The SUV formed are larger than the minimal size formed by sonication.

10.4.3 Large unilamellar vesicles

These vesicles have diameters in the size range 100–1000 nm. Because of the large size of the vesicles attainable, a high entrapment can often be achieved. If properly made there is no reason why fusion and/or growth of these vesicles should occur. However, depending upon their lipid composition, these vesicles may be more leaky to entrapped substances than the equivalent MLV preparation.

Reverse-phase evaporation

Since its inception the reverse-phase evaporation (REV) protocol has until recently been the preferred method of preparing LUV if high solute entrapment is required (Szoka and Papahadjopoulos, 1978; Szoka et al., 1980). This was despite the inherent problems associated with hydration of lipid from organic phases and the need for column chromatography or dialysis to remove traces of solvent from vesicle preparations. In the REV method multicompartment vesicles are prepared by the introduction of an aqueous buffer into an organic solution of surfactant. A water-in-organic solvent emulsion is produced by sonicating the two phases. The organic solvent is then removed by evaporation to collapse the reverse micelles to leave a gel which subsequently collapses to form vesicles. Vesicular dispersions prepared in this manner are mixtures of LUV contaminated with MLV and LOV.

There are number of critical parameters/stages in the preparation of vesicles by this method, namely: the solvent/buffer ratio (volume ratios and lipid concentrations are critical factors in the protocol) and the formation of the gel (which is dependent upon the surface area/volume ratio of the preparation during the evaporation procedure).

Vesicles prepared by this method have four times larger aqueous compartments than those formed by the simple swelling of surfactants (Szoka and Papahadjopoulos, 1978). It is thought that entrapment of a hydrophobic solute is optimized in a reverse-phase procedure where the solute is cosolubilized (with surfactant) in the organic phase (Hope et al., 1986). In addition to its high trapping efficiencies, it is able to produce vesicles at high lipid concentrations. The homogeneity of the preparation can be increased by filtration through polycarbonate filters of defined size (Olson et al., 1979; Szoka et al., 1980). The disadvantages are that the components are exposed to organic solvents and sonication which may be damaging, and that the amount of amphiphile required is relatively large. In addition the preparation conditions may be too severe for some labile materials.

Extrusion

As indicated above, repeated extrusion of MLV under moderate pressures $(260\,N/m^2)$ through two stacked polycarbonate filters of 0.1 μm pore size results in a relatively homogeneous population of LUV (LUVETS) with a mean diameter of approximately 90 nm (Hope et al., 1985). The considerable advantages of this technique are that it is rapid (with preparation times of the order of 10 min), works directly from MLV, and can be used for a variety of surfactant molecules. Mayer et al. (1986) extended this work and showed that, using similar procedures, homogeneously sized unilamellar and plurilamellar vesicles (VET) could be produced by utilizing filters with pore sizes ranging from 30 to 400 nm. Prior to extrusion, these vesicles underwent freeze–thawing (Mayer et al., 1985). The unilamellarity and trapping efficiencies of these vesicles were significantly enhanced by the freeze–thawing. Filters with larger pore sizes give rise to larger systems with higher trapping efficiencies and increasingly mutlilamellar character. An important feature of the procedure is the fact that it is particularly suitable for high lipid concentrations (up to $400\,mg\;ml^{-1}$).

Although the technique is limited to producing LUV in the diameter range 40–150 nm, the most commonly used LUV systems fall within this size range.

The above method of preparation also has the advantage of being able to generate large unilamellar vesicles from long-chain saturated lipids (Nayar *et al.*, 1989). Other procedures have serious problems in the production of large unilamellar vesicles composed of unsaturated lipids, e.g. the limited solubility of long-chain saturated lipids in many common organic solvents. In fact most vesicle studies have employed unsaturated lipids in the liquid-crystal (fluid) state. Nayar *et al.* (1989) showed that it was possible to use the extrusion procedure to produce LUV of defined size from saturated phosphatidylcholines of chain lengths from C_{14} to C_{20}, provided they were extruded at temperatures above their phase transition (i.e. they were in the liquid-crystal state).

Recently, MacDonald *et al.* (1991) have developed a hand-driven extrusion apparatus, which does not require gas under pressure and is easily made at low cost. Although it has a capacity of less than 1 ml, it readily meets the needs of investigators wanting to prepare small quantities of large unilamellar vesicles.

Infusion methods

Hydration of lipids from a non-polar organic phase can result in the formation of large unilamellar vesicles (diameter 150–250 nm). Typically solvents such as ether (Deamer and Bangham, 1976), or petroleum ether (Schieren *et al.*, 1978) are used, although solvents such as pentane (Deamer, 1984) have been used, and methanol is sometimes added to increase the solubility of the lipids in the solvents (Deamer, 1984). In the infusion method, the lipid dissolved in the immiscible organic solvent is injected very slowly into an aqueous solution through a narrow bore needle at a temperature high enough for rapid evaporation of the solvent (Deamer and Bangham, 1976). The mechanism whereby LUV are formed is not clearly understood. Since the solvent is removed at the same rate as it is introduced, there is no limit to the final concentration of lipid that can be achieved. Vesicle size is independent of initial lipid concentration. The resulting preparation is relatively dilute and the size distribution heterogeneous. The technique also has the advantage of being fairly gentle in that no sonication is required.

Formation from templates

Lasic (1988) has reported a simple method for the preparation of homogeneous populations of unilamellar vesicles with a diameter of around 1 μm. The vesicles are formed spontaneously upon the swelling of slightly charged phospholipid films deposited on special (insoluble) supports in excess water. Upon addition of water, vesicles are formed directly, and without any additional physical or chemical treatment. The formation of multilamellar structures is prevented by inducing a surface charge on the bilayers. The size distribution can be controlled by the topography of the surface upon which the phospholipid film is deposited.

10.4.4 Giant vesicles

These vesicles have diameters in the region of 1000–10 000 nm. Only a few methods of producing giant vesicles (GV) have been reported. As a result of their large size and the presence of a single lipid membrane their mechanical stability and retention of solutes may not be high.

Slow hydration

In a variation of the method of Bangham *et al.* (1965), Reeves and Dowben (1969) hydrated the dried lipid film, first with wet nitrogen gas that had passed through a water phase, followed by addition of water and gentle swirling. The slower hydration was found to produce very large vesicles up to several hundred micrometers in diameter, but only in the absence of ions or protein. Hub *et al.* (1982) produced GV by hydrating a thin lipid film without agitation for several hours. This procedure was carried out at a relatively high temperature (70°C, higher than the phase transition temperature of the lipid). During lyophilization, the lipid film detaches and forms vesicles of <1.0 μm diameter. These vesicles grow over a few hours to 5–10 mm. Reshaking for less than 5 s allows formation of vesicles of 1–50 μm size. The majority of vesicles formed in this manner are unilamellar, and are virtually all uncontaminated by multilamellar vesicles, although a few contain smaller unilamellar vesicles. The method described is simple and applicable to a range of lipid materials, producing a high yield of single bilayer vesicles. Since the original authors intended the vesicle preparation for electrical measurements, they do not record any encapsulation data. However, because of their large size, vesicles of this type would be expected to exhibit a high percentage capture.

Reverse-phase evaporation

Kim and Martin (1981) described a method for preparing cell-size vesicles with a high captured volume and defined size. The method, which is much more complicated than the method of Hub *et al.* (1982), is similar in idea to the reverse-phase evaporation method of Szoka and Papahadjopoulos (1978) and involves removing organic solvents from microscopic chloroform spherules containing smaller water droplets within. The average diameter of the vesicles was 9.2 μm, with a standard deviation of 3.0 μm. The authors reported a high trapped volume, almost an order of magnitude greater than that reported by Szoka and Papahadjopoulos (1978). Depending upon the lipid(s) and the nature of the entrapped material, encapsulations of over 80% have been reached.

Freeze–thawing

If SUV are freeze–thawed in the presence of a high concentration of electrolytes and subsequently dialysed against a lower concentration, unilamellar vesicles with a diameter of more than 10 μm are formed. The formation of these giant liposomes (GV) is caused by the influx of water due to differences in osmotic pressure (Oku and MacDonald, 1983a, b).

10.4.5 Miscellaneous

There are a number of methods of vesicle preparation that are specific to a particular type or combination of lipids. For example, MLV can be induced to reassemble 'spontaneously' into unilamellar vesicles by simply changing pH (Hauser and Gains, 1982). This process, termed 'pH induced vesiculization', is an electrostatic phenomenon and obviously depends upon the presence of a charged lipid. Oleic acid vesicles (ufasomes) are also prepared by changing solution pH (Gebicki and Hicks, 1973). Small unilamellar vesicles composed of acidic phospholipid can be made to aggregate in the presence of calcium to produce large unilamellar vesicles (Papahadjopoulos *et al.*, 1975).

10.5 Industrial preparation

To date most work examining vesicles has concentrated on overcoming the problems of *in vitro* and *in vivo* stability. Practical extension of this work has been limited by the lack of any technology that might allow the reproducible manufacture of liposomes of predictable, uniform size distribution. Many research and industrial groups are now at the stage of scaling up from laboratory procedures to large volume industrial processes. Consequently, cost and scale-up factors have begun to take on increased significance as well as quality control problems such as consistency and reproducibility on a commercial scale are required. While a number of different methods have been described for preparing liposomes most tend to be impractical for commercial purposes. For example, the methods tend to be cumbersome, poorly reproducible, or limited in production batch size, employ harsh procedures (such as sonication) or involve organic solvents. Although some industrial applications, notably diagnostics, have not been hampered by the scale-up problem (200 mg of lipid can produce 2000 tests), the problem of large-scale production remains.

An ideal method of preparation should be simple, standardized, reproducible, and cost effective, and the yield (on both micro- and macroscales) should be homogeneous and stable for sufficient periods of time. Moreover, the size of vesicles should be controllable within wide limits. Although several methods are now available, as yet no method encompasses all of these objectives.

10.5.1 Detergent dialysis

The commercially available Liposomat (Dianorm, Munich, Germany) and Lipoprep (Bachofer, Reutlinger, Germany) are both based on the method of detergent removal from detergent–lipid comicelles by fast and controlled flow dialysis. The method relies on passing the solution containing the comicelles over a flat dialysis membrane at the same time as the dialysis fluid is passed over the other side of the membrane. The comicellar solution is repeatedly passed over the dialysis membrane until all the detergent is removed. Indeed the vesicles are formed within a few minutes as only one passage across the dialysis membrane is required to produce them. However, after one passage through the dialyser the vesicles formed contain about 50–60% of residual detergent. Further cycles are necessary to reduce the detergent content. After about 60 min a residual detergent concentration of about 2–3% is achieved, while 90 min is required for detergent concentrations less than 1%. Continual dialysis leads to a practically detergent-free vesicle preparation. As the procedure has to be performed at constant temperature the temperature of the dialysis unit is accurately controlled.

As the Liposomat is designed for the rapid preparation of 5–10 ml of vesicular suspension (lipid concentration greater than $20 \, mg \, ml^{-1}$) large volumes take considerably longer to prepare. For example, a production volume of 200 ml at a flow rate of $0.5 \, ml \, min^{-1}$ takes 400 min for a single pass through the dialyser. Recycling of the solution to reduce detergent concentrations down to a sufficiently low concentration takes hours. In order to avoid recycling through the Liposomat, the company has designed a second dialysis system which can remove the residual detergent within 30 min, such that 200 ml of vesicle suspension can be produced within 7–8 h. The company has recently announced the availability of a pilot-size plant (Liposomat-Plant) which is capable of producing up to thousands of litres a day.

10.5.2 Extrusion

Another commerically available vesicle device is the Lipex Extruder (Lipex Biomembranes Inc., Vancouver, Canada). Unfortunately the present 10 ml maximum capacity is too low to be of use from an industrial viewpoint. At the moment the company does not produce a pilot plant or industrial-size extruder. When preparing vesicles using the extruder, the dry lipid is first hydrated and dispersed by vortexing to produce MLV. The manufacturers recommend freeze–thawing the vesicle preparation before introduction into the extruder. Depending on the size of the vesicles required, the extruder is assembled using polycarbonate filters of the appropriate size. The use of two stacked filters is recommended. The vesicle preparation may then be introduced into the extruder at a temperature above the phase transition of the lipid. Unfortunately, the standard model does not come with the temperature-control unit required for extruding lipids with high phase transitions. To avoid any degradation of the lipid the extruder is pressurized with nitrogen gas. The vesicles are then repeatedly passed (at least 10 times) through the barrel of the extruder to produce vesicles of the required size.

10.5.3 Microfluidizer

The microfluidizer (Microfluidics Corp., Newton, MA, USA) is a relatively new device that provides a practical and convenient means of preparing liposomes continuously and reproducibly, in large quantities and with excellent encapsula-tion efficiency (Mayhew *et al.*, 1984, 1985). Presently pilot-production size has the capability of producing about 5 500 l/h. Larger scale production units are under development. The microfluidization process exploits the dynamic interaction of two fluid streams in precisely defined microchannels. Fluid is pumped under high pressure (3 300 n/m^2) through a filter (5 μm) into the interaction chamber where it is separated into two streams which interact at extremely high velocities in the dimensionally defined microchannels. The suspension can be used after a single pass or recycled through a reservoir. When using the laboratory-scale model, sample volumes typically of 15 ml with a flow rate of 90 ml min^{-1} and a rate of four cycles per minute are used. Under these conditions a vesicle preparation can be completed within 10 min.

By altering lipid composition and concentration and recycling time it is possible to produce vesicles of controlled size and narrow size distribution. For example, using soy lecithin and cholesterol and processing conditions of 300 N/m^2 and 10 recycles results in a clear, stable liposome preparation of 44 nm. The SUV formed by this process are larger than the minimal size formed by sonication and significant amounts of larger particles are also present. Encapsulation efficiencies as high as 75% have been obtained (Mayhew *et al.*, 1985). The technique also has the advantage of being able to process samples with a very high proportion of lipid (>20%). Such concentrated suspensions can not be handled by other more conventional methods of preparing MLV.

10.5.4 French press

Although commercially available from SLM-Aminco Inc. (Urbana, IL, USA) the main drawback with the use of the French press as a means of vesicle production is the high initial cost of purchasing a full new system which consists of an electric hydraulic press and pressure cell. In addition, while there are two pressure cells available, the larger one only holds up to 40 ml. Furthermore, when using the French press, one is essentially working with an open system. Therefore, when using radioactive or otherwise hazardous material, utmost care must be taken to avoid spilling and splashing during the filling and extrusion steps. Under most circumstances the French press would not be considered suitable for large-scale production of vesicles.

10.6 Vesicle stability

In spite of their thermodynamic instability, vesicles once formed are kinetically stable for weeks, even months. The slow leaving rates of the monomers from vesicles are responsible for their kinetic stabilities. However, on prolonged standing, vesicles frequently undergo aggregation or fusion to form larger and more polydisperse entities and ultimately form the lamellar phases from which they were initially formed.

Vesicle stability is a function of a number of different parameters, in particular their composition and method of preparation (de Gier *et al.*, 1968; Gebicki and Hicks, 1976). It is generally accepted that the size stability of MLV is greater than that of 'minimal' size SUV, although some workers have variously attributed the reduced stability of SUV, to preparation below the T_c, to pH changes, or to the presence of contaminants. With respect to the composition of the external medium it should be noted that large vesicles, in contrast to SUV, exhibit a marked osmotic sensitivity. Vesicles without a net charge, or polymeric hydrophilic groups, can be expected to aggregate readily, due to attractive van der Waals interaction. In contrast, vesicles with a sufficiently high surface charge will have little tendency to aggregate. As yet, only a few studies on the stability of surfactant vesicles have been published. MLV and SUV prepared by sonication of dioctyldimethylammonium chloride appear to be stable at ambient temperature for several weeks (Romero *et al.*, 1978). Similarily, MLV produced from non-ionic surfactants of the polyoxyethylene[17]glycerol-1,2-dialkyl ether type are stable with respect to size for several months (Lawrence, 1992).

Retention of any encapsulated material on storage is important if vesicles are to be widely used. Continuous and rapid leakage of a variety of materials from vesicles is well documented in the literature. Solute retention during storage is dependent on the nature of the encapsulated compound(s), the type of vesicle, the composition of the bilayer and the external medium. By careful manipulation of the components of the vesicle, for example, incorporation of cholesterol or long-chain saturated phospholipids, it is possible to reduce solute leakage, particularly of water-soluble compounds (Demel *et al.*, 1968). Single-chain polyglycerol surfactant vesicles appear to be leaky with respect to encapsulated water-soluble material, although incorporation of cholesterol into the vesicles reduces loss (Rogerson, 1986). In contrast, vesicles made from triethoxycholesterol (Patel *et al.*, 1984) gave an initial loss of 15% of the encapsulated material, but thereafter remained stable.

If vesicles are to be used for drug delivery purposes they have the added limitation that they must be stable *in vivo* as well as *in vitro*. Again the composition and type of vesicle to a large extent determines the stability. For example MLV are removed from the circulation at a much quicker rate than SUV. Vesicles composed of phospholipids containing long, saturated 'solid-phase' chains are more stable in plasma than those prepared from shorter, unsaturated 'fluid-phase' lipids (Senior and Gregoriadis, 1982; Allen and Chonn, 1987).

In order to overcome the problems associated with vesicle stability, considerable effort has been placed on producing either a reconstitutable form of vesicle, for example by freeze drying (Fransen *et al.*, 1986), or a stable vesicle by the use of polymerizable amphiphiles (Regen *et al.*, 1982; ishiwatari and Fendler, 1984).

10.7 Characterization of vesicle preparations

The *in vitro* and *in vivo* behaviour of vesicles is determined to a large extent by factors such as physical size and surface properties, although the chemical composition, membrane permeability, quantity of entrapped solutes, as well as the quality and purity of the starting materials are also of consequence. It is important therefore to have as much information as possible regarding these parameters. Vesicular preparations can be characterized by a number of techniques. Table 10.3 gives some of the more common techniques; however, the list is not exhaustive. As most of the biophysical techniques for measuring the size of vesicles require a homogeneous population with a well-defined shape, techniques such as light scattering, nuclear magnetic resonance (NMR), or ultracentrifugation are most often used to follow size changes.

Table 10.3 *Physicochemical characterization of vesicles*

Technique	Comments
Solute entrapment (Kunitake *et al.*, 1985)	Entrapment of water-soluble substances, such as glucose and sucrose, is often used as a means of proving the existence of closed vesicles
Osmotic shock techniques (Bittman and Blau, 1986)	Shrinkage in vesicle size in the presence of a hypermolar solution is often used to provide evidence for the presence of closed vesicles
Fluorescence quenching (Bittman and Blau, 1986)	The fluorescence quenching of carboxyfluorescien upon entrapment in vesicles is useful for proving the formation of closed vesicles
Transmission electron microscopy (Kunitake *et al.*, 1981)	Freeze fracture and negative staining are used to determine aggregate morphology. Need care in interpreting negative staining data (Talmon, 1983; Lukac and Perovic, 1985)
Video-enhanced differential interference contrast microscopy (Brady *et al.*, 1984; Kachar *et al.*, 1984)	Technique allows rapid, direct characterization of the vesicles. It enables the analysis of the vesicles in real time, freeze frame, slow motion or time lapse
Fluorescence probes (Kano *et al.*, 1979)	Used in structural studies of vesicle fluidity, phase transition, mobility and lateral diffusion, polarity and surface potential
Nuclear magnetic resonance (Mason and Huang, 1978; Gabriel and Roberts, 1986)	Yields information about, the ratio of lipids molecules on the outside to the inside of the vesicle, the number of bilayers present, and the orientation of the lipid molecules within the vesicle
Differential scanning calorimetry (Malliaris, 1988)	Used to examine the thermotropic (reversible gel to liquid crystalline) phase transition of the vesicle
Small angle X-ray scattering (Pidgeon *et al.*, 1987)	Used to gain information about the number and width of the bilayers in the vesicle

Table 10.3   *(Continued)*

Technique	Comments
Intrinsic viscosity (Mason and Huang, 1978; Kano *et al.*, 1979)	The intrinsic viscosity of a vesicular preparation supplies information about vesicle shape and hydration
Static and dynamic light scattering (Kremer *et al.*, 1977; van Zanten and Monobouquette, 1992)	Together these techniques yield the anhydrous and hydrodynamic radii of the vesicles
Gel exclusion chromatography (Lesieur *et al.*, 1991)	Quick and convenient method for both fractionation and average size determination
Turbidity (Barrow and Lentz, 1980)	Simple method for estimating the level of large contamination of small unilamellar vesicle preparations
Analytical ultracentrifugation (Ruf *et al.*, 1989)	Supplies the hydrodynamic radius of the vesicles
Sedimentation field flow fractionation (Kirkland *et al.*, 1980)	Determination of size distribution of vesicular preparations
Eletrophoresis (Bangham *et al.*, 1974)	Surface charge density of vesicles

10.8 Conclusion

It is possible to prepare a wide range of vesicle types using a number of different methods. The choice between different methods will depend to a large extent on the intended application of a vesicular preparation. However, it is usually possible to find one method of preparation that will provide a satisfactory approximation to the desired characteristics.

References

Allen, T. M. and Chonn, A. (1987) Large unilamellar liposomes with low uptake into the reticuloendothelial system. *FEBS Lett.*, **223**, 42–46

Ballie, A. J., Florence, A. T., Hume, I. R., Muirhead, G. T. and Rogerson, A. (1985) The preparation and properties of niosomes–non-ionic surfactant vesicles. *J. Pharm. Pharmacol.*, **37**, 863–868

Assadullahi, T. P., Hider, R. C. and McAuely, A. J. (1991) Liposome formation from synthetic polyhydroxyl lipids. *Biochim Biophys. Acta*, **1083**, 217–276

Bangham, A. D., Standish, M. M. and Watkins, J. C. (1965) Diffusion of univalent ions across the lamellae of swollen phospholipids. *J. Mol. Biol.*, **13**, 238–252

Bangham, A. D., Standish, M. M., Watkins, J. C. and Weissmann, G. (1967a) The diffusion of ions from a phospholipid model membrane. *Protoplasma*, **63**, 183–187

Bangham, A. D., de Gier, J. and Greville, G. D. (1967b) Osmotic properties and water permeability of phospholipid crystals. *Chem. Phys. Lipids*, **1**, 225–246

Bangham, A. D., Mill, M. W. and Miller, N. G. A. (1974) Preparation and use of liposomes as models of biological membranes. *Methods Member. Biol.*, **1**, 1–68

Barenholz, Y., Gibbes, D., Litman, B. J., Goll, J., Thompson, T. E. and Carlson, F. D. (1977) A simple method for the preparation of homogeneous phospholipid vesicles. *Biochemistry*, **16**, 2806–2810

Barenholtz, Y., Amselem, S. and Lichtenberg, D. (1979) A new method for preparation of phospholipid vesicles (liposomes). *FEBS Lett.*, **99**, 210–214

Barrow, D. A. and Lentz, B. R. (1980) Large vesicle contamination in small unilamellar vesicles. *Biochim. Biophys. Acta*, **597**, 92–99

Batelle Memorial Institute (1979) *British Patent Application 2001929A*

Batzri, S. and Korn, E. D. (1973) Single bilayer liposomes prepared without sonication. *Biochim. Biophys. Acta*, **298**, 1015–1019

Bittman, R. and Blau, L. (1986) Permeability behaviour of liposomes prepared from fatty acids and fatty acid methyl esters. *Biochim. Biophys. Acta*, **863**, 115–120

Brady, J. E., Evans, D. F., Kachar, B. and Ninham, B. W. (1984) Spontaneous vesicles. *J. Am. Chem. Soc.*, **106**, 4279–4280

Brockerhoff, H. and Ramsammy, L. S. (1982) Preparation and structural studies of cholesterol bilayers. *Biochim. Biophys. Acta*, **691**, 227–232

Brunner, J., Skrabal, P. and Hauser, H. (1976) Single bilayer vesicles prepared without sonication: physico-chemical properties. *Biochim. Biophys. Acta*, **455**, 322–331

Cantu, L., Corti, M., Musolino, M. and Salina, P. (1991) Spontaneous vesicle formation from a one-component solution of a biological surfactant. *Progr. Colloid Polym. Sci.*, **84**, 21–23

Chauhan, S. (1993) The synthesis and characterization of vesicle forming nonionic surfactants. PhD Thesis

Chauhan, S. and Lawrence, M. J. (1989) The preparation of polyoxyethylene containing nonionic surfactant vesicles. *J. Pharm. Pharmacol.*, **41**, 6P

Cornell, B. A., Middlehurst, J. and Separovic, F. (1980) The molecular packing and stability within highly curved phospholipid bilayers. *Biochim. Biophys. Acta*, **598**, 405–410

Cullis and de Kruijff, B. (1979) Phosphorous-31 NMR studies of unsonicated aqueous dispersions of neutral and acidic phospholipids. Effects of phase transitions, p^2H and divalent cations on the motion in the phosphate region of the polar headgroup. *Biochim. Biophys. Acta*, **559**, 399–420

Deamer, D. W. (1984) Preparation of solvent vaporization liposomes. In: *Liposome Technology*, Vol. 1 (ed. Gregoriadis, G.) CRC Press, Boca Raton, FL

Deamer, D. W. and Bangham, A. D. (1976) Large volume liposomes by an ether injection vapourisation method. *Biochim. Biophys. Acta*, **443**, 629–634

de Gier, J., Mandersloot, J. G. and van Deenen, L. L. M. (1968) Lipid composition and permeability of liposomes. *Biochim. Biophys. Acta*, **150**, 666–675

Demel, R. A., Kinsky, S. C., Kinsky, C. B. and van Deenen, L. L. M. (1968) Effects of temperature and cholesterol on the glucose permeability of liposomes prepared with natural and synthetic lecithins. *Biochim. Biophys. Acta*, **150**, 655–665

Estep, T. N., Mouncastle, D. B., Biltonen, L. and Thompson, T. E. (1978) Studies on anomalous thermotropic behaviour of aqueous dispersions of dipalmitoyl phosphatidlycholine–cholesterol mixtures. *Biochemistry*, **17**, 1984–1989

Estep, T. N., Mouncastle, D. B., Barenholz, Y., Biltonen, L. and Thompson, T. E. (1979) Thermal behaviour of synthetic sphingomyelin–cholesterol dispersions. *Biochemistry*, **18**, 2112–2117

Evans, E. and Needham, D. (1986) Giant vesicle bilayers composed of mixtures of lipids, cholesterol and polypeptides. *Faraday Discuss. Chem. Soc.*, **81**, 267–281

Fendler, J. H. (1980) Surfactant vesicles as membrane mimetic agents: characterisation and utilisation. *Acc. Chem. Res.*, **13**, 7–13

Fendler, J. H. (1982) *Biomimetic Membrane Chemistry*, Wiley-Interscience, New York

Forge, A., Knowles, P. F. and Marsch, D. (1978) Morphology of egg phosphatidylcholine–cholesterol single-bilayer vesicles studies by freeze–etch electron microscopy. *J. Membr. Biol.*, **41**, 249–263

Fransen, G. J., Salemink, P. J. M. and Crommelin, D. J. A. (1986) Critical parameters in freezing of liposomes. *Int. J. Pharm.*, **33**, 27–35

Gabriel, N. E. and Roberts, M. F. (1986) Interaction of short-chain lecithin with long-chain phospholipids: characterisation of vesicles that form spontaneously. *Biochemistry*, 2812–2821

Gebicki, J. M. and Hicks, M. (1973) Ufasomes are stable particles surrounded by unsaturated fatty acid membranes. *Nature*, **243**, 232–234

Gebicki, J. M. and Hicks, M. (1976) Preparation and properties of vesicles enclosed by fatty acid membranes. *Chem. Phys. Lipids*, **16**, 142–160

Gregoriaids, G. (1988) In *Lipososmes As Drug Carriers* (ed. Gregoriadis, G.) Wiley, Chichester

Gregoriadis, G., da Silva, H. and Florence, A. T. (1990) A procedure for the efficient entrapment of drugs in dehydration–rehydration liposomes (DRVs). *Int. J. Pharm.*, **65**, 235–242

Gruner, S. M., Lenk, R. P., Janoff, A. S. and Ostro, M. J. (1985) Novel multilamellar lipid vesicles: comparison of physical characteristics of multilamellar liposomes and stable plurilamellar vesicles. *Biochemistry*, **24**, 2833–2842

Handjani-Vila, R. M., Ribier, A., Rondot, B. and Vanlerberghie, G. (1979) Dispersion of lamellar phases of non-ionic lipids in cosmetic products. *Int. J. Cos. Sci.*, **1**, 303–314

Hamilton, Jr, R. L., Goerke, J., Guo, L. S. S., William, M. C. and Havel, R. J. (1980) Unilamellar liposomes made with the French pressure cell. *J. Lipid Res.*, **21**, 981–992

Hargreaves, W. R. and Deamer, D. W. (1978) Liposomes from ionic, single-chain amphiphiles. *Biochemistry*, **17**, 3759–3768

Hauser, H. and Gains, N. (1982) Spontaneous vesiculation of phospholipids: A simple and quick method of forming unilamellar vesicles. *Proc. Natl. Acad. Sci. USA.*, **79**, 1683–1687

Helfrich, W. (1973) Elastic properties of lipid bilayers: theory and possible experiments. *Z. Naturforsch, Teila*, **28**, 693–703

Ho, R. J. Y. and Huang, L. (1988) Immunoliposome assays: perspectives, progress and potential. In *Liposomes As Drug Carriers* (ed. Gregoriadis, G.) Wiley, New York

Hope, M. J., Bally, M. B., Webb, G. and Cullis, P. R. (1985) Production of large unilamellar vesicles by a rapid extrusion procedure. Characterisation of size distribution, trapped volume and ability to maintain a membrane potential. *Biochim. Biophys. Acta*, **812**, 55–65

Hope, M. J., Bally, M. B., Mayer, L. D., Janoff, A. S. and Cullis, P. R. (1986) Generation of multilamellar and unilamellar phospholipid vesicles. *Chem. Phys. Lipids*, **40**, 89–107

Huang, C. (1969) Studies of phosphatidylcholine vesicles, formation and physical characteristics. *Biochemistry*, **8**, 344–352

Hub, H. H., Zimmermann, U. and Ringsdorf, H. (1982) Preparation of large unilamellar vesicles. *FEBS Lett.*, **140**, 254–256

Hundscheid, F. J. A. and Engberts, J. B. F. N. (1984) Aggregation behaviour of a series of structurally related single-chain amphiphiles. Structural effects on aggregate morphology as studied by electron microscopy *J. Org. Chem.*, **49**, 3088–3091

ICI (Imperial Chemicals Industries Ltd) (1978) *Belgian Patent 866697*

Ishiwatari, T. and Fendler, J. H. (1984) Polymerized surfactant vesicles: effects of polymerization on intravesicular ester hydrolysis. *J. Am. Chem. Soc.*, **106**, 1908–1912

Israelachvili, J. N., Mitchell, D. J. and Ninham, B. W. (1976) Theory of self-assembly of hydrocarbon amphiphiles into micelles and bilayers. *J. Chem. Soc., Faraday Trans. 2*, **72**, 1525–1568

Israelachvili, J. N., Mitchell, D. J. and Ninham, B. W. (1977) Theory of self-assembly of lipid bilayers and vesicles. *Biochim. Biophys. Acta*, **470**, 185–202

Israelachvili, J. N., Marcelja, S. and Horn, R. G. (1980) Physical principles of membrane organisation. *Q. Rev. Biophys.*, **13**, 121–200

Jousma, H., Talsma, H., Spies, F., Joosten, J. G. H., Junginger, H. E. and Crommelin, D. J. A. (1987) Characterisation of liposomes. The influence of extrusion of multilamellar vesicles through polycarbonate membranes on particle size, particle size distribution and number of bilayers. *Int. J. Pharm.*, **35**, 263–274

Kachar, B., Evans, D. F. and Ninham, B. W. (1984) Rapid characterisation of colloidal systems by video-enhanced light microscopy. *J. Colloid Interface Sci.*, **99**, 593–596

Kaler, E. W., Murthy, A. K., Rodriguez, B. E. and Zasadzinski, J. A. N. (1989) Spontaneous vesicle formation in aqueous mixtures of single-tailed surfactants. *Science*, **245**, 1371–1374

Kaneko, T. and Sagitani, H. (1992) Formation of homogeneous liposomes with high trapping efficiency by the surface chemical method. *Colloid. Surfaces*, **69**, 125–133

Kano, K., Romero, A., Djermouni, B., Ache, H. J. and Fendler, J. H. (1979) Characterisation of surfactant vesicles as membrane mimetic agents. *J. Am. Chem. Soc.*, **101**, 4030–4037

Kasahara, M. and Hinckle, P. C. (1977) Reconstituted and purification of the D-glucose transporter from human erythrocytes. *J. Biol. Chem.*, **252**, 7384–7390

Kato, A., Arakawa, M. and Kondo, T. (1984) Preparation and stability of liposome-type artificial red blood cells stabilized with carboxymethylchitin. *J. Microencapsulation*, **1**, 105–112

Kim, S. and Martin, G. M. (1981) Preparation of cell-size unilamellar with high captured volume and defined size distribution. *Biochim. Biophys. Acta*, **646**, 1–9

Kirby, C. J. and Gregoriadis, G. (1984a) A simple procedure for preparing liposomes capable of high encapsulation efficiency under mild conditions. In *Liposome Technology*, Vol. 1 (ed. Gregoriadis, G.) CRC Press, Boca Raton, FL

Kirby, C. J. and Gregoriadis, G. (1984b) *Biotechnology*, **2**, 979

Kirkland, J. J., You, W. W., Doerner, W. A. and Grant, J. J. (1980) Sedimentation field flow fractionation of macromolecules and colloids. *Anal. Chem.*, **52**, 1944–1954

Kiwada, H., Niimura, H., Fujisaki, Y., Yamada, S. and Kato, Y. (1985) Application of synthetic alkyl glycoside vesicles as drug carriers. I. Preparation and physical properties. *Chem. Pharm. Bull.*, **33**, 753–759

Kondo, T. (1988) Liposomes as carriers for haemoglobin. In *Liposomes As Drug Carriers* (ed. Gregoriadis, G.) Wiley, Chichester

Kremer, J. M. H., v. d Esker, M. W. J., Pathmamanoharan, C. and Wiersema, P. H. (1977) Vesicles of variable diameter prepared by a modified injection method. *Biochemistry*, **16**, 3932–3935

Kunieda, H., Makino, S. and Ushio, N. (1991a) Anionic reversed vesicles. *J. Colloid Interface Sci.*, **147**, 286–288

Kunieda, H., Nakamura, K. and Evans, D. F. (1991b) Formation of reversed vesicles. *J. Am. Chem. Soc.*, **113**, 1051–1052

Kunitake, T. and Okahata, Y. (1977) A totally synthetic bilayer membrane. *J. Am. Chem. Soc.*, **99**, 3860–3861

Kunitake, T. and Okahata, Y. (1978) Synthetic bilayer membranes with anionic head groups. *Bull. Chem. Soc. Jpn.*, **51**, 1877–1879

Kunitake, T. and Higashi, N. (1985) Bilayer membranes of triple-chain fluorocarbon amphiphiles. *J. Am. Chem. Soc.*, **107**, 692–696

Kunitake, T. and Okahata, Y. (1980) Formation of stable bilayer assemblies in dilute aqueous solution from ammonium amphiphiles with the diphenylazomethine segment. *J. Am. Chem. Soc.*, **102**, 549–553

Kunitake, T., Okahata, Y., Shimomura, M., Yasunami, S. and Takarabe, K. (1981) Formation of stable bilayer assemblies in water from single-chain amphiphiles. Relationship between the amphiphile structure and the aggregate morphology. *J. Am. Chem. Soc.*, **103**, 5410–5413

Kunitake, T., Okahata, Y. and Tawaki, S.-I. (1985) Bilayer characteristics of 1,3-dalkyl- and 1,3-diacyl-rac-glycero-2-phosphocholines. *J. Colloid Interface Sci.*, **103**, 190–201

Lai, M.-Z., Duzgunes, N. and Szoka, F. C. (1985) Effects of replacement of the hydroxyl group of cholesterol and tocopherol on the thermotropic behaviour of phospholipid membranes. *Biochemistry*, **24**, 1646–1653

Lasic, D. D. (1988) The spontaneous formation of unilamellar vesicles, *J. Colloid Interface Sci.*, **124**, 428–435

Lasic, D. D., Kidric, J. and Zagorc, S. (1987) A simple method for the preparation of small unilamellar vesicles. *Biochim. Biophys. Acta*, **896**, 117–122

Lawaczeck, R., Kainosho, M. and Chan, S. I. (1976) The formation and annealing of structural defects in lipid bilayer vesicles. *Biochim. Biophys. Acta*, **443**, 313–330

Lawrence, S. M. (1992) Characterization and computer modelling of novel non-ionic surfactant. Ph.D. Thesis, University of London

Lawrence, S. M., Tipple, S. C., Barlow, D. J. and Lawrence, M. J. (1991) Structure of novel nonionic surfactant vesicles (NVS's). *Pharm. Res.*, **8**, PDD 7082

Lesieur, S., Grabiello-Madelmont, C., Paternostra, M.-T. and Ollivon, M. (1991) Size analysis and stability of lipid vesicles by high-performance gel exclusion chromatography, turbidity and dynamic light scattering. *Anal. Biochem.*, **192**, 334–343

Litman, B. J. (1974) Determination of molecular asymmetry in phosphatidylcholine surfactant distribution in mixed phospholipid vesicles. *Biochemistry*, **13**, 2844–2848

Lukac, S. and Perovic, A. (1985) Effects of stains and staining procedures on transmission electron microscopy. *J. Colloid Interface Sci.*, **103**, 586–589

MacDonald, R. C., MacDonald, R. I., Menco, B. Ph. M., Takeshita, K., Subbarao, N. K. and Hu, L. (1991) Small-volume extrusion apparatus for preparation of large unilamellar vesicles. *Biochim. Biophys. Acta*, **1061**, 297–303

Malliaris, A. (1988) Thermotropic phase transitions in totally synthetic surfactant vesicles and bilayers. *Progr. Colloid Polym. Sci.*, **76**, 176–182

Mason, J. T. and Huang, C. (1978) Analysis of egg phosphatidylcholine vesicles. *Ann. N.Y. Acad. Sci.*, **308**, 29–48

Mayer, L. D., Hope, M. J., Cullis, P. R. and Janoff, A. S. (1985) *Biochim. Biophys. Acta*, **817**, 2833–2843

Mayer, L. D., Hope, M. J. and Cullis, P. R. (1986) Vesicles of variable sizes produced by a rapid extrusion procedure. *Biochim. Biophys. Acta*, **858**, 161–168

Mayhew, E., Lazo, R., Vail, W. J., King, J. and Green, A. M. (1984) Characterisation of liposomes prepared using a microemulsifier. *Biochim. Biophys. Acta*, **775**, 169–174

Mayhew, E., Nikolopoulos, G. T., King, J. J. and Siciliano, A. A. (1985) A practical method for the large-scale manufacture of liposomes. *Pharm. Manufacturing*, **Aug.** 18–22

McDaniel, R. V., McIntosh, T. J. and Simon, S. A. (1983) Nonelectrolyte substitution for water in phosphatidylcholine bilayers. *Biochim. Biophys. Acta*, **731**, 97–108

Menger, F. M., Lee, J.-J., Aikens, P. and Davis, S. (1989) Vesicle size dependence on experimental parameters. *J. Colloid Interface Sci.*, **129**, 185–191

Murakami, Y., Nakano, A. and Ikeda, H. (1982) Preparation of stable single-compartment vesicles with cabionic and zwitterionic amphiphiles involving amino acid residues. *J. Org. Chem.*, **47**, 2137–2144

Müller, C. C., Gaymann, H.-J. and Hamann (1991) Pharmacosomes – multilamellar vesicles consisting of pure drug. *Eur. J. Pharm. Biopharm.*, **37**, 113–117

Nagamura, T., Takeyama, N. and Matsuo, T. (1985) Self-assembling and photochemical properties of ammonium amphiphiles containing naphthyoxy group. *J. Colloid Interface Sci.*, **103**, 202–209

Nayar, R., Hope, M. J. and Cullis, P. R. (1989) Generation of large unilamellar vesicles from long-chain saturated phosphatidylcholines by extrusion technique. *Biochim. Biophys. Acta*, **986**, 200–206

New, R. R. C. (1990) *Liposomes: A Practical Approach*, IRL Press, Oxford

Ninham, B. W., Evans, D. F. and Wei, G. J. (1983) The curious world of hydroxide surfactants. Spontaneous vesicles and anomalous micelles. *J. Phys. Chem.*, **87**, 5020–5025

Nusselder, J.-J. H. and Engberts, J. B. F. N. (1989) Surfactant structure and aggregate morphology. The urge for aggregate stability. *J. Am. Chem. Soc.*, **111**, 5000–5002

Okahata, Y., Tanamachi, S., Nagai, M. and Kunitake, T. (1981) Synthetic bilayer membranes prepared from dialkyl amphiphiles with nonionic and zwitterionic head groups. *J. Colloid Interfae Sci.*, **82**, 401–417

Oku, N. and MacDonald, R. C. (1983a) Differential effects of alkali metal chlorides on formation of giant liposomes by freezing and thawing and dialysis. *Biochemistry*, **22**, 855–863

Oku, N. and MacDonald, R. C. (1983b) Formation of giant liposomes from lipids in chaotropic ion solutions. *Biochim. Biophys. Acta*, **734**, 54–61

Olson, F., Hunt, A. H., Szoka, F. C., Vail, W. J. and Papahadjopoulos, D. (1979) Preparation of liposomes of defined size distribution by extrusion through polycarbonate membranes. *Biochim. Biophys. Acta*, **557**, 9–23

Ostro, M. J. (1987) Liposomes. *Sci. Am.*, **256**, 90–99

Papahadjopoulos, D., Vail, W. J., Jacobsen, K. and Poste G. (1975) Cocleate lipid cylinders: formation by fusion of unilamellar vesicles. *Biochim. Biophys. Acta*, **394**, 483

Patel, K. R., Li, M. P., Schuh, J. R. and Baldeschwieler, J. D. (1984) The pharmacological efficacy of a rigid non-phospholipid liposome drug delivery system. *Biochim. Biophys. Acta*, **797**, 20–26

Payne, N. I., Browning, I. and Hynes, C. A. (1986a) Characterisation of proliposomes. *J. Pharm. Sci.*, **75**, 330–333

Payne, N. I., Timmins, P., Ambrose, C. V., Ward, M. D. and Ridgeway, F. (1986b) Proliposomes: a novel solution to an old problem. *J. Pharm. Sci.*, **75**, 325–329

Perrett, S., Golding, M. and Williams, W. P. (1991) A simple method for the preparation of liposomes for pharmaceutical applications: characterization of the liposomes. *J. Pharm. Pharmacol.*, **43**, 154–161

Pick, U. (1981) Liposomes with a large trapping capacity prepared by freezing and thawing of sonicated phospholipid mixtures. *Arch. Biochem. Biophys.*, **212**, 186–194

Pidgeon, C., Hunt, A. H. and Dittrich, K. (1986) Formation of multilayer vesicles from water, organic-solvent (w/o) emulsions. Theory and practice. *Pharm. Res.*, **3**, 23–34

Pidgeon, C., McNeely, S., Schmidt, T. and Johnson, J. E. (1987) Multilayered vesicles prepared by reverse-phase evaporation: liposome structure and optimum solute entrapment. *Biochemistry*, **26**, 17–29

Reeves, J. P. and Dowben, R. M. (1969) Formation and properties of thin-walled phospholipid vesicles. *J. Cell. Physiol.*, **73**, 49–54

Regen, S. L., Singh, A., Oehme, G. and Singh, M. (1982) Polymerized phosphatidylcholine vesicles. Synthesis and characterisation. *J. Am. Chem. Soc.*, **104**, 791–795

Ribier, A., Handjani-Vila, R. M., Bardez, E. and Valeur, B. (1984) Bilayer fluidity of non-ionic vesicles. An investigation by differential polarized phase fluorometry. *Colloids and Surfaces*, **10**, 155–161

Rhoden, V. and Goldin, S. M. (1979) Formation of unilamellar lipid vesicles of controllable dimensions by detergent dialysis. *Biochemistry*, **18**, 4173–4176

Rogerson, A. (1986) Ph.D. Thesis, University of Strathclyde

Romero, A., Tran, C. D., Klahan, P. L. and Fendler, J. H. (1978) Drug entrapment in surfactant vesicles. *Life Sci.*, **22**, 1447–1450

Ruf, H., Georgalis, Y. and Grell, E. (1989) Dynamic laser light scattering to determine size distributions of vesicles. In *Methods in Enzymology* (ed. Fleischer, S. and Fleischer, B.) Academic Press, New York

Safran, S. A., Pincus, P. A. and Andelman, D. A. (1990) Theory of spontaneous vesicle formation in surfactant mixtures. *Science*, **248**, 354–356

Safran, S. A., MacKintosh, F. C., Pincus, P. A. and Andelman, D. A. (1991) Spontaneous vesicle formation by mixed surfactants. *Progr. Colloid Polym. Sci.*, **84**, 3–7

Schieren, H., Rudolph, S., Finkelstein, M., Coleman, P. and Weissmann, G. (1978) Comparison of large unilamellar vesicles prepared by a petroleum ether vapourization method with multilamellar vesicles. *Biochim. Biophys. Acta*, **542**, 137–153

Schmidt, D. (1981) *United States Patent 4271196*

Schneider, M. and Lamy, B. (1979) *British Patent Application 2002319A*

Seltzer, S. E. (1988) Liposomes in diagnostic imaging. In *Liposomes as Drug Carriers* (ed. Gregoriadis, G.) Wiley, Chichester

Senior, J. and Gregoriadis, G. (1982) Stability of small unilamellar vesicles in serum and clearance from the circulation. *Life Sci.*, **30**, 2123–2136

Shew, R. L. and Deamer, D. W. (1985) A novel method for encapsulation of macromolecules of liposomes. *Biochim. Biophys. Acta*, **816**, 1–8

Six, H. R., Anderson, C. and Kasel, J. A. (1988) Further studies of liposomes as carriers of purified protein vaccines. In *Liposomes as Drug Carriers* (ed. Gregoriadis, G.) Wiley, Chichester

Szoka, F. and Papahadjopoulos, D. (1978) Procedure for preparation of liposomes with large internal aqueous space and high capture by reverse-phase evaporation. *Proc. Natl. Acad. Sci.*, **75**, 4194–4198

Szoka, F. and Papahadjopoulos, D. (1980) Comparative properties and methods of preparation of lipid vesicles (liposomes). *Ann. Rev. Biophys. Bioeng.*, **9**, 467–508

Szoka, F. and Papahadjopoulos, D. (1981) Liposomes: preparation and characterisation. In *Liposomes: From Physical Structure To Therapeutic Applications* (ed. Knight, C. G.) Elsevier, Amsterdam

Szoka, F., Olson, F., Heath, T., Vail, J., Mayhew, E. and Papahadjopoulos, D. (1980) Preparation of unilamellar liposomes of intermediate size (0.1–0.2 μm) by a combination of reverse phase evaporation and extrusion through polycarbonate membranes. *Biochim. Biophys. Acta*, **601**, 559–571

Talsma, H., Jousma, H., Nicolay, K. and Crommelin, D. J. A. (1987) Multilamellar or multivesicular vesicles? *Int. J. Pharm.*, **37**, 171–173

Talmon, Y. (1983) Staining and drying-induced artifacts in electron microscopy of sulfactant dispersions. *J. Colloid Interaces Sci.*, **93**, 366–382

Vanlerberghe, G., Ribier, A. and Handjani-Vila, R. M. (1978) *Communication au Collegue du CNRS sur la Physicohemie des Composes Amphiphiles*, Bordeaux, 303–311

van Zanten, J. H.and Monobouquette, H.G. (1992) Characterization of vesicles by classical light scattering. *J. Colloid Interface Sci.*, **146**, 330–336

Chapter 11

Control of solid particle comminution

D. J. Wedlock

11.1 Introduction

In essence there are two procedures for production of solid particulates of controlled size: condensation mechanisms and comminution mechanisms. The former infers growth to the final particle-size distribution, the latter some process of energy input which gives rise to a finer than original particle-size distribution by particle fission or erosion.

This chapter will seek to outline the processes by which controlled size reduction of solid particulates is effected using the comminution approach. Condensation processes are covered elsewhere (see Chapter 8). Comminution processes involving liquid-in-liquid dispersions (emulsions) are also covered elsewhere (see Chapter 7) and therefore will not be reviewed here.

Milling processes are many and varied, but may be divided simply into two principal groups; wet milling and dry milling. The theoretical considerations have some similarities for the two cases, although there is also sufficient divergence under extremes of process conditions that it will be worth dealing with theoretical aspects in conjunction with specific examples.

11.2 Solid-in-liquid (wet) milling

11.2.1 General considerations

In this approach, particle-size reduction is obtained while having the initial solid feed mixture and the finally desired particle-size distribution dispersed in the same liquid continuous phase. This necessarily requires that the disperse phase (the particle) should be wetted by and have a relatively low solubility in the continuous phase. The continuous phase might equally be polar (aqueous) or apolar (oil based). Depending on the preferred choice of continuous phase in relation to disperse phase, particle–liquid interfacial effects may be of some significance, especially for particles in the $1-10$ μm size range, and this will have an influence on the process design for comminution.

The choice of wet milling may be purely for safety considerations (e.g. explosion risk minimization) during the size reduction process. In this case the comminuted particles would be separated from the continuous phase post-milling by filtration or centrifugation. Under such circumstances a relatively higher particle solubility in the continuous phase might be tolerated than would be permissible were the suspension produced the final product. In the latter case the necessarily polydisperse particle-size distribution is more susceptible to the chronic effects of Ostwald ripening (mean size enlargement) upon storage.

Particle ripening processes

Ostwald ripening (Hunter, 1987) is a process which opposes the effect of particle-size reduction, by increasing mean particle size as a function of time. It is a completely general phenomenon, dictated by thermodynamic considerations but is probably of most significance for solid-in-liquid and liquid-in-liquid dispersions because of overall kinetic considerations. Essentially, particles below a critical radius dissolve and the solute supersaturation caused by this is relieved by precipitation onto particles above the critical particle radius (see Chapter 3).

This occurs as a result of the higher Laplace pressure of the smaller particles in comparison with the bigger particles in the complete particle-size distribution. The Laplace pressure ΔP for our purposes may be regarded as the vapour pressure of the particulate:

$$\Delta P = \frac{-2\gamma_{SL}}{r} \tag{11.1}$$

where γ_{SL} is the interfacial tension, and r is the particle radius. Therefore it may be seen that the smallest particles will have a higher solubility than the larger particles in the continuous phase. As the smallest particles dissolve and the largest particles become larger, the particle-size distribution skews to a larger mean size, negating the effect of comminution. This effect can be minimized by judicious matching of particle and continuous phase to minimize average particle solubility. This will minimize the solute flux through the continuous phase, conferring some degree of kinetic stability on the particle size distribution. Similarly, a sensible choice of surfactants can have a significant controlling influence on the ripening process, if they are required for the overall processing. Anionic surfactant micelles have greater difficulty approaching (generally) net negatively charged surfaces than do non-ionic micelles, due to charge repulsion. Thus, anionic surfactants will contribute less to micellar-enhanced solubilization and transport of solute in the overall regrowth process.

For crystalline (as opposed to amorphous) milled suspensions where the microcrystallites can have a number of growth faces, each with significantly different interfacial excess free energies, differential growth rates at preferred faces can lead to a somewhat similar mean particle size enlargement. This is distinct from the true Ostwald ripening process, but must be recognized as of equal importance. This aspect of crystal growth is comprehensively dealt with in Chapter 4. It is worth restating that the process of particle regrowth is not limited purely to the liquid mediated state, but is also a solid-state phenomenon. It is simply that the kinetics of the solid-state growth process are generally slower.

Dispersion of a particle in a continuous phase of different polarity

In the process of preparation of solid-in-liquid dispersions one encounters a significant increase in the surface area per unit mass of disperse phase as a function of reducing particle size. This increase in interfacial area may not be of great significance where interfacial tensions are low, for example where the solid and liquid phase are of similar chemical type or more correctly have a similar polarity. However, when there are large mismatches in the polarity of the continuous and disperse phase, interfacial tensions or the equivalent excess Gibbs' interfacial free energies must be considered and catered for.

The following general considerations apply whether for comminution of a hydrophobic particle in an aqueous continuous phase, say a pharmaceutical suspension, or for any other combination of chemically mismatched interfaces. Before comminution can occur in a wet milling process, the solid particle must be effectively and fully wetted by the continuous phase. In order for this to occur the wetting process must be spontaneous.

The wetting of a solid surface by a liquid may be described in terms of the equilibrium contact angle and the appropriate interfacial tension using the Young equation:

$$\gamma_{SV} - \gamma_{SL} = \gamma_{LV} \cos \theta \tag{11.2}$$

or

$$\cos \theta = \frac{\gamma_{SV} - \gamma_{SL}}{\gamma_{LV}} \tag{11.3}$$

where γ represents the interfacial tension and the subscripts S, L and V refer to the solid, liquid and vapour phases respectively, and θ is the solid–liquid contact angle at a smooth surface.

For wetting of a perfect cubic particle it can be shown (Patton, 1964) that the work of dispersion W_d is given by the sum of liquid adhesion to one surface of the particle, penetration of the liquid by four surfaces and, finally, spreading of the liquid to enclose the particle in dispersion. It may easily be shown that for this idealized cubic particle

$$W_d = -6\gamma_{LV} \cos \theta \tag{11.4}$$

In this case it is assumed the particle surfaces are perfectly planar. The wetting of this ideal particle will thus be determined by two relatively easily measurable quantities: the contact angle θ, and the surface tension. The process of wetting for real (i.e. irregular-shaped) particles can also be described essentially by three processes (Patton, 1964):

(a) work of adhesion, W_a, in which the solid and liquid phases are brought into molecular contact
(b) work of penetration, W_p, in which the continuous phase penetrates the pores and irregularities of the solid
(c) work of spreading, W_s, in which there is formation of a film of significant thickness with distinct interfaces (solid–liquid and liquid–vapour), and distinct character on either side of those films.

A factor i, the rugosity, relating to the irregularity of the surface may be defined as A/A_0, where A is the effective (molecular) surface area and A_0 is the simple geometric planar surface area. Hence,

$$W_a = -\gamma_{LV} (i \cos \theta + 1) \tag{11.5}$$

$$W_p = -\gamma_{LV} (i \cos \theta) \tag{11.6}$$

$$W_s = -\gamma_{LV} (i \cos \theta - 1) \tag{11.7}$$

The essential conclusions of these equations are:

(a) adhesional wetting is always spontaneous when $\theta < 90°$, but when $\theta > 90°$ adhesional wetting may only occur when the surface has an appreciable roughness
(b) spontaneous penetration of pores and irregularities occurs when $\theta < 90°$
(c) spreading over a planar surface is spontaneous only for $\theta = 0°$, but happens in practice for small contact angles when there is a finite rugosity to the surface.

Rate of wetting of particles into dispersion

Assuming the rate of wetting is limited by the expulsion of air, an understanding of the process variables of significance can be ascertained from the Washburn–Rideal equation (Washburn, 1921; Rideal, 1922), in its differential form eqn (11.8), assuming the wetting process can be represented by the wetting of a capillary:

$$\frac{dl}{dt} = \frac{r\gamma_{LV} \cos \theta}{4\eta l} \tag{11.8}$$

where l is the length wetted of the capillary in time t, r is the capillary radius, and η is the viscosity of the continuous phase. The rate of wetting can be seen to be proportional to the same factor, $\gamma_{LV} \cos \theta$ which controls the work of dispersion.

Stabilization against particle reaggregation

In making a fine particulate suspension by wet milling, the first stage is necessarily the mixing of a polydisperse mix of primary feed particles and coarse agglomerates of the powder, with the liquid, free of entrapped air. Until the air is displaced, effective wetting and suspending of the particle cannot occur. The wetting may be assisted by surface active agents – in particular a low molecular weight 'wetter' which can diffuse to the interface rapidly by virtue of its relatively high translational diffusion coefficient. The wetter is usually subsequently replaced by a higher molecular weight surfactant which will impart long-term colloidal stability to the particle (Cosgrove, 1987; Wedlock, 1992) and prevent reaggregation of the freshly formed interface/particle due to attractive van der Waals forces. The exchange of the small for large surface active species is driven largely by the reduced probability of desorption of the larger dispersant molecule which will generally have multiple adsorption points or, alternatively, some large chemical 'block' with an intrinsically higher affinity for the surface. The colloidal forces are only of significance for particle sizes less than about 1–5 μm. In some cases even feed particles can be of these dimensions initially.

Exclusion of bubbles

Since many industrial particulate dispersions will be of very high particle volume fraction Φ_p, the relative viscosity η_{rel} of the dispersion may be too high for easy and rapid displacement of the entrained and entrapped air, since for high Φ_p, Φ_p^2 dominates:

$$\eta_{rel} \approx 1 + 2.5\Phi_p + 6.2\Phi_p^2 \tag{11.9}$$

and in this case some advantage can be gained from the use of antifoams to assist bubble collapse. The mechanism of action of antifoams is complex and varied, but in essence they tend to be surface active species of very low solubility in the continuous phase which necessarily locate at the air–liquid interface and assist bubble coalescence through enhanced interbubble film drainage.

11.2.2 Specific considerations in wet milling

Media charges

Bead wet milling is a universally utilized approach to achieve relatively fine particle-size distributions in dispersion (about 1–5 μm). The ball wet mill, as its name implies, employs larger diameter grinding media than the bead mill, and is used for milling to a larger mean particle size and also when grinding harder materials. Ball mills are used as wet mills when size reduction of very hard materials such as minerals is required, say >4.0 on the Moh hardness scale. Such a combination is used because the alternative dry milling techniques would tend to result in too much mill erosion. The grinding media variants of bead and ball milling work under essentially the same principles. Ball milling is an operation that is routinely also performed without a liquid phase present, but this would be quite unusual for bead milling. The grinding media are usually either glass or ceramic spheres, but may sometimes be hardened steel.

There is an empirical rule relating the size ratio of the initial feed mean particle size to the media size for bead wet milling that should be adhered to for most effective processing. The initial size ratio should be in the range 1:10 to 1:20. If this is likely to limit the final particle size achievable then a premilling stage may be required using either an attritor or colloid mill (Williams and Simons, 1992) (see below).

The grinding media charge will typically occupy around 50–80% of the chamber volume. Obviously there is a balance to strike between the rate of size reduction required, which will increase with the amount of grinding media used (number of bead–bead collisions) and the total quantity of suspension that can be milled in a single batch. Sand instead of beads is sometimes used as the milling medium, but really only offers the advantage of a cheaper milling medium where the medium cost in relation to the product cost is of significance. The mean size of bead and sand media are comparable.

Mill design and process variables in wet mills

The form of this type of mill occurs in many variants. These include vertical and horizontal milling chambers on the manufacturing scale to very simple bead mills for use in the laboratory which in essence are no more than a single disc rotating at high speed on a vertical axis in an open chamber. Milled preparations on this very small scale are useful for such things as dispersant screening, since visual observations may be made of unacceptable thickening processes or the propensity for air incorporation in the suspension during milling (Wedlock, 1992).

A typical bead wet milling process is shown schematically in Figure 11.1. Essentially, a horizontal high speed impeller and agitator shear the beads and suspension giving rise to (usually) turbulent flow in the mill chamber. The mill chamber is jacketed to allow water cooling, particularly for organic compounds, pigments etc., since substantial frictional and viscous heat dissipation may ensue. This heat must be removed to allow efficient milling. If the heat build up is too great then several problems may be encountered. In the extreme, the melting point of the milled particle may be approached. More likely though, a high mill temperature will give rise to a change in the performance of the dispersant leading to catastrophic thickening of the suspension through colloidal instability. An example of the sensitivity of suspension viscosity to temperature is shown in Figure 11.2 and discussed in more depth elsewhere (Wedlock, 1992). Under conditions of higher suspension viscosity and lower mill speeds where flow may be non-turbulent, shear processes may be of some significance in the rate of size reduction, but this is inherently more complex to model and explain since many more processes are involved. In general, wet mills will not be operated under such process conditions very often and so this will not be discussed further here.

It is usual to circulate continuously the suspension through the mill chamber and the recirculation circuit is shown in Figure 11.1. This is useful to allow sampling of the suspension during the milling process and permits monitoring of size reduction during comminution. To effect this, a filter screen or ring slit constriction is inserted in the suspension flow to allow separation of the recirculating suspension from the grinding media.

Energy considerations in wet milling

Comminution occurs primarily by particle impact with the beads, bringing about size reduction mainly by breaking up agglomerates and shattering the primary particles. Particle–particle impacts are considered to be of relatively less importance in this type of mill. We now consider the relationship between energy input and the particle-size distribution obtained. The relationship between numbers of particle–bead impacts and the rate of particle-size reduction is discussed later.

During the process of size reduction there is a substantial overall increase in particle surface area per unit mass of particles, with a corresponding increase in excess surface-free energy. Both the equilibrium and kinetic aspects of the 'surface effect' have been dealt with above. However, the change in surface-free energy is relatively much smaller, by approximately two orders of magnitude in

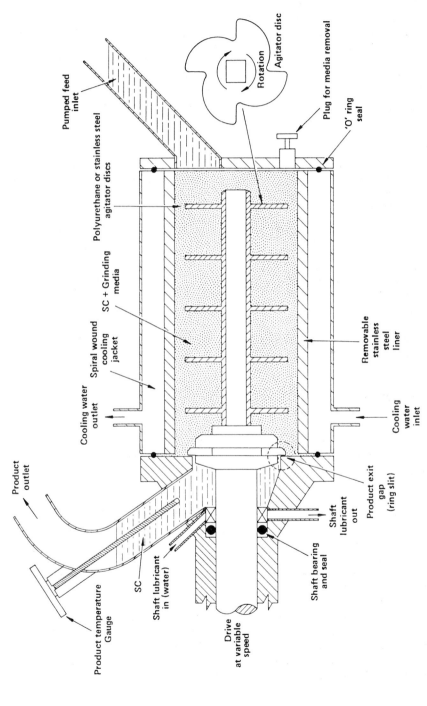

Figure 11.1 *A schematic representation of a bead milling process (with permission of W. A. Bachofen AG, Switzerland)*

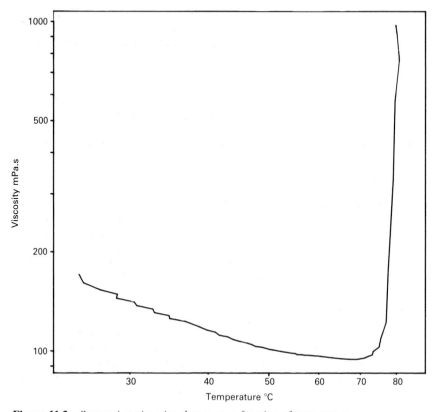

Figure 11.2   *Suspension viscosity change as a function of temperature*

general, compared with the energy required for the actual particle fracture process. Some relationships between the energy required for particle fracture and the resultant particle size have been developed.

One of the first attempts at deriving such a relationship (Rittinger, 1867) stated that the energy expended in the size reduction process is proportional to the new surface area formed:

$$\overline{E}_a = K_1 \Delta S \tag{11.10}$$

where $\overline{E}_a$ is the statistical average mass-related net specific energy, ΔS is the increase in mass-related specific surface area, and K_1 is a constant. Later, Kick (1885) stated that the energy required to produce a reduction ratio for a given mass is constant and independent of the original particle size. Hence,

$$\overline{E}_a = K_2 \log (d_1/d_2) \tag{11.11}$$

where d_1 is the particle size of the feed, d_2 is the final particle size, and K_2 is a constant. Much later, a compromise between these two extreme relationships was evolved by Bond (1952); it stated that the energy required for size reduction was proportional to the geometric mean of d^2 (proportional to surface area) and d^3 (proportional to volume), that is $d^{2.5}$. Thus the specific energy requirement is given by

$$\overline{E}_a = K_3(1/\sqrt{d_2} - 1/\sqrt{d_1}) \tag{11.12}$$

where K_3 is a constant. This empirical relationship has proven very useful for size reductions down to about 10 μm, where surface area considerations are of rather less importance.

Charles (1952) incorporated a variable exponent, n, into the energy equation,

$$\overline{E}_a = K_4 d^{-n} \tag{11.13}$$

However, when attempting size reduction to particle sizes much less than 10 μm it became apparent that n varied as the size of the particles reduced. This may be a manifestation of the relative increase in the importance of the surface excess free energy term. In other words, a lack of colloidal stability of the final particle size, leading to particle agglomeration post-fracture. The net result will be energy dissipation as heat without significant particle-size reduction.

Particle-size reduction rates in wet milling

In the bead wet milling process, crystalline dispersions can be manufactured to particle sizes in the supermicrometre to submicrometre range. Studies show that the rate determining factor in size reduction is usually the number of collisional encounters between the beads. Since the number of collisions between beads is effectively constant with time, for a given set of mill process parameters, the rate of size reduction will necessarily decrease with time. This is because of the considerable increase in the number of particles as comminution progresses, the number of particles increasing approximately cubically with mean particle-size reduction. Figure 11.3 shows the quite reasonable correspondence between the model and experimental reduction in particle size, assuming each collision between beads leads to the same number of fractured particles being created. The model relationship is cubic since, for example, halving the particle size in all three dimensions leads to an increase in the particle number concentration by a factor of 8.

In the sense that the number of bead–bead collisions is constant for a given bead charge volume and for a given rotation rate of the bead mill, then the only way to increase the number of collisions between beads is to increase their number. This broadly entails reducing the size of the beads within the constraints of the ratio of the initial feed particle size to bead size, as previously mentioned. However, it is both possible and usual to mix media of differing sizes to achieve this. Some efficiency advantage may be gained from the increased total charge achievable in the mill by big bead–small bead interstitial packing. Figure 11.4 shows that the effect of reducing the size of beads is indeed to increase the rate of particle size reduction.

There is an obvious limit to how far this approach may be invoked in reducing the time to achieve a certain particle-size for a given set of process conditions in

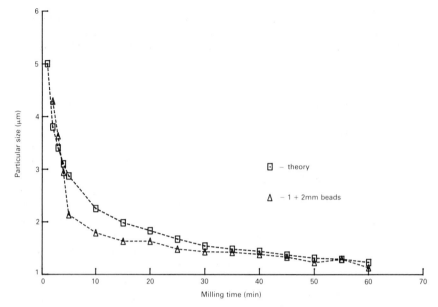

Figure 11.3   *Experimental and theoretical rates of particle size reduction. Theory assumes a cubic rate law*

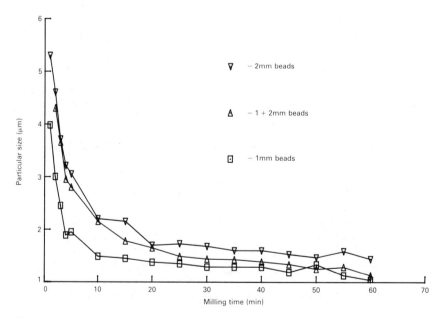

Figure 11.4   *Effect of mixing media of different size (1 and 2 mm) on rate of size reduction*

the mill. Eventually, the separating screen dimensions would have to become unacceptably small, giving rise to a too high resistance to circulation of the product, heat build up and added difficulty in separating the product from the milling media.

The colloid mill

The colloid mill is a form of process equipment which indirectly finds applications for size reduction of solid-in-liquid dispersions, more for making the dispersion initially, through high shear mixing processes rather than for its effectiveness in particle-size reduction. It is essentially a high shear homogenization device which can serve to homogenize liquid-in-liquid with size reduction but for solids-in-liquids is largely restricted to a dispersion role.

It functions by forcing a large volumetric liquid throughput into small channels or restrictions. Typically it comprises a conforming rotor and stator assembly with an adjustable clearance (about 1 mm to about 100 μm). The rotor and stator have channels to allow entrance and exit of the dispersion. This is shown in Figure 11.5. Rotor speeds can be very high, approaching 15 000 r.p.m., and this can result in significant viscous heat dissipation. Thus the colloid mill is generally cooled. It is an ideal device to use in the formation of combined suspension–emulsion systems.

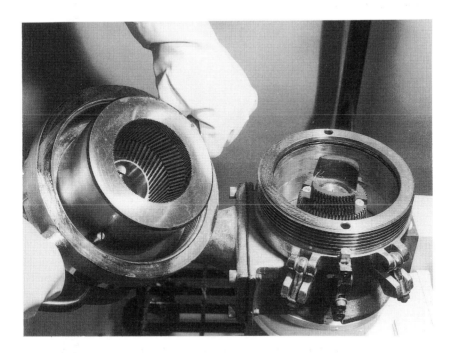

Figure 11.5   *A typical colloid mill rotor and stator (FRYMA-Maschinen AG)*

11.3 Dry milling

11.3.1 General considerations

Perhaps the largest proportion of comminution processes, be it food production, mineral processing or even size reduction of crystalline organic materials, involves the use of some form of dry milling. The range of particle-sizes routinely achievable is almost as great as that produced by wet milling processes. For example, air milling (see later) can relatively easily be applied to produce micrometer-sized organic particles.

However, a number of complicating factors influence the use of dry milling. For example, where organic materials are involved, melting points are generally low and risks of explosion significant. Inertization procedures such as full or partial inert blanketing must be used, but with the generally high gas flows involved in air milling procedures this produces the complicating factor of a potential asphyxiation risk to operators unless special precautions are taken. Furthermore, cooling procedures for low-melting particulates (cryogenic milling) are not simple, particularly at the pilot or manufacturing scale.

Other advantages and disadvantages could be highlighted, but it is probably sufficient to summarize by saying that, while dry milling processes are apparently simple and intuitively attractive, final process design can be at least as complex as wet milling processes.

11.3.2 Roll milling

The roll mill is a highly energy-efficient means of particle-size reduction that achieves comminution by a combination of shear and compression. In fact the highest utilization of energy in the comminution of solids by mechanical means is obtained by the breakage of single particles under slow compressional loading (Stairmand, 1975; Schoenert, 1986; Kapur *et al.*, 1990). Single-particle comminution under rigidly held rollers is only marginally less efficient (Kerber, 1984). It has been shown that the process energy efficiency of single-particle breakage in roll mills is up to five times that of ball/bead mills (Fuerstenau *et al.*, 1990). The median particle-size obtained by roll milling is generally inversely proportional to the net energy expended per unit mass of solids reduced. The comminution energy itself depends on the physical properties of the material – in particular the compression strength and, more specifically, the size of the feed material in relation to the roller gap.

Roll milling is effected by passing the pre-mix through the nip of a pair of rollers (generally spring loaded for safety) moving at varying speeds with respect to each other. This angular velocity differential is typically of the order of $1.1:1$ to $2:1$.

A special type of roll mill arrangement used for the production of quite narrow particle-size distributions of pigment materials and resin formulations is that manufactured by Bauermeister Gmbh (Hamburg, Germany). The rollers are not smooth but have a surface engraved with long pitched spiral-like teeth which serve to assist the grinding process by gripping the solid material. The relative speed of the rollers may be changed by altering the diameter of the pulley wheels which drive the rollers. Furthermore, the roller gap and hence the effective nip angle may be changed by an automated movement of the gap between the roller axes.

A cascade of such pairs of rollers (typically one, two or three pairs) may be arranged vertically above each other. Such an arrangement would allow an initial pregrind to produce a feed more suited to the second and third roller pair which may be set up to achieve the required throughput and particle-size cut. Figure 11.6a shows a schematic representation of a gravity-fed cascade of three roller pairs. Figure 11.6b is a schematic diagram showing how the pressure profile, $P(\alpha)$ or force profile, $F(\alpha)$ on a particle, represented by an equivalent sphere,

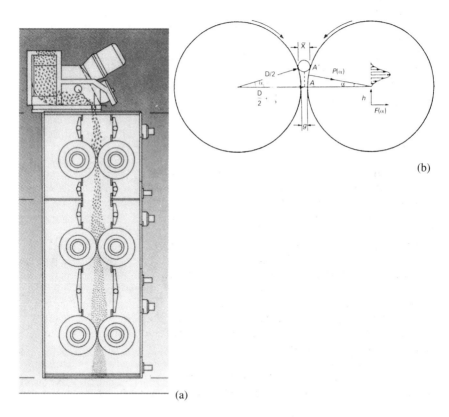

(b)

(a)

Figure 11.6   (a) *Schematic representation of a cascade of roller mills (courtesy of Bauermeister GmbH). (b) Pressure profile on a particle as a function of passage through a roll mill (after Kapur* et al. *(1990) courtesy of Elsevier Science Publishers BV)*

changes as a function of passage through a roller mill. The angle α_i represents half the nip angle; α_i is the angle subtended by the centre of the rollers with the horizontal and the radius touching the particle at the point where the pressure on the particle begins to rise; x is an equivalent sphere diameter and g the roller gap. As the particle is caught in the nip, the pressure will increase as a function of rotation angle α in some manner until at an intermediate angle α_0 the particle breaks at pressure P_0 and the pressure decays away.

Hence, the situation can be described as a power function of $P(\alpha)$ of the type

$$P(\alpha) = \left(\frac{\alpha_i - \alpha}{\alpha_i - \alpha_0} \right)^n P_0 \tag{11.14}$$

for the rising limb, $\alpha_0 \leqslant \alpha \leqslant \alpha_i$, and

$$P(\alpha) = \left(\frac{\alpha}{\alpha_0} \right)^m P_0 \tag{11.15}$$

for the falling limb, $0 \leqslant \alpha \leqslant \alpha_0$, where m and n are exponents in the power functions. Thus the pressure profile will almost certainly not be symmetrical.

Kapur *et al.* (1990) showed empirically that the size-reduction ratio obtained in a simple roller operation was proportional to the energy expended irrespective of the feed size and roll gap combination used. This rather simple relationship is similar to the finding of Kick (1885) for wet media milling operations, but appears to be more universally applicable in roll milling.

Roll milling is generally used on dry mixes but can also be adapted as another form of wet milling and is especially useful for concentrated organic pigment dispersions and inks where it is generally reserved for making relatively coarse dispersions.

11.3.3 Air and jet milling

Alternative methods of particle comminution that do not rely on the use of either impacting (bead) or compressive (roller) media, but rather on the exploitation of particle–particle collisions, are air milling or jet milling.

Air milling

Air milling is rather a misnomer since any appropriate gas may be used as the accelerating fluid to bring about particle–particle collisions. In particular, inert gases such as nitrogen may be required when grinding potentially explosive materials such as organic compounds.

A schematic diagram for an air mill is shown in Figure 11.7. A tangential air flow is introduced into the mill chamber. Particles of coarse material are introduced through a central orifice and accelerated in the gas swirl. Their increased velocity in turbulent flow leads to a large number of high-energy particle–particle collisions and a size-reduction process comes about. This kind

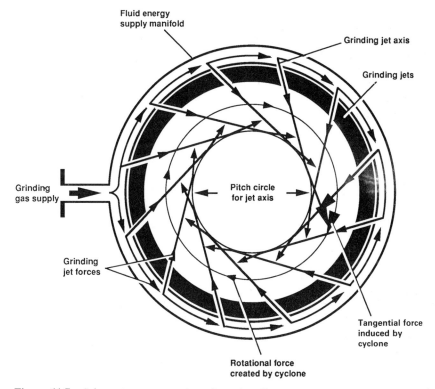

Figure 11.7   *Schematic representation of an air mill*

of milling process might typically be used for preparation of a dry mix for some subsequent particle-growth process such as agglomeration or granulation, where it would be inappropriate to wet mill initially.

The particle contacting elements of the mill are constructed of a hardened steel, but the comminution process causes erosion of the mill surface where particle velocities are greatest. Product contamination with the eroded steel is a consequence. In some cases, air mills may be lined with renewable polymer-based liners (e.g. PTFE) that are resistant to mill erosion. An alternative approach to this is to isolate the high-energy particle–particle collisions to a region of the mill where such an erosion process cannot occur. An example of this is in the process of opposed jet milling.

Opposed jet milling

In opposed jet milling particle–particle collisions can be isolated in a defined volume element by the convergence of gas jets. An example is in the fluidized bed opposed jet mill of the type produced by Hosokawa Micron. Figure 11.8 shows the arrangement schematically. The feed material is conveyed into the jet mill by a suitable feeder. A fluidized bed of the particles to be milled is created in the mill chamber. Any of those particles that are undersize already are

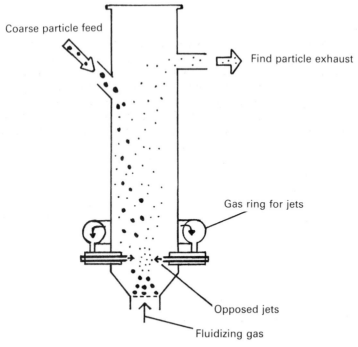

Coarse particle feed

Find particle exhaust

Gas ring for jets

Opposed jets

Fluidizing gas

Figure 11.8   *Schematic representation of an opposed jet mill*

immediately fed to the air classifier system above the fluidized bed. A ring-pipe circling the fluidized bed can feed a number of jets with high-pressure gas. The jets converge in a single volume element where the comminution process takes place by collisions between particles with directly opposed velocities, unlike the largely parallel particle velocities of the air mill. The fluidized bed acts to continuously renew the comminution volume with oversized particles, and those milled to the lower size cut can be removed by the air classifier. The fluidized bed acts to buffer the most highly energetic particle collisions and ensure they do not take place with the chamber surface. The outer part of the fluidized bed contains the slowest moving particles. This ensures minimal product contamination with eroded mill particles. Furthermore, the fluidized bed acts as a partial sound absorbent medium, since jet milling is an inherently noisy process.

This milling approach is appropriate for making fairly narrow particle-size distributions since the air classifier can be tuned to return oversize particles and those smaller can be removed almost immediately they fall into the correct size range.

This kind of jet mill is relatively efficient in terms of specific energy requirements compared to other jet mill techniques – and particularly classic air mills, since the particle collisions are from opposed flows and not purely reliant on turbulent conditions to give rise to the relative particle–particle collision velocities.

11.3.4 Impact mills

'Impact milling' is the generic term for mills designed to bring about particle-size reduction by particle impact with a solid surface and is generally reserved for the production of quite fine dry particulates/powders (about 10 μm). The solid surfaces take the form of rapidly moving plates, hammers (anvils), studs or pins. Hammer mills usually have a static grinding track in the periphery and a sieve insert for classifying the comminuted particles. Pin mills on the other hand require no sieves, relying on rotational speed, product throughput and airflow to regulate particle size. These mills have a turbine-like effect and give rise to substantial air flow. In some cases additional suction may be applied to enhance or control throughput rates.

In general, materials used in impact mills should not be too abrasive, and materials with a Moh's hardness index of greater than 3.5 are generally uneconomical to grind on mechanical impact and mechanical classifier mills because mill erosion becomes too significant. Very hard materials are better ground in for example fluid bed jet mills of the type described above.

References

Bond, F. C. (1952) *Min. Eng. Trans. A.I.M.E.*, **193**, 484–494

Charles, R. J. (1952) Energy-size reduction relationships in comminution. Trans. A.I.M.E., **208**, 80–88

Cosgrove, T. (1987) *Solid/Liquid Dispersions* (ed. Tadros, Th. F.) Academic Press, London, 131

Fuerstenau, D. W., Kapur, P. C., Schoenert, K. and Marktsheffel, M. (1990) Comparison of energy consumptions in breakage of single particles in rigidly mounted rolls with ball-mill grinding. *Int. J. Mineral Proc.*, **28**, 109–125

Hunter, R. J. (1987) *Foundations of Colloid Science*, Vol. 1, Oxford University Press, Oxford, 98

Kapur, P. C., Schoenert, K. and Fuerstenau, D. W. (1990) Energy-size relationship for breakage of particles in a rigidly mounted roll-mill. *Int. J. Mineral. Proc.*, **29**, 221–233

Kerber, A. (1984) Einfluss von beanspruchungsgeschwindingkeit Profielierung und Rauhigkeit auf die Einzelkorn Druckzerkleinerung. Dr.-Ing. Thesis, University of Fridericiana, Karlsruhe

Kick, F. (1885) *Das gesetz der proportionaten widerstanole und seine anwendungen*, Arthur Felix, Leipzig

Patton, T. C. (1964) *Paint Flow and Pigment Dispersion*, Wiley Interscience, New York

Rideal, E. K. (1992) On the flow of liquids under capillary pressure. *Phil. Mag.*, **44**, 1152

Rittinger, von P.R. (1867) *Lehrbuch der auflereiteingshunde*, Berlin

Schoenert, K. (1986) Limits of energy saving in milling. In *1st World Congress on Particle Technology* (Nurenburg), *Part II, Comminution*, 21

Stairmand, C. J. (1975) The energy efficiency of milling processes. *DECHEMA Monogr.*, No. **79**, 1–17

Washburn, E. W. (1921) The dynamics of capillary flow. *Phys. Rev.* **17**, 273

Wedlock, D. J. (1992) Practice of product formulation. In *Colloid and Surface Engineering: Application in the Process Industries* (ed. Williams, R. A.) Butterworth-Heinemann, Oxford, 112–139

Williams, R. A. and Simons, S. J. R. (1992) Handling colloidal materials. In *Colloid and Surface Engineering: Application in the Process Industries* (ed. Williams, R. A.) Butterworth-Heinemann, Oxford, 55–111

Chapter 12

On-line measurement and control of particle production and handling processes

R. A. Williams and S. J. Peng

12.1 Introduction

The control of any manufacturing or processing plant handling colloids or fine particles relies upon the acquisition of reliable mass flow, particle concentration and particle size data, upon which operating decisions are based. Recently the development of suitable sensors has blossomed in step with the increased availability of powerful computational aids coupled with developments in transducer technology which have resulted in a plethora of instrumentation that monitors the chemical and physical composition of a dispersion. In the following introductory discussion the term 'particle' is defined as a discrete body that is small in relation to its surroundings, hence the definition includes liquid-, gas- and solid-state dispersions.

Prior to these developments it has to be stated that manufacture of colloidal products has been performed largely without the use of advanced instrumental technology, instead relying upon previous experience and empirical testing, perhaps based on *ex situ* analysis (for example, of rheological properties). The incentive for adopting new measurement technology is that it allows products to be formulated in a consistent manner with proven product quality. The same plant can be run to produce batches that have to meet slightly different specifications, hence effective process control is necessary. Careful control is also required to produce speciality materials for the electronics and ceramics industries by liquid-based (e.g. crystallization) or gas-based (e.g. aerosol) routes. Indeed some of these materials can only be produced in conjunction with precision monitoring of the production stage using advanced sensor technologies.

Process streams that contain more than one type of solid component of different sizes and compositions in multiple phases (gas/solid/liquid) still present a number of challenging instrumental problems. For instance, sometimes the property being sensed (mass flow rate, solid concentration, or composition) may vary widely with time at a given position in the process route – this is unlike conventional liquid–vapour systems encountered in organic chemical processing whose properties can be defined uniquely from measurements of chemical composition, temperature and pressure. In short, it is still difficult to perform measurements on ultrafine particulate systems.

This chapter focuses on the importance of instruments that can operate on-line, ideally, satisfying one or more of the following criteria:

(a) without recourse to sampling the process stream(s) (which can introduce substantial errors), or if this is not possible

(b) without diluting the particulate mixture (which introduces further sampling errors and other unwanted effects)
(c) without changing the environment of the particulates (e.g. changing the nature of the prevailing fluid flow, shear etc.).

The value of this type of *in situ* measurement approach and its industrial significance are highlighted. In addition some of the common pitfalls associated with interpretation of data that have been abstracted on-line for process modelling purposes are illustrated. Some specific case studies based on particle-size measurement are presented to allow more detailed discussion of some practical applications which have implications for modelling, design and control of particulate processes. The discussion concentrates on solid–fluid systems, although many of the principles apply equally well to liquid–liquid and gas–liquid systems. Firstly, practical constraints encountered in the industrial production environment are considered before listing some of the sensing systems that can be used to probe ultrafine particulate systems.

12.2 Industrial requirements for on-line measurement

Considering the simplest case of a fine powder (containing a single type of solid) being transported in a pneumatic pipe, it is possible to identify five types of measurement required to characterize the operation:

(a) concentration of the solid phase (e.g. solid volume fraction)
(b) velocity of the solid phase (which may differ from the velocity of the gas phase)
(c) solids mass flow rate, gas mass flow rate (or volumetric flow rates)
(d) the particle-size distribution
(e) solids concentration profile within a given cross-section of the pipe (to assess homogeneity of powder distribution due to saltation etc.).

The product of the measured values of (a) and (b) is often used to deduce the mass flow rate of solids (c). Technology for routine measurement of (a)–(c) is in widespread use, whereas (d) and (e) are generally more difficult (unless the particulate system can be observed using simple optical devices).

An additional level of complexity is introduced if the particulate mixture contains more than one type of solid component, since information on (a)–(e) is then desired for *each* component. In particular, it is important to know the mixture composition (a). There are cases, such as in the processing of mineral ores, where composite particles exist – quantification of powder flow in these circumstances is extremely difficult.

12.2.1 Uses for on-line measurements

On-line measurements allow improved control and monitoring during manufacturing operations and transportation of particulate products. Several key functions can be identified:

- particle size and concentration measurement within process reactors for process control (decision-making) and a quality monitoring (checking product specification) functions
- mass flow measurements in product streams, for process control and auditing purposes
- alarm functions (e.g. to detect blockages)
- to enable development of process models and simulators.

The design, modelling and simulation of many unit operations and flow sheets involving the treatment of ultrafine dispersions continues to be impeded by the lack of appropriate *models*. Whilst some particle computational simulation packages exist as part of chemical process simulators, or in their own right (e.g. ASPEN, CAMP, GSIM, JKSimMet, METSIM, MICROSIM, MINDRES, MODSIM, PROCESS, Speedup, SysCAD and USIM-PAC) they tend to be severely constrained by the lack of any mechanistic models and consequently rely upon empirical 'input–output'-type models based on experimental data for existing plant. Hence 'process design from scratch' may not be possible, except for a well-known process scenario and making use of extensive solids properties databases which are contained within some simulators (e.g. METSIM). Generally, the simulators that exist are restricted to solids larger than the micrometer size range. For submicrometer materials very few simulators are applicable, unless the macroscopic properties of the mixture can be modelled, for instance, as a non-Newtonian continuum.

Without models it is difficult to *simulate* particulate processes, and the absence of suitable models is due largely to three factors. Firstly, the general lack of suitable and robust on-line measurement instrumentation that can operate in a truly *in situ* sense in flowing solid–fluid mixtures. Secondly, the poor mathematical procedures used to reconcile and interpret experimental data abstracted from multicomponent streams (since, in practice, the input and output stream measurements rarely balance). Thirdly, in some cases, over-reliance on existing computational fluids dynamics packages or particle-simulation packages which may generate useful information (e.g. sensitivity analysis) but tend to mask or de-emphasize inherent long-standing problems (fundamental mechanisms, handling of measurement errors etc.).

The first aspect is essentially one of the major driving forces for development of on-line instrumentation and will be surveyed in more detail below. The second and third aspects will provide the focus for one of the case studies presented later.

12.3 Sensor and instrument selection

The selection of an instrument depends upon identifying an appropriate means of sensing the properties of interest. Few generalizations can be made since the exact choice will depend on the needs of each individual particulate system and the measurement requirements, since all instruments have operational limitations. In general, most sensor systems involve generation of an excitation signal from an appropriate source (e.g. radiation) and a distant detector which measures the arriving signal and hence the mediating effect of the particulate mixture (e.g. absorbence of part of the signal) can be assessed and quantified (e.g. the solids concentration can be deduced via calibration). The most simple instruments use a single signal source and one or more detectors.

Some of the most critical factors affecting sensor selection are related to:

- the concentration of solids and fluctuations of this value
- size, density, shape, hardness of particles
- velocity of particles and fluctuations of this value
- dimensions of the desired sensing volume and its geometry (e.g. narrow pipe cross-section, large diameter crystallizer)
- temperature, pressure, opacity and chemical composition of the suspending fluid
- safety considerations (possibility of explosive hazard, fluid leakage, use of radiation, sterility etc.)
- frequency at which measurements are required (e.g. continual, once per day)
- response time required
- accessibility to plant and ease of maintenance and cleaning of instrument
- capital and operating costs.

Possible methods of meeting the on-line measurement requirements (a)–(e) listed in Section 12.2 will now be considered.

12.3.1 Phase concentration, velocity and mass flow measurement

Table 12.1 provides a summary of the principal types of instrumentation amenable to quantifying the flow rate of materials being transported as in liquid or gas fluids in pipelines based on measurements of phase concentration, velocity and mass flow. Some of these methods can also be used as the basis for 'probes' to place inside reactors or other vessels to monitor particulate properties. Further details of the techniques cited in the table have been described by Huskins (1982), McAdie (1986), Iinoya *et al.* (1988), Allen (1990), Williams *et al.* (1991), Williams (1992) and Cooper (1992).

One of the most demanding tasks for any powder or suspension sensor is the measurement of *in situ* properties of an optically opaque mixture (e.g. at a high solids concentration and/or where the suspending fluid itself is opaque), or for a system where optical access is not possible (e.g. particles flowing within a thick-walled corrosion-resistant pipe). Using conventional technology developed over the last three decades, this would have necessitated the use of penetrative radiation (e.g. γ-ray, X-ray, or neutron sources) from which (after calibration) an indication of, say, local solids concentration in a specific zone over a period of time could be obtained. Hence the mass flow rate could be obtained provided that an independent measurement of stream velocity was available. Alternatively, in some situations it is possible to measure mass flow rate directly based on mechanical phenomena (e.g. a Coriolis mass flow meter, Table 12.1) but,

Table 12.1 *Sensor selection for multiphase flow measurement techniques*

Phase volume/number concentration
Ultrasonics
Wave resonance
Capacitance/dielectric methods
Conductivity
Electromagnetic
Radiation attenuation (various)
Electrostatic charging
Particle counters
Laser radar
Light scattering
Light absorption
Mechanical resonance
Centrifugal differentiation

Phase velocity
Cross-correlation (various): e.g. nuclear and nuclear magnetic resonance; capacitance, conductance, electrodynamic; microwave; optical; ultrasonic
Doppler velocimetry (various): e.g. laser; microwave; ultrasound
Laser (time-of-flight) and optical microscopy
Tracer and vector tomography techniques (various): e.g. fluorescence; electromagnetic; conductivity; radioactivity; pulsed neutron activation

Mass flow
Mechanical devices (various): sensitive load cells; Coriolis force; turbine flowmeter; venturi/orifice-type meters; target (drag force) flowmeters
Ultrasonics
Electromagnetic flowmeter
Microwave resonance
Correlation flowmeters (see phase-velocity methods)
Heat emission and absorption methods (various)

---

although this is an on-line method, it may still require sampling or splitting of the process stream. For colloidal mixtures which carry entrained gas, methods based on mechanical oscillation can be problematic.

12.3.2 Particle-size measurement and concentration profiling

Easy solutions to the problem of finding a versatile instrument for on-line particle-size measurement remain elusive, unless the particle size is very coarse. Possible solutions could include the use of nucleonic methods; however, these tend to be hazardous and may not yield the required time response for dynamic processes. In general, most particle-size sensing techniques fall into one the following two categories:

(a) virtual sensing methods, where the measurement signals are analysed and fitted to a model from which the size distribution is obtained (e.g. laser light diffraction and other spectroscopic-based methods) or inferred (e.g. from centrifugal fractionation of the stream)
(b) counting methods, in which properties of *individual* particles are sensed and the information is accumulated to enable a statistical analysis of the particulate system (e.g. optical counting).

In essence, the first class of instruments tend to provide an instantaneous 'snapshot' of the size distribution whilst the second class may require some time (a few seconds or in some cases minutes) to produce a result. A variety of commercial instruments are available, but most are suited to *dilute* mixtures. A detailed discussion of these methods pertaining to coarse particulate and finer colloidal mixtures has been given by Allen (1990) and Williams (1992), respectively.

Two useful methods appropriate to both dilute *and* concentrated solid–liquid mixtures are *ultrasound spectroscopy* and *scanning laser microscopy* (described below).

Ultrasound methods exploiting absorption, scattering, spectroscopic and 'time-of-flight' techniques are being investigated intensively by several research groups, since they offer the possibility of on-line applications (Allen, 1990; Hoyle and Xu, 1993). Figure 12.1 shows a commercial on-line spectroscopic-based sizing device reported to be capable of sizing in the size range 5–3000 μm at solids volume concentrations up to 25% (Riebel and Löffler, 1989). This instrument is appropriate for use on crystallizers but cannot resolve submicrometre solid particles. Pendse (1993) and Alba *et al.* (1993) have also described the performance of other prototype on-line instruments that can size in the range 0.01–1000 μm in dispersions containing up to 70% solids by volume. In due course it may also be able to extract information on particle *shape* from such measurements (Pendse, 1993).

Ultrasound methods can be also be applied to follow rapid transients, such as aggregation kinetics, of submicron materials (Challis *et al.*, 1990; Holmes and Challis, 1993), but these techniques are not yet sufficiently developed for routine on-line use. Figure 12.2 shows the experimental cell of a broad bandwith (60 MHz) ultrasonic transmission spectrometer for rapid measurement (65 ms) of wave velocity and attenuation as a function of frequency. In the form shown in this figure, the instrument performs measurements on a 25 cm^3 volume sample of a concentrated colloidal suspension. The dynamic response of the dispersed phase to changes in hydrodynamics (e.g. by stirring the sample) or reagent additions can be sensed (Holmes and Challis, 1993). Other applications for monitoring droplet sizes in submicrometer emulsion systems using ultrasound techniques has also been reported (Allen, 1990). Preliminary industrial testing of *acoustic emission* sensors have also been reported to be successful in monitoring the crystallization and mixing processes (Parkinson, 1992).

Figure 12.1 *On-line particle-size analyser based on ultrasound spectroscopy of slurry flowing through a process pipeline, for sizing in the range 5–3000 μm, 1–25% (v/v) solids (OPUS, Sympatec GmbH)*

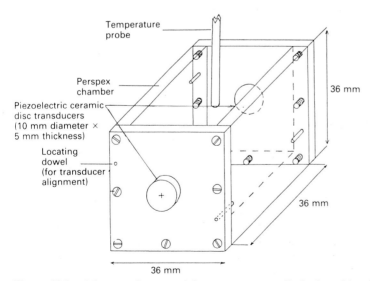

Figure 12.2 *Schematic diagram of the measurement cell of a broad bandwidth ultrasound transmission spectrometer for* in situ *measurement on concentrated colloidal suspensions (after Challis* et al. *(1991))*

Similarly, the area of dielectric spectroscopy and other electrically based sensors are receiving attention, since they too may also enable rapid fingerprinting of the size and composition of complex polydisperse and multicomponent mixtures (Russel *et al.*, 1989; Shi *et al.*, 1993).

The task of obtaining solids concentration profiles implies extension of the measurement information from being a simple one-dimensional 'average value' to two- or three-dimensional (spatial) information. This requires the use of multiple sensors, both signal emitters and detectors. This area of technology is sometimes referred to as *flow imaging* or *tomography* and has its origins in the design of medical body scanners, which first utilized X-rays and, more recently, nuclear magnetic resonance. A number of sensing methods can be used, such as electrical conductivity, electrical capacitance, ultrasound, optics, neutrons, X-rays, γ-rays or tracer-based methods involving positrons or electromagnetic radiation (Beck *et al.*, 1993; Beck and Williams, 1994). One sample result is presented in a case study given below.

12.4 Industrial examples

Five examples illustrating several types of on-line instrumentation will now be considered briefly:

(a) control of grinding mills
(b) control of crystallizers
(c) control of aggregation processes
(d) transportation of friable aggregates
(e) modelling and control of particle hydrocyclone classifiers.

These examples embrace two distinct measurement protocols. The first two studies involve on-line size measurement techniques that require *sampling* and *pretreatment* of the sample before analysis. The last three methods use wholly *in situ* methods of size analysis. The last case involves *in situ* particle size measurement coupled with on-line mass flow rate measurements and solids concentration profiling using electrical tomographic techniques to deduce a model for a particle separator.

12.4.1 Control of grinding mills

Particle formation through comminution consumes large amounts of energy and has long been a test-bed for on-line size analysers, since substantial costs savings are to be gained through better control to avoid overgrinding.

Early reviews of the subject area (Hinde and Lloyde, 1975) identified the potential of on-line techniques but their implementation, on an industrial scale, has not met with great success. The primary reason being the feasibility of using techniques that are on-line but essentially *ex situ*, in the sense that they require sampling and dilution (Hobbel *et al.*, 1991) and these processes are prone to malfunction (blockage or leakage). The poor reliability of some of the early on-line size analysers has not encouraged implementation of more recently developed and robust technology (Salter, 1991).

Alternatively, control of grinding circuits can be so designed to obviate the need of complex size analysis instrumentation through careful control strategies based on other process measurements coupled with process models (Alba and Herbst, 1987; Herbst *et al.*, 1988, 1992). Figure 12.3 shows an example of a typical milling circuit consisting of an open-circuit rod mill followed by two ball mills (5 m × 7.3 m) operated in parallel and closed using two banks of hydrocyclones (in this case, each containing four cyclones). The control circuit

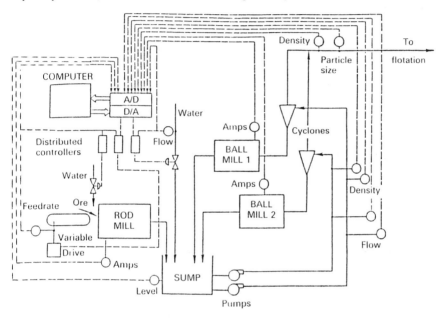

Figure 12.3 *Schematic layout of a grinding circuit and associated instrumentation required for model-based control of rod and ball mills (after Herbst* et al. *(1992))*

is based on a model for the various unit operations that is designed to require a small number of variables to be measured. Typically this demands measurement of stream densities, sump level, water addition rates, pumping rates, power drawn by the mills, hydrocyclone cut-size and a particle-size characteristic (e.g. the percentage by mass greater than a given size (150 μm) in the hydrocyclone overflow product) (Figure 12.3). Herbst *et al.* (1992) have demonstrated that by using an integrated approach it is possible to optimize the control strategy to yield improved energy efficiency and product quality. Ideally this type of control scheme should be operated using on-line size analysers. In practice, single-point size analysis is often more amenable to routine implementation and automation in process plant (as described below), although for many types of comminution operations it would also be desirable to measure the full size distribution. More complicated instrumental techniques are required to measure over the complete size-distribution range.

Some uses have been reported of commercial ultrasound attenuation devices (e.g. Autometrics PSM System-100 and Debex Debmeter instruments) which rely on acquiring a sample of the particulate slurries (Allen, 1990; Salter, 1991). Great care has to be taken in siting the installation within the process plant and to ensure that any entrained air is removed from the sample before analysis. Automatic deaeration systems can be deployed; one example of such a system is shown in Figure 12.4. Early versions of these instruments provided a single point

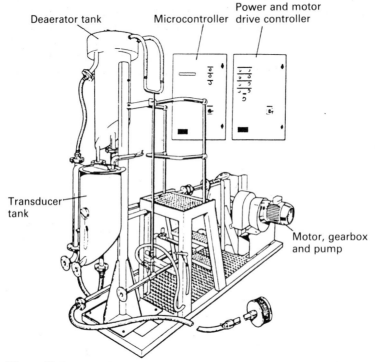

Figure 12.4   *Example of the general features of an on-line size analyser (Debex Debmeter) based on ultrasound attenuation of a diluted, deaerated slurry sampled from the process line (after Salter (1991))*

on the cumulative size-distribution curve, which was a useful parameter for grinding circuit control (Figure 12.3). Later modifications enabled several size fractions to be quantified, but with definite limitations on the mean size (typically 20–1000 μm) and the width and modality of the size distribution. The complexity of these instruments, arising from the sampling processes involved, contrasts greatly with more recent wholly on-line instruments based on ultrasound spectroscopic sensors (e.g. Figure 12.1). Advances in the use of ultrasound methods are likely to emerge in the near future.

12.4.2 Control of industrial crystallizers

Of all the particle formation processes, modelling and control of crystallizers have been the focus of extensive international interest (Rawlings *et al.*, 1992; Jager *et al.*, 1992). It is evident from Chapter 6 that successful control of agglomerative precipitation processes demands detailed knowledge of particle-size distribution and particle number density, in addition to chemical factors, such as the extent of supersaturation and reactor hydrodynamics. The ultimate objective is to produce a product that meets a given specification with regard to crystal-size distribution (CSD), shape and purity. An example based on a forward light-scattering instrument based on the results of De Wolf (1990), Jager (1990) and Boxman (1992) as part of one major study, the Universal Instrumentation and Automation of Crystallizers (UNIAK) project, will be highlighted here. Figure 12.5 shows one configuration of the crystallizer plant, which is based on a draft tube baffled evaporative crystallizer having a volume of 970 dm^3, equipped with a fines removal loop. It may be operated in a batch or continuous mode.

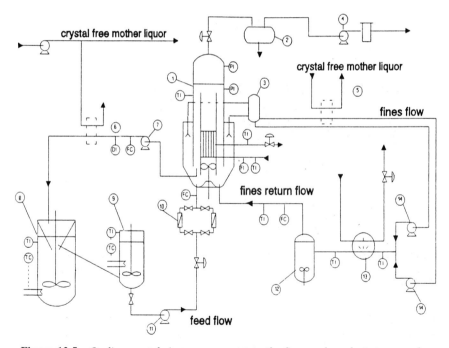

Figure 12.5 *On-line crystal size measurement on the fines and product streams of the UNIAK 970 dm^3 evaporative crystallizer using forward light scattering (after Boxman (1992)). 1, Crystallizer; 2, condenser; 3, fines collection vessel; 4, vacuum pump; 5, fines measurement unit; 6, product dilution and measurement unit; 7, product removal pump; 8, feed regeneration vessel; 9, feed supply vessel; 10, filters; 11, feed pump; 12, fines dissolving vessel; 13, fines heat exchanger; 14, fines removal pumps*

Several on-line sizing principles could be utilized for dynamic CSD measurements based on conventional forward light scattering, back scattering (discussed later) and acoustic methods (described above). The application of laser light diffraction developed by the UNIAK team represents an advanced on-line method. Forward scattering offers a reasonably fast and reliable measurement method, but is limited to very dilute slurries. Consequently, much effort has to be expended in ensuring that a representative sample of crystal magma is removed for analysis and that this is diluted in a consistent and reliable fashion without changing the degree of supersaturation (temperature, liquor composition etc.). The size analysis was based on a modified commerical instrumentation (Malvern 2600c) employing carefully designed sampling and flow arrangements for measurement of fines (location 5, Figure 12.5) and the crystal product (location 6, Figure 12.5), described in detail by Jager *et al.* (1990), Jager *et al.* (1992) and Boxman (1992).

Considering measurements of mean size based on mass and area during start-up of the crystallizer, good agreement has been demonstrated between conventional off-line sieving and sampled on-line forward light scattering (Figure 12.6). However, close examination of all the raw measurement

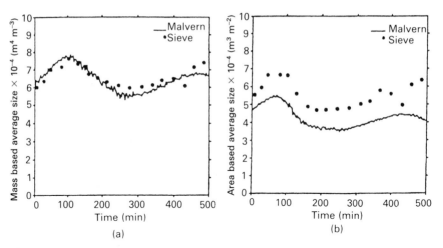

Figure 12.6 *Comparison of mass* (a) *and area* (b) *based average sizes for on-line forward light scattering and off-line sieve analysis during crystallizer start-up (after Jager (1990))*

information coupled with careful statistical analysis and correlation of data interpreted with respect to fundamental principles of light scattering and associated inversion algorithms also allows other indicators to be devised (Boxman, 1992). For example, trends in the spread of measured data pertaining to certain size intervals can be used to judge the reliability of the measurements – this being of critical importance if the on-line measurements are to be used for *control* purposes, since unrecognized rogue data could cause serious problems in crystallizer control. This serves to demonstrate that the measurement of size alone is not the only objective of performing measurements on-line, since the quality of the measurement process must be verifiable in some way and often time-varying characteristics can be utilized for this purpose. The quality of measurement may also be dependent on the form of the CSD.

Figure 12.7 shows the effect of the fines removal rate on the oscillatory nature of the CSD in the product during start-up of a continuous evaporative crystallizer. The figure shows the dynamic behaviour for four different CSD volume fractions measured at point 6 (Figure 12.5) over a 40 h period for the crystallizer operated in continuous mode. Oscillation is a common problem in the industrial operation of these types of crystallizers. In this case the period of oscillation is approximately 6 h. The upper trace for each of the two sets of measurements shows the response of median size with time. The period of oscillation is similar for both fines removal rates (1 and 2.2 dm^3 s^{-1}), but it is clear that the oscillations are less damped at the higher rates of fines withdrawal, which is consistent with other observations (Randolph and Larson, 1962). This is generally attributed to variations in the level of supersaturation in the crystallizer, which cause periodic bursts of new nuclei, and due to generation of new surface from crystal fragmentation.

12.4.3 Control of aggregation processes

Two types of direct on-line and *in situ* measurement methods have found application in following aggregation processes: *scanning laser microscopy* and *ultrasound spectroscopy*. Unlike the sizing techniques discussed in the two preceding examples, these methods do not require splitting of the process stream or dilution. In this sense they are truly on-line techniques, but in fact do still involve what might be termed 'convective sampling', in the sense that the result they display may be influenced by the hydrodynamics of the process being investigated. A specific example is discussed later. Nevertheless, the methods are of considerable interest since they can be employed with minimum disruption to the particulate process.

A specific case study based on laser microscopy is considered in detail below. Riebel and Löffler (1989) have described the basis of on-line ultrasound spectroscopy and possible emerging applications for obtaining information on aggregation and aggregate structure (Riebel, 1992).

Scanning laser microscopy

The scanning laser microscope (SLM) is available as a commercial instrument in the form of a probe (25 mm diameter) which is inserted into the process stream or reactor (Lasentec Corporation, USA). It operates by rapidly scanning a high intensity and sharply focused laser beam (to form an elliptical spot 0.8 μm × 2 μm) over the subject particle passing through the focal point and measuring the time duration of the back scattered pulse of light (Figure 12.8). The focal point can be adjusted from a distance of zero up to a few millimetres away from the sapphire window. The pulses produced from individual particles going through

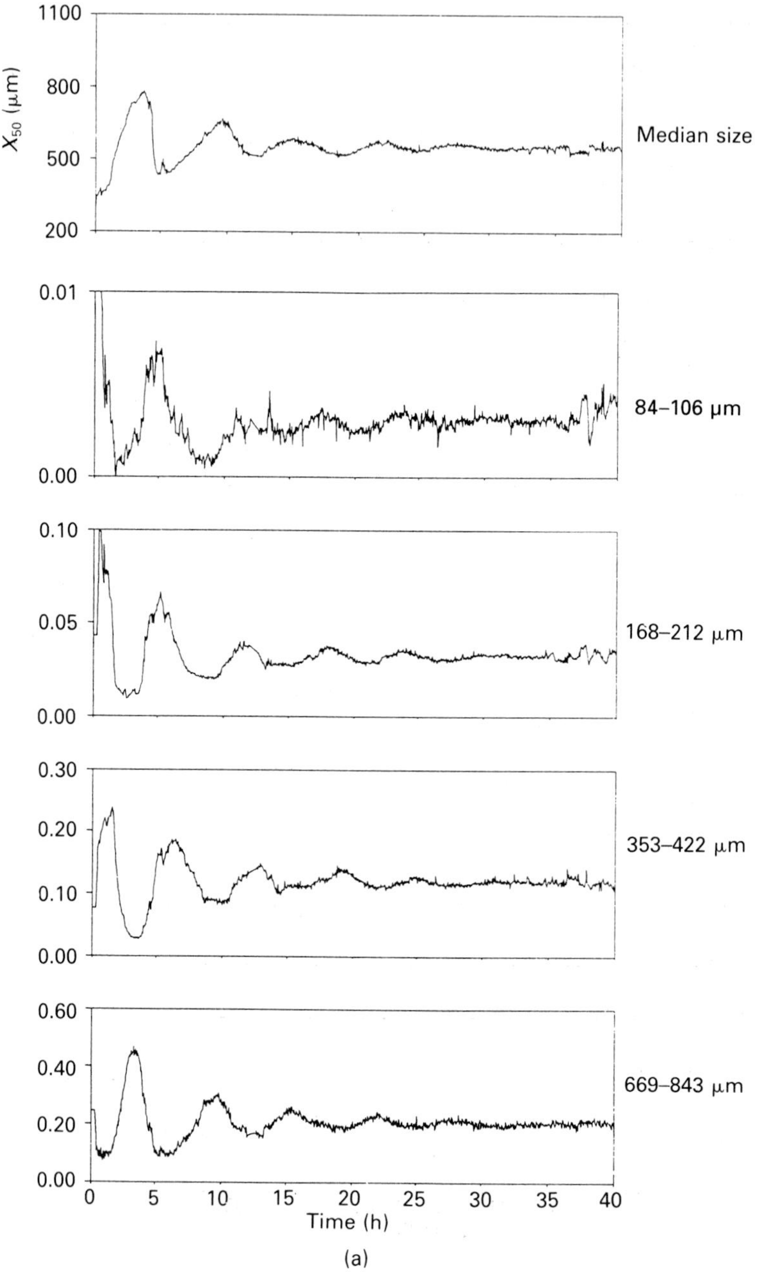

Figure 12.7   *Start-up response of the median (x_{50}) size and four crystal volume fractions in a continuous evaporative crystallizer for fines withdrawal rates of 1.0 dm^3 s^{-1} (a) and 2.2 dm^3 s^{-1} (b). (After Broxman (1992))*

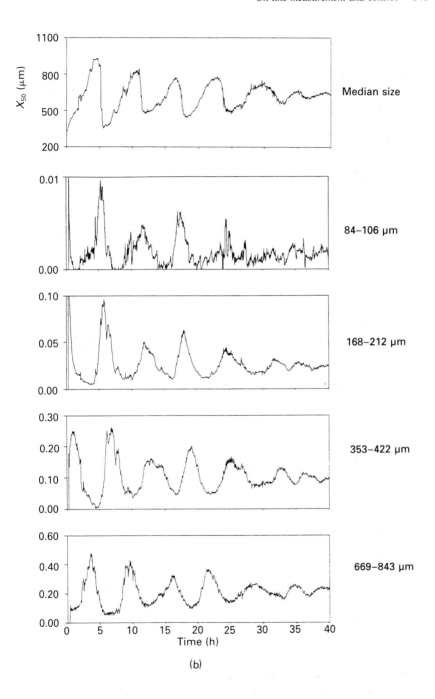

(b)

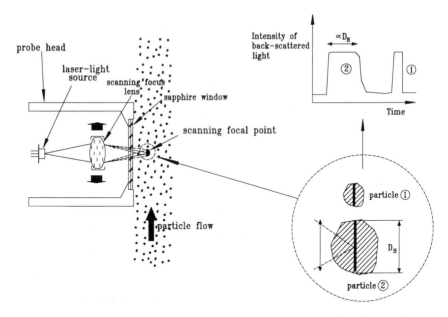

Figure 12.8 *Schematic diagram of scanning laser microscope probe, showing the size measurement principle based on the duration of back-scattered light pulses*

the focal point represent the measurement of individual particles – in this sense the probe can be considered to be a particle-counting device which may be useful in providing information for population balance purposes. In reality this method measures a chord length across the particle (the intersection of the scan plane with the projected area of the particle in the focal plane) which must be corrected based on a probability density function which predicts the chance of encountering a given chord length for a particle of a given shape. After this correction a number–size frequency distribution is statistically determined from the total number of counts. The device can measure a size distribution between 1.2 and 1000 μm every 1.6 to 8 s (depending on the number concentration of particles) in 38 channels.

Some results from an experimental investigation of the formation of silica flocs and their subsequent breakage in a stirred tank, as a function of agitation rate, solids concentration and polymer flocculant dosage will be considered (Williams *et al.*, 1992). In this arrangement the evolution and disintegration of aggregates (in the size range 1–500 μm) in 'concentrated' aqueous particulate slurries (up to 5% solids by volume) was measured using the SLM positioned in a vertical orientation in 5 dm³ experimental baffled stirred tank equipped with a six-bladed Rushton turbine impeller. It is assumed that the flocs are approximately spherical in shape.

Figure 12.9a shows a time sequence of measurements performed on a silica slurry on pulse addition of a flocculant (at the end of region I, Figure 12.9d), the

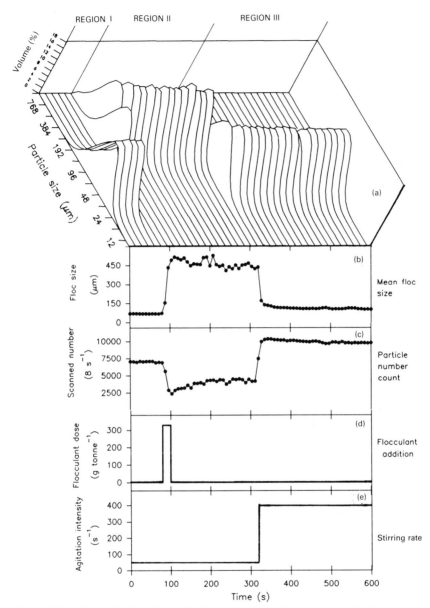

Figure 12.9   (a) *Evolution of size distribution measured using a scanning laser microscope on adding a polymer to a primary polydisperse silica suspension (1% (v/v)) (region I) showing floc growth (region II) and subsequent breakage on increasing the rate of agitation from a nominal velocity gradient of 50 to 400 s^{-1} (region III). (b) Mean size (by mass). (c) Number of particles counted during 8 s counting period. (d) Profile of flocculant addition. (e) Agitation rate, expressed as nominal velocity gradient*

growth of flocs during agitataon at a constant rate (region II), and the breakage of flocs on suddenly increasing the rate of agitation (region III). Figure 12.9b traces the change in mean particle size (by mass), from which it is evident that the floc breakage process on increasing the agitation rate (Figure 12.9e) is rapid.

It is interesting to note the variation in the number of scanned counts (Figure 12.9c) since this reveals information about the *number density concentration* of particles. On addition of the flocculant the number counts fall off immediately, increase slightly to a plateau and then increase dramatically when the stirring rate is increased. Increased stirring causes generation of new particles, but it also would be expected to increase the number of particles analysed in a given analysis time. The latter factor may introduce additional 'sampling' errors. Not only do convective sampling effects exist, but the very fact that when the particle-size distribution coarsens and is broad then the analysis of one larger particle is performed to the exclusion of being able to analyse several smaller particles. Hence some further bias may be introduced due to the form of the size distribution itself.

The floc formation and breakage kinetics can be studied in some detail by varying the rate of agitation and the manner in which the system is flocculated. For example, Figure 12.10a shows the effect of polymer dosage on the final equilibrium floc size (reported here as a dimensionless mean size compared with primary unflocculated particle size) for two different rates of stirring. The influence of the magnitude of the step change in agitation intensity (reported here as the increase in 'nominal velocity gradient', ΔG, having units of reciprocal seconds) to break up the flocs is shown in Figure 12.10b. The equilibrium floc size is inversely proportional to log (ΔG). These types of experiment allow practical measurements to be performed to characterize and quantify the formation and strength of aggregates. Similarly, the effects of shear-induced electrostatic coagulation of colloidal particles can be investigated (Peng and Williams, 1994).

Using these methods it is possible to measure and model the formation and breakage *kinetics*, and therefore to undertake simulations of floc behaviour under process conditions. This allows the process engineer to probe the sensitivity of flocculated systems to changes in real process variables (pH, shear rate, reagent addition etc.) which other methods based on sampling and removal of fragile flocs from their environment would not allow.

An example of such a procedure is shown in Figure 12.11. On subjecting a floc population to a step change in agitation rate in a stirred tank (Figure 12.11c) there may be a spontaneous reduction in the mean size (Figure 12.11a). The solid line (I) corresponds to flocs that degrade due to rupture mechanisms producing an ensemble of daughter particles. The semi-solid line (II) represents the other extreme case where degradation arises from erosion or chipping of individual particles from the surface of the aggregate producing a large number of small particles. These different mechanisms are reflected in the number count data (Figure 12.11b). In practice, both types of mechanism occur during breakage of flocs (Tambo and Hozumi, 1979; Tambo and Francois, 1991; Bache and Hossain, 1991; Torres *et al.*, 1991; Williams *et al.*, 1992). The breakage process can be

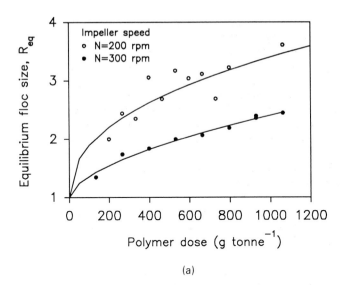

(a)

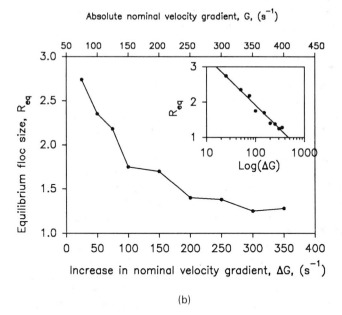

(b)

Figure 12.10 (a) *Effect of flocculant dosage on mean silica floc size for two different stirring conditions (size is normalized with respect to mean size of primary, unflocculated particles).* (b) *Effect of the magnitude of the step change in agitation (expressed as change in nominal velocity gradient ΔG) to induce breakage of flocs on the final mean floc size obtained*

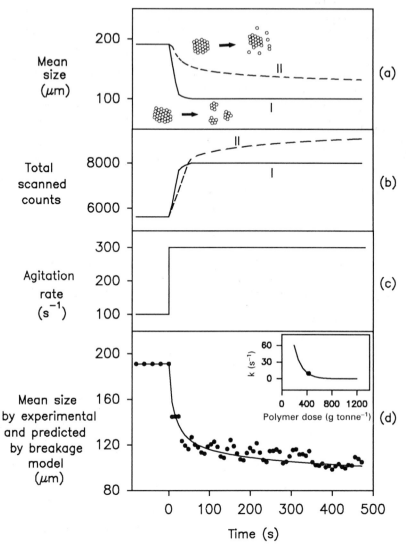

Figure 12.11 *Monitoring breakage kinetics of polymer flocculated silica on increasing the stirring rate (expressed as nominal velocity gradient, s^{-1}) in a batch mixing tank. (a) Change in mean size for floc breaking due to predominance of (I) rupture and (II) erosion mechanisms. (b) Corresponding change in number of particles counted per unit time for cases (I) and (II). (c) Step change in stirring speed (expressed as nominal velocity gradient, s^{-1}). (d) Comparison of experimentally measured change in mean size due to shear perturbation with model prediction based on breakage model. Inset diagram shows variation in breakage constant K (inversely related to floc strength) with polymer dose used in forming the floc population, indicating data point corresponding to condition in main figure*

modelled in an analogous fashion to comminution using a simplified analytical solution of the mass balance equation (Peng and Williams, 1993). Breakage constants (K) can be computed to quantify the breakage characteristics of flocs that have been formed in different ways and hence their in-process performance can be predicted. A typical result is shown in Figure 12.11d for a polymer flocculated silica and the inset shows the role of polymer dosage on the breakage constant. The extent of breakage is reduced with increasing polymer dosage addition. A further example is given later in relation to transport of colloids and flocs in pipelines.

In situ methods such as SLM and ultrasound spectroscopy should, in due course of time, allow better quantitative modelling of particle formation and disruption dynamics and lead to improved industrial practices, for instance, in the area of quality assurance and process control.

12.4.4 Transportation of friable aggregates

The SLM probe (Figure 12.8) can also be mounted on process pipelines to monitor the size distribution of particulates. This can be useful to assess the extent of particle breakage in passing a slurry through a pump or to follow the shift in particle size during prolonged transport of a solids in a closed circuit hydraulic loop (Roldan-Villasana, 1992). In an analogous manner to the quantification of floc strength in a stirred tank by means of the breakage constants embodied in conventional comminution modelling (Figure 12.11), the breakage properties of flocs flowing in a pipe can be described using a standard breakage test procedure.

Figure 12.12 shows a schematic diagram of a floc testing apparatus in which floc are formed in a stirred chamber and then driven under pressure through a pipe section of desired length and bore. The floc size distributions in the feed reservoir and the floc population emerging from the pipe are measured using SLM devices. By comparing the two size distributions it is possible to compute the breakage coefficients. This procedure is performed continuously while increasing the flow rate through the pipe (e.g. in small steps or as a ramp function). If necessary, the apparatus can be operated in closed circuit, although this is not generally desirable for flocculated systems.

Figure 12.13a shows how the mean size of polymer-flocculated silica (at a solids concentration of 1% (v/v)) decreases as the flow velocity in the tube increases. When the 'mean size (ratio)' reaches unity the flocs are deemed to be totally broken and primary particle size is recovered. Experiments can be performed on flocs that have been formed in different ways (and hence having different mechanical strengths) for which values of the breakage constant K can be computed as a function Reynolds number (Re) in the pipe (Peng and Williams, 1994). Figure 12.13b gives a typical result for a 0.5% (v/v) silica suspension. Hence decisions can be made using such data on the critical or largest acceptable breakage constant for a given floc transportation process.

Since K is strongly dependent on Re, the extent of floc breakage can be controlled by careful choice of pipeline diameter, pumping rate and pipe run

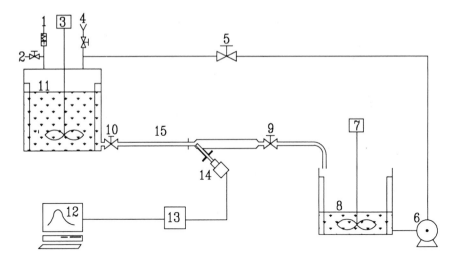

Figure 12.12 *Floc-strength flow apparatus. 1, Pressure relief valve; 2, pressurized gas inlet; 3, adjustable speed motor; 4, flocculant inlet; 5, return flow valve; 6, return flow pump; 7, stirrer motor; 8, floc collection tank; 9, adjustable flow valve; 10, ball valve (fully open during all analyses); 11, floc reservoir; 12, PC control; 13, laser control; 14, scanning laser microscope probe; 15, interchangeable pipe measurement section*

lengths. The same methodology can be used to investigate floc formation and breakage using in-line mixers. However it should be borne in mind that errors will arise when non-isotropic flocs are considered, due to the nature of the laser scanning measurement method which normally assumes that actual size distribution constructed using the measured chord length statistics is for measurements on spherical particulates. This limitation can be surmounted for some non-spherical shapes by incorporating shape-factor corrections in the algorithm which reconstructs the size distribution.

12.4.5 Modelling and control of particle hydrocyclone classifiers

Modelling classification performance
Hydrocyclones can be used to classify particles according to their size and/or density with respect to the continuous fluid phase. They are used commonly to close grinding circuits (Figure 12.3), to classify crystals leaving continuous crystallizers or as part of a fines removal circuit and for thickening applications. The feed slurry enters the hydrocyclone body tangentially and larger particles within the feed population tend to move to the wall of the conical body and down towards the apex (or underflow), while smaller particles move to the centre of the body and leave via the overflow orifice (see inset, Figure 12.14a).

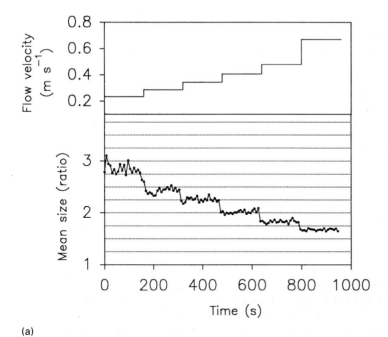

(a)

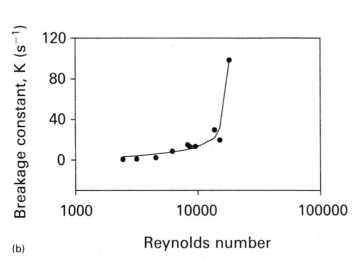

(b)

Figure 12.13 (a) *Effect of increasing flow velocity on mean size of polymer-flocculated silica exiting measurement pipe section (2 mm diameter, 800 mm length) for 1% (v/v) solids. Size data are shown in reduced units where ratio corresponds to multiples of the mean size of the primary disperse silica particles;* (b) *computed breakage constant versus Reynolds number for 0.5% (v/v) silica flocs (after Peng and Williams (1994))*

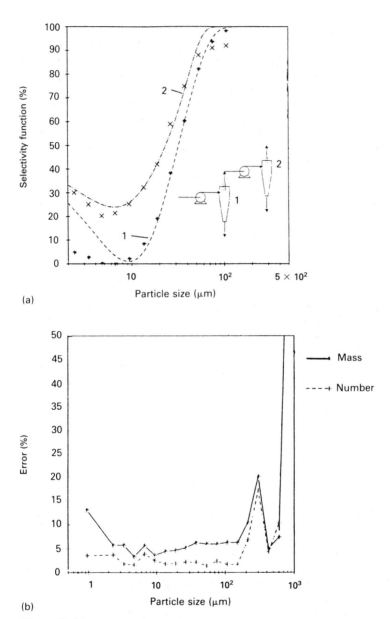

(a)

(b)

Figure 12.14 (a) *Particle classification function (% by mass if a given size reporting to underflow stream) for reconciled measurements (data points) and model-fitted prediction (continuous lines) corresponding to performance if a simple network of two hydrocyclones in series (labelled 1 and 2, see inset). (b) Measured errors in particle-size channels using on-line scanning laser microscopy based on mass and number statistics*

This case-study concerns measurements of the classification efficiency, i.e. the probability that a particle of a given size will report to the coarse (underflow) product. An example of such curves is given in Figure 12.14a, for two hydrocyclones placed in series, whereby the overflow stream (containing fine particles and water) from the first unit is reprocessed to clarify further the effluent. It can be seen that large particles are recovered at high efficiency (approaching 100%), while smaller ones are not. The task is to measure such effects experimentally and model their form as a function of key process variables. In this context the problem of formulating an accurate population balance is particularly troublesome.

For convenience, many workers modelling particle separation in hydrocyclones (and other similar operations) choose to measure one input stream (e.g. the feed) and one output stream (e.g. the underflow) and to obtain the other stream (i.e. the overflow) by *difference* – thereby eliminating all residual errors. This can lead to considerable uncertainties and propagation of errors and is not to be recommended. To acquire reliable models for the separation function it is necessary to make as many experimental measurements as is practically possible and then to reconcile any apparent duplication of information in a rigorous statistical manner.

To give an example of this procedure, Table 12.2 reports measurements on a single 50-mm diameter hydrocyclone performed on-line using an automated apparatus in which the mass flow rates were measured using Coriolis flowmeters (see Table 12.1) and the size distributions of streams analysed with the SLM described in the preceding example. Hence information for the *size distribution* and *mass flow rate* in all three streams was obtained together with measurements of other operational conditions (pressure drop, slurry density etc.). The first step in reconciling the experimental data is to measure (or estimate) the measurement errors of each instrument (e.g. by replicate testing). Hence the likely error for the mass flowmeter and the particle-size analyser can be deduced. Figure 12.14b shows the errors in the size analysis which is seen to be size dependent. This information can be put into a suitable data-reconciliation algorithm which will find an optimum solution and then proceed to fit the reconciled data to a model for the separation (or 'selectivity') function itself (Roldan-Villasana, 1992; Roldan-Villasana *et al.*, 1993). This involves consideration of the size-distribution data (on a mass basis) and the mass flow rates of the process streams.

Table 12.2 summarizes the results for measurements for the hydrocyclone unit 2 in Figure 12.14a – the figure shows the data fitted to a separation function model. The cut-size for unit 2 (the size which has an equal chance of reporting to the underflow or overflow products) is lower than for unit 1. The tabulated data show the importance of the data reconciliation procedure around a single hydrocyclone. For example, a simple mass balance for the solids in the 13th (coarsest) size interval, gives the values for the selectivity function shown in Table 12.3, depending on which of the two streams is chosen to perform the calculations. It is evident that if the streams were not reconciled the model parameters obtained would have depended upon the streams selected for analysis, and that even taking the average value (1.06) of the three possible

Table 12.2 *Particle size and mass flow data for hydrocyclone unit 2 (Figure 12.14a) derived from raw experimental measurements and after applying data-reconciliation procedures*

	Mass flow (kg s^{-1})					
	Feed		Overflow		Underflow	
	Exp.	Recon.	Exp.	Recon.	Exp.	Recon.
Liquid	0.3771	0.3773	0.1969	0.1959	0.1594	0.1814
Solids	0.0055	0.0056	0.0025	0.0026	0.0026	0.0030

	Particle-size distribution (mass fraction)					
	Feed		Overflow		Underflow	
Size interval	Exp.	Recon.	Exp.	Recon.	Exp.	Recon.
1	0.0000	0.0000	0.0000	0.0000	0.0000	0.0000
2	0.0002	0.0002	0.0003	0.0003	0.0001	0.0001
3	0.0011	0.0014	0.0022	0.0022	0.0006	0.0006
4	0.0044	0.0058	0.0104	0.0102	0.0022	0.0022
5	0.0215	0.0277	0.0487	0.0475	0.0113	0.0109
6	0.0616	0.0733	0.1216	0.1199	0.0347	0.0341
7	0.1292	0.1401	0.2088	0.2075	0.0839	0.0833
8	0.2089	0.2049	0.2601	0.2599	0.1584	0.1586
9	0.2665	0.2431	0.2139	0.2181	0.2572	0.2641
10	0.2109	0.1948	0.1054	0.1070	0.2633	0.2688
11	0.0779	0.0841	0.0226	0.0220	0.1411	0.1365
12	0.0163	0.0219	0.0047	0.0043	0.0416	0.0367
13	0.0015	0.0027	0.0013	0.0011	0.0056	0.0041

combinations is erroneous. Without proper data-reconciliation procedures the scatter of data in Figure 12.14 would have prohibited the fitting of a selectivity model with any confidence. This serves to demonstrate that installation of on-line instrumentation in itself will not suffice for control and monitoring of complex separation processes involving particulate materials and multiple process streams – measured results still require careful interpretation.

Internal modelling of flow structure
In situ measurements on the behaviour of the hydrocyclone unit itself rather than the process pipes servicing the separator are becoming feasible. One promising

Table 12.3 *Comparison of raw and reconciled data for a specific size class (the coarsest size, interval 13 in Table 12.2) used to estimate the classification efficiency, (the raw experimental results are based on consideration of only two of the three process streams)*

Streams chosen as basis for calculation of classification efficiency of coarsest particle-size fraction	Experimental data	Reconciled data
Feed and underflow	1.76	0.81
Overflow and underflow	0.83	0.81
Feed and overflow	0.61	0.81
Average	1.06	–

technique utilizes electrical resistance tomography to monitor radial fluctuations in the resistance of the process fluid at a given cross-section in the separator. Although individual particles cannot be visualized, it is possible to measure local changes in resistivity which is related to the local solids concentration.

Figure 12.15a shows an arrangement for performing such measurements, involving the installation of a number (16) of electrodes in the wall of the hydrocyclone. A small alternating electrical current is passed via one pair of neighbouring electrodes and the resulting distribution of electrical potential is measured (as a small oscillating voltage signal) at all the other electrodes. This procedure is repeated for all possible combinations of electrodes and the resulting voltage measurements are then reconstructed to produce a map showing the variations of conductivity within the cross-section. In practice this is a quite a difficult non-linear inverse problem, but some short-cuts (e.g. using a linear back-projection method) can be used if semiquantitative images are required, whereas quantitative reconstruction may require access to advanced computational facilities. Nevertheless, images can be acquired at high rates (up to 100 frames per second) since the nature of the measurement is all electronics based.

An example of a single image through a hydrocyclone separating a silica slurry is shown in Figure 12.15b, which shows a significant enhancement of resistivity near the cyclone walls and near the central air core which is present within the hydrocyclone. The increase in resistance is due to the presence of a higher local concentration of solids, which can be interpreted in a quantitative manner in order to deduce solids concentration profiles (Williams and Dickin, 1993). This type of measurement allows the dynamic response of the separator to be observed and is helpful in refining and validating more fundamental mechanistic models based on computational fluid dynamics techniques. Ultimately, image-based information could be used for control purposes, either using a reduced data set (e.g. following the response of a subset of pixels in a predetermined location) or pattern recognition (e.g. using a neural net).

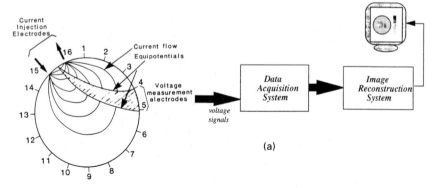

Figure 12.15   (a) *Principle of electrical resistance tomography using 16 wire electrodes inserted through the wall of a hydrocyclone. Oscillating current is injected through adjacent electrodes generating voltages at other sensing electrodes (e.g. shaded region depicts non-linear path of voltage equipotentials relevant to sensors 4 and 5). (b) Relative particle concentration isometric plots and two-dimensional contour map for a cross-section located 15 mm below the feed point on a 50 mm diameter hydrocyclone treating a silica slurry (broken circle drawn on lower figure denotes position of central air core)*

It is also possible to measure the complex impedance of the mixture (i.e. the real and imaginary components) which for particulate and colloidal mixtures may be dependent on the frequency of the excitation signal. The use of spectroscopic techniques (by sweeping the frequency) appears to offer considerable potential for the future, since any frequency dependencies could be used to discriminate between components (e.g. particle types) within a slurry. This may provide a means to enable mapping of *component* concentrations within slurries in pipes and separation vessels.

By using more than one plane of electrodes, three-dimensional images can be obtained and vectoral information could be abstracted. Similar measurements can be made on unit operations other than hydrocyclones. In particular, liquid–liquid mixing processes have received some attention (Ilyas, 1992; Dickin *et al.*, 1993) and non-conducting particle–gas systems using dielectric detection methods (Halow *et al.*, 1990; McKee *et al.*, 1993; Xie *et al.*, 1993).

12.5 Conclusion

The development of advanced on-line measurement technology is of critical importance for the control of particulate processes. Without appropriate tools to enable measurement of velocities, size and flow rates of particulate components to produce viable models for particle processing operations and opportunities for simulation remains severely impeded. The availability of more measurement information helps to improve understanding of mechanistic processes (e.g. in

HIGH RESISTIVITY DUE TO
SOLIDS AT PERIPHERY
OF AIR CORE

HIGH RESISTIVITY DUE TO
CONCENTRATION OF
SOLIDS AT WALLS

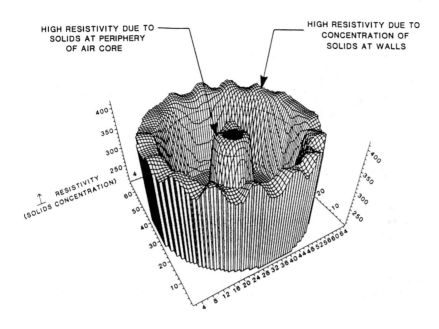

LOCATION OF
FEED POINT
(ABOVE IMAGED
PLANE)

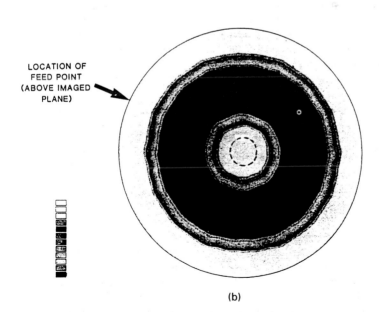

(b)

particle separation and transport) and, therefore, our ability to control these processes to ensure optimum productivity in industrial or manufacturing operations.

It has been demonstrated that having *more information* about a particulate process also presents new problems. Firstly, the need to reconcile experimental data when performing a population balance. Secondly, knowing how to use the information content of the measured information for the purpose of improved *control* of the process. Such issues must be addressed if intelligent use is to be made of the emerging on-line instrumentation and associated technology.

References

Alba, F. J. and Herbst, J. A. (1987) Particulate and multiphase processing. In *Colloidal and Interfacial Phenomena* (ed. Ariman, T. and Verizoglu, T. N.) Hemisphere, Houston, 297–311

Alba, F., Davies, R., Allen, T. and Boxman, A. (1993) Use of acoustic attenuation for measuring the particle size distribution of slurries. In *Control of Particulate Processes III* (Santa Barbara) Engineering Foundation, New York

Allen, T. (1990) *Particle Size Measurement*, 4th edn, Chapman & Hall, London

Bache, D. H. and Hossain, M. D. (1991) Optimum coagulation conditions for coloured water in terms of floc properties, *J. Water SRT-Aqua*, **40**, 170–178

Beck M. S. and Williams, R. A. (eds) (1994) *Process Tomography – Principles, Techniques and Applications*, Butterworth-Heinemann, Oxford

Beck, M. S., Campogrande, E., Morris, M., Waterfall, R. W. and Williams, R. A. (eds) (1993) *Tomographic Techniques for Process Design and Operation*, Commission of European Communities/BriteEuram/Computational Mechanics Publications, in press

Boxman, A. (1992) Particle size measurement for the control of industrial crystallizers. Ph.D. Thesis, Delft University of Technology, The Netherlands

Challis, R. E., Holmes, A. K., Harrison, J. A. and Cocker, R. P. (1990) Digital signal processing applied to ultrasonic absorption measurements. *Ultrasonics*, **28**, 5–15

Cooper, H. R. (1992) Advances in on-line particulate composition analysis. *Powder Technol.*, **69**, 93–99

De Wolf, S. (1990) Modelling, system identification and control of an evaporative continuous crystallizer. Ph.D. Thesis, Delft University of Technology, The Netherlands

Dickin, F. J., Williams, R. A. and Williams, R. A. (1993) Determination of composition and motion of multi-component mixtures in process vessels using electrical impedance tomography: (I) Principles and process engineering applications. *Chem. Eng. Sci.*, **48**, 1883–1897

Halow, J. S., Fasching, G. E. and Nicoletti, P. (1990) Three dimensional capacitance imaging of fluidised beds. *AIChE Symp. Ser. No. 276*, **86**, 41–50

Herbst, A., Alba, F., Pate, W. T. and Oblad, A. E. (1988) Significance of the efficiency index in evaluating the economics of benification of coal fines by the oil agglomeration technique. *Int. J. Mineral Process.*, **22**, 275–290

Herbst, A., Pate, W. T. and Oblad, A. E. (1992) Model-based control of mineral processing operations. *Powder Technol.*, **69**, 21–32

Hinde, A. L. and Lloyd, P. J. D. (1975) Real time particle size analysis in wet closed-circuit milling. *Powder Technol.*, **12**, 37–50

Hobbel, E. F., Davies, R., Rennie, F. W., Allen, T., Butler, L. E., Waters, E. R., Smith, J. T. and Sylvester, R. W. (1991) Modern methods of on-line size analysis for particulate process streams. *Part. Part. Syst. Charact.*, **6**, 29–34

Holmes, A. K. and Challis, R. E. (1993) Ultrasonic scattering in concentrated colloidal suspensions. *Colloids Surfaces: Physicochem. Eng. Aspects*, **77**, 65–74

Hoyle, B. S. and Xu, L. A. (1993) Ultrasonic sensors. In *Process Tomography, Principles, Techniques and Applications* (ed. Beck, M. S. and Williams, R. A.) Butterworth-Heinemann, Oxford, in press

Huskins, D. J. (1982) *Quality Measuring Instruments in On-Line Process Analysis*, Ellis Horwood, Chichester

Iinoya, K., Masuda, H. and Watanabe, K. (1988) *Powder and Bulk Solids Handling Processes: Instrumentation and Control*, Marcel Dekker, New York

Ilyas, O. M. (1992) Interrogation of liquid–liquid and solid–liquid processes using electrical impedance tomography. M.Sc. Thesis, University of Manchester Institute of Science and Technology, Manchester

Jager, J. (1990) Control of industrial crystallizers: the physical aspects. Ph.D. Thesis, Delft University of Technology, The Netherlands

Jager, J., Kramer, H. J. M., de Jong, E. J., de Wolf, S., Bosgra, O. H., Broxman, A., Merkus, H. G. and Scarlett, B. (1992) Control of industrial crystallizers. *Powder Technol.*, **69**, 11–20

Jager, J. S., Kramer, H. J. M., de Jong, E. J. and de Wolf, S. (1990) On-line particle size measurement in dense slurries. *Powder Technol.*, **62**, 155–162

McAdie, H. G. (ed.) (1986) On-line Particle Size Analysis: Industrial Applications. Fine Particle Technology Group, Toronto

McKee, S. L., Dyakowski, T., Bell, T. A., Allen, T. and Williams, R. A. (1993) Solids flow imaging and attrition studies in a pneumatic conveyor. *Powder Technol.*, in press

Parkinson, N. (1992) Acoustic emission monitoring. In *M62 Sensors Club* (Nov. 1992) A V Technology Limited, Cheadle, Stockport

Pendse, H. P. (1993) Development of ultrasound sensors to monitor particle size distribution of industrial pigment slurries. In *Control of Particulate Processes III* (Santa Barbara) Engineering Foundation, New York

Peng, S. J. and Williams, R. A. (1993) Control and optimisation of mineral flocculation and transport processes using on-line particle size analysis. *Mineral Eng.*, **6**, 133–153

Peng, S. J. and Williams, R. A. (1994) Direct measurement of floc breakage in flowing suspensions. In press

Randolf, A. D. and Larson, M. A. (1962) Transient and steady state size distributions in continuous mixed suspension crystallizers. *AIChE J.*, **8**, 639–645

Rawlings J. B., Witkowski, W. R. and Eaton, J. W. (1992) Modelling and control of crystallizers. *Powder Technol.*, **69**, 3–9

Riebel, U. (1992) Characterization of agglomerates and porous particles by ultrasound spectrometry. In *5th European Symposium on Particle Characterization* (Nürnberg, Germany), Vol. 2, 545–559

Riebel, U. and Löffler, F. (1989) The fundamentals of particle size analysis by means of ultrasonic spectroscopy. *Part. Part. Syst. Charact.*, **6**, 135–143

Roldan-Villasana, E. J. (1992) Simulation and modelling of hydrocyclone networks for fine particle processing. Ph.D. Thesis, university of Manchester Institute of Science and Technology

Roldan-Villasan, E. J., Dyakowski, T., Lee, M. S. and Williams, R. A. (1993) Design and modelling of hydrocyclones and hydrocyclone networks for fine particle processing. *Mineral. Eng.*, **6**, 41–54

Russel W. B., Saville, D. A. and Schowalter, W. R. (1989) *Colloidal Dispersions*, Cambridge University Press, Cambridge

Salter, J. D. (1991) Mill control using particle size monitors. *Mineral. Eng.*, **4**, 707–716

Shi, T. M., Simons, S. J. R. and Williams, R. A. (1993) Electrical sensing of dispersions. *Colloids Surfaces: Physicochem. Eng. Aspects*, **77**, 9–27

Tambo, N. and Francois, R. J. (1991) Mixing, breakup and floc characteristics. In *Mixing in Coagulation and Flocculation* (ed. Amirtharajah, A., Clark, M. M. and Trussel, R. R.) American Water Works Association Research Foundation, Denver, CO, 256–281

Tambo, N. and Hozumi, H. (1979) Physical characteristics of flocs, II. Strength of floc. *Water Res.*, **13**, 421–427

Torres, F. E., Russel, W. B. and Schowalter, W. R. (1991) Floc structure and growth kinetics for rapid shear coagulation of polystyrene colloids. *J. Colloid Interface Sci.*, **142**, 554–574

Williams R. A. (1992) Characterisation of process dispersions; handling colloid materials. In *Colloid and Surface Engineering: Applications in the Process Industries* (ed. Williams, R. A.) Butterworth-Heinemann, Oxford, 3–111

Williams, R. A. and Dickin, F. J. (1993) Recent developments in the use of dynamic tomographic imaging for the design and control of mineral processing plant. In *Proceedings XVIII International Mineral Processing Congress* (Sydney, May), Australasia Institution of Mining and Metallurgy, Parkville, Vol 2, p. 443–452

Williams, R. A., Xie C. G., Dickin F. J., Simons S. J. R. and Beck M. S. (1991) Multi-phase flow measurements in powder processing. *Powder Technol.*, **66**, 203–224

Williams, R. A., Peng, S. J. and Naylor, A. (1992) *In situ* measurement of particle aggregation and breakage kinetics in a concentrated suspension. *Powder Technol.*, **73**, 75–83

Xie, C. G., Huang, S. M., Hoyle, B. S., Lenn, C. P. and Beck, M. S. (1993) Transputer-based electrical capacitance tomography for real-time imaging of oil-field flow pipelines. In *Tomographic Techniques for Process Design and Operation* (ed. Beck, M. S., Campogrande, E., Morris, M. A., Waterfall, R. W. and Williams, R. A.) Commission of European Communities/BriteEuram/Computational Mechanics Publications, Southampton, p. 333–346

Author index

Subject index

Absorption, and aerosol coagulation, 147
Acacia gum, coacervation, 264
6'-Acetamido-6-bromo-2-cyanodiethylamino-4-
    nitroazobenzene, packing coefficient,
    101–2
Acetic acid, fracture tensions, 171
Acid-base reactions, bubble formation, 172
Acoustic aerosol coagulation, 142, 151–3
Acoustic bubble formation, 170
Acoustic emission sensors, 332
Acrylamide, in addition polymerization, 219
Acrylates, in addition polymerization,
    218, 219, 226
Acrylic acids, alkyl esters, in addition
    polymerization, 217
Acrylics, in emulsion polymerization, 218,
    249–51
Addition polymerization, 217–29
Additives, effect on particle shape, 121–31
Adhesion, in wetting, 311–12
Adhesives, dextrine, 224–5
Adipic acid precipitation, 80
Adsorption
    of additives, 123–4
    of colloids, 225
    of surfactants, 210
Aerogels, 32
Aerosil, scattering studies, 21–2
Aerosols, 137–58
Agglomeration, 62
    crystals, 61, 62, 72–4
    and particle size, 88
    see also aggregation; coagulation
Agglomerative precipitation processes, 61–94
Aggregated particle systems, 10, 11, 17–22
    gelation processes, 23–9
Aggregates, friable, transportation, 347–8
Aggregation, 19–20
    control of, 339–47
    droplets in emulsions, 199, 210–15
    in emulsion polymerization, 236–2, 247
    orthokinetic, 76–8
    perikinetic, 74–6

in precipitate formation, 41, 61
reaggregation, in wet milling, 312–13
and secondary crystal agglomeration, 72–4
and supersaturation, 80
vesicles, 299, 300
see also agglomeration; coagulation
Aggregation kernels, 74, 75, 76
Agitation
    in addition polymerization, 221
    and agglomeration/aggregation, 73, 76, 342,
        344
    bubble formation, 176–7
    and emulsion droplet size, 191
    and particle size, 88
Air and jet milling, 322–4
Air nebulizers, aerosol formation, 139
L-Alanine, morphological predictions, 111,
    115–16
Alarm functions of measurement, 329
n-Alkanes, morphological predictions, 117–19
Alkoxides
    gel formation, 2
    as precursors, 44
Alkyl esters, in addition polymerization, 217
Alkyl glycosides, in vesicle preparation, 292
Alkyl phenyl ethoxylates, in addition
    polymerization, 222
Alumina
    crystallization, 80, 83
    pyrogenic, scattering studies, 20–2, 27–9
Aluminium compounds, particle formation, 41,
    44, 46, 48
Aluminium salts, solutions, polycations, 6
Amines, in in-situ polymerization, 269
Ammonium salts
    in addition polymerization, 220, 221, 222
    vesicles, 300
Amorphous precipitates, 62
Amphipathic behaviour, 221
Amphiphiles, vesicle formation, 278–80, 300
Amphiphilic behaviour, vesicles, 277, 278
Amphoteric surface activity, 222
Analytical ultracentrifugation, vesicle studies,

Measurement
  particle processes, 327–56
  size of particles, 236–8
Mechanical bubble formation, 160
Mechanical microencapsulation, 259, 271–2
Mechanical multiphase flow measurement, 331
Mechanical resonance measurement, multiphase
    flow, 331
Mechanical vesicle preparation, 282, 293
Media charges, wet milling, 313
Medical imaging, use of vesicles, 278
Membrane emulsification, 203–4, 214
Membrane filtration, sol formation, 7
Membrane mimesis, vesicles, 278
Membranes, sequential extrusion, vesicle
    preparation, 287
Meniscus formation, 29–30
Mercury, fracture tensions, 171
Metal chelates, decomposition, 45
Metal hydroxides, precipitation with urea, 46
Metal oxalates, particle formation, 48
Metal oxides, dispersed particle formation, 40,
    44, 52
Metal phosphates, particle formation, 48
Metal selenides, particle formation, 46, 50
Metal sulphates, particle formation, 48
Metal sulphides, particle formation, 46, 50
Metastable limit, 70
Methacrylates, in addition polymerization, 218,
    219, 226
Methacrylic acids, in addition polymerization,
    217, 219
Methacrylics, in addition polymerization, 218
Methanol, in vesicle preparation, 295
Methyl methacrylate, in addition polymerization,
    226
Micelles
  critical micelle concentration (CMC), 222–4
  formation,
    phospholipids, 278
    and surfactants, 222
  and particle formation, 231
  in precipitation, 50
  in vesicle preparation, 292, 297
Microattrition, 79
Microbubbles, 159
Microemulsions, 192
Microencapsulation, 259–73
Microfluidization, vesicle preparation, 287–9,
    298–9
Microfluidizers, 202
Micropore filtration, particle size measurement,
    238
Microscopy

aggregation measurement and control, 339–47
emulsions, 212–13
microsphere assessment, 272–3
multiphase flow measurement, 331
vesicle studies, 301
Microspheres
  preparation, 264–72
  release from, 261, 262–4
  size and morphology, 272–3
  structure, 260–1
  surface analysis, 273
Microstructure, gels, 10
Microwave Doppler velocimetry, multiphase flow,
    331
Mie scattering, 213
  microsphere assessment, 273
Milling
  solid particles, 309–25
    measurement and control, 334–6
Mini emulsion process, 230, 249
Mini-homogenizers, 200
Minimum film-forming temperature (MFT), 241
Miscible solvents, water displacement, 30
Mixed-suspension, mixed-product removal
    (MSMPR) crystallizers, population
    balance, 67–8, 69, 95
Mixers, emulsification by, 202
Mixing
  and particle growth, 84–6
  and particle size, 88
    in microencapsulation, 267–9
  and precipitation, 72, 73
MNDO computer software, 113
Models and simulations, 329
  flow structure, 353–5
Molar enthalpy, and bonds, 104
Molecular crystals, morphology, 95–135
Molecular interactions *see* intermolecular;
    intramolecular
Molecular materials, crystal chemistry, 99–104
Molecular shape, and crystal morphology,
    113–18
Monodisperse aerosols, 138, 140
Monodisperse bubbles, 160
Monodisperse colloids, preparation, 6, 193
Monodisperse emulsions, 203, 214, 248
Monodisperse particles
  emulsion polymers, 248
  inorganic, formation, 39–59
Monomers, addition polymerization, 217–18
MORANG computer software, 106
Morphology
  molecular crystals, 95–135
    prediction examples, 110–21